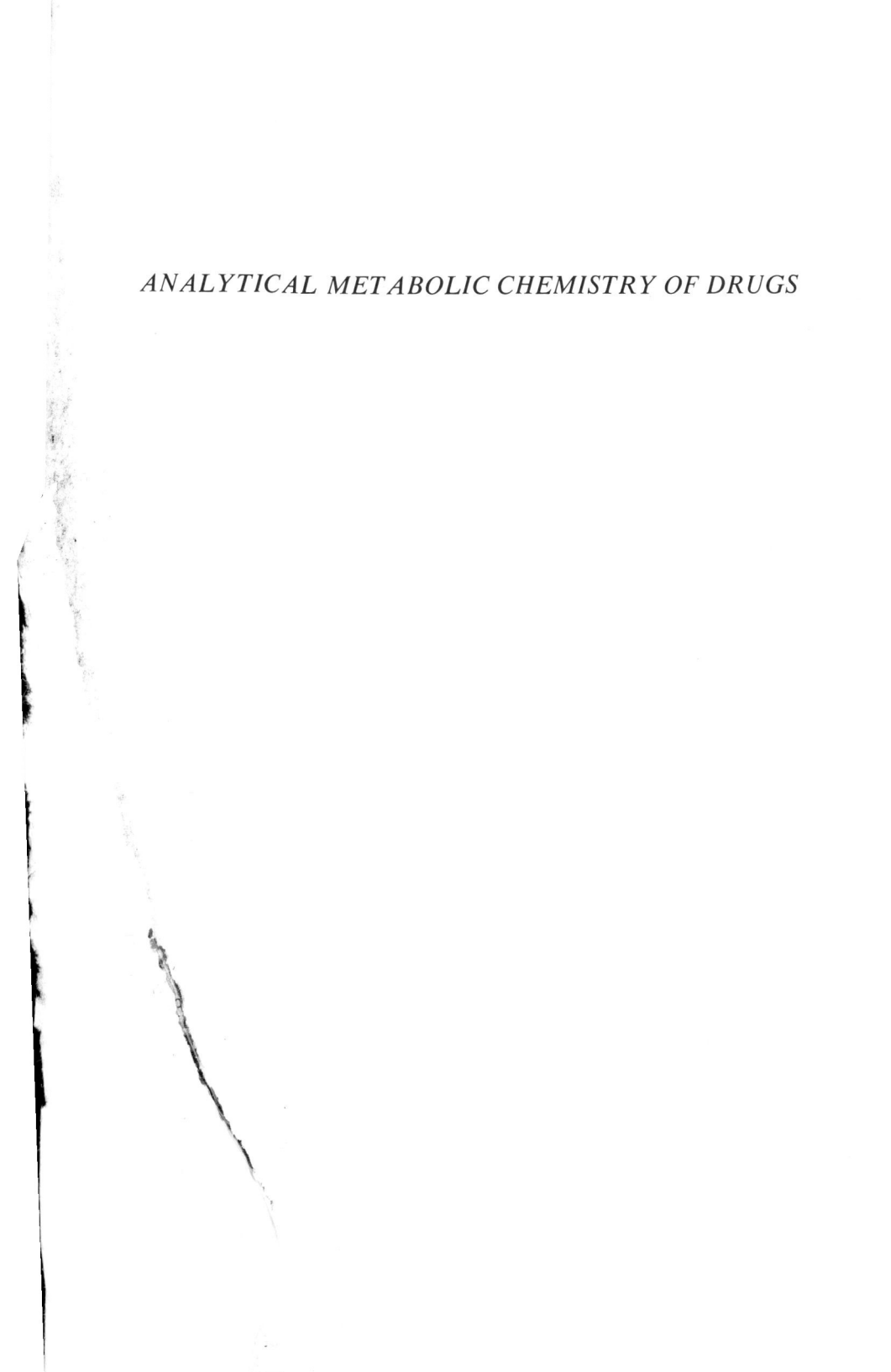

ANALYTICAL METABOLIC CHEMISTRY OF DRUGS

MEDICINAL RESEARCH:
A SERIES OF MONOGRAPHS

Consulting Editor: GARY L. GRUNEWALD

VOLUME 1: *Drugs Affecting the Peripheral Nervous System*, edited by Alfred Burger
VOLUME 2: *Drugs Affecting the Central Nervous System*, edited by Alfred Burger
VOLUME 3: *Selected Pharmacological Testing Methods*, edited by Alfred Burger
VOLUME 4: *Analytical Metabolic Chemistry of Drugs*, by Jean L. Hirtz
VOLUME 5: *Biogenic Amines and Physiological Membranes in Drug Therapy,
Edited by John H. Biel and Leo G. Abood*

In Preparation:

VOLUME 6: *Search for New Drugs, Edited by Alan D. Rubin*

ANALYTICAL METABOLIC CHEMISTRY OF DRUGS

JEAN L. HIRTZ
DIRECTOR
BIOPHARMACEUTICAL RESEARCH CENTER
CIBA-GEIGY
RUEIL-MALMAISON, FRANCE

Translation Editor

EDWARD R. GARRETT

This book was first published in France by Masson & Cie in 1968, under the title LES METHODES ANALYTIQUES DANS LES RECHERCHES SUR LE METABOLISME DES MEDICAMENTS, in the series "Monographies de Pharmacie," under the direction of Prof. M. M. Janot, and with a preface by Prof. J. R. Boissier.

1971

MARCEL DEKKER, INC., NEW YORK

COPYRIGHT © 1971 by MARCEL DEKKER, INC.

ALL RIGHTS RESERVED

No part of this work may be reproduced or utilized in any form or by any means, electronic or mechanical, including xerography, photocopying, microfilm, and recording, or by any information storage and retrieval system, without the written permission of the publisher.

Notice Concerning Trademark or Patent Rights: The listing or discussion in this book of any drug in respect to which patent or trademark rights may exist shall not be deemed, and is not intended as, a grant of, or authority to exercise, or an infringement of, any right or privilege protected by such patent or trademark.

MARCEL DEKKER, INC.

95 Madison Avenue, New York, New York 10016

LIBRARY OF CONGRESS CATALOG CARD NUMBER: 75-157838

ISBN NO.: 0-8247-1308-7

Printed in the United States of America

FOREWORD

The major intent of the author, Jean Hirtz, was to present the analytical techniques used in the identification and quantitative analysis of metabolites transformed from drugs in biological systems. He anticipated that it would be a proper companion to R. T. Williams' classic work on <u>Detoxication Mechanisms</u> (John Wiley & Sons, Inc., New York, 1959), delineating the metabolic patterns of various drugs. The work and effort involved in the pursuit and unearthing of this analytical metabolic chemistry from the hidden corners of manifold publications is obviously prodigious and has not been heretofore accomplished. Since I knew of the need of this book in our country, I was happy to volunteer to "Americanize" the translation. Incidentally, I have found that the "American" language is quite different than that spoken and written by the English.

In addition to enjoying the fruit of these Herculean labors, I have had other interests in seeing that the American language edition of this book was made available. Although it was not a major intent of the author, he has provided us with the most recent compilation of metabolic products derived from drugs and presented this information in an appropriate context. I have had occasion often to use it as a detailed exposé of the metabolism of drugs and drug classes for our own pharmacokinetic and biopharmaceutic studies. It has served us as a readily available summary of the significant work on delineation of metabolites, the advantages and limitations of the analytical methods used, and as a ready bibliography. I do know that Dr. Hirtz is collaborating with a special group in Paris, so his efforts in this field are continuing. I am hopeful that in the future reviews of analytical metabolic chemistry will be continued from where they end in this book.

The need for, and appreciation of, analytical metabolic chemistry of body fluids and tissue is vital to the appraisal of modern drug action, the design of dosage regimens, and the evaluation of bioavailability of drugs from their dosage forms.

In a practical sense, no new drug application can expect to be approved without knowledge and quantification of the metabolic pathways.

Preclinical and clinical studies to determine drug and metabolites demand these analytical methods. Some of the vital uses and needs of such methodology are summarized in the following.

The metabolites of a drug are present in and isolable from the blood, urine, and feces of the animal organism. Too frequently, the concentration in the blood may be too low to permit characterization of drug derivatives whose structure or chemical properties are unknown. The feces may be too difficult as an isolation source. The urine serves to concentrate physiologically those products of the body's molecular alterations in quantities sufficient for isolation and characterization.

The United States Food and Drug Administration has urged that isolation and characterization of metabolites be effected in several animal species. If at all possible, consistent with ethical considerations, metabolites in man should be considered at the same time. An accounting of 60-70% of the drug and its metabolites in the excreta would seem to be a reasonable initial aim.

Initial studies with radiochemically labeled drugs are advantageous. Chromatographic methods (paper, thin layer, column, gas, mass) can be employed for separation directly from the biological fluids and tissues, or from extracts made at various pH values with selected solvents of varying polarities. The separated components can be differentiated and quantified by radiochemical detection methods, which may also include autoradiography or thin-layer scraping, paper cutting, elution, combustion and monitoring, etc. Thus, "metabolic profiles" or "metabolic fingerprints" can be devised for each of the various animal species, even when the exact chemical structures of the metabolites are unknown. A minimum of three different solvent, column, etc., systems should be used for each chromatographic method.

The labeled original drug and labeled potential metabolites can be used as standards. The expert knowledge of the biochemist can be brought into play to predict most plausibly how the drug would be modified by the body. Such potential metabolites are considered on the basis of potential oxidations (aromatic and alkyl group hydroxylations, carboxylic acid formation from alkyl groups, oxidative deamination, O-dealkylation, N-dealkylation, sulfoxidation, and sulfur replacement by oxygen), reductions (of aldehydes and ketones to hydroxyls, of nitro and azo groups to amines, and disulfide conversion to sulfhydrils), hydrolyses (of some esters and amides), and conjugations of the drug and its other derivatives (glucuronidation, acetylation, methylation, and conjugation with sulfate, glycine, and glutamine). The total output of radioactivity can provide estimates of drug recovery. Subsequent to the controlled extractions with organic solvents, the aqueous phases can be heated with strong acids or subjected to incubations with splitting enzymes such as β-glucuronidase. The resultant solutions can be reextracted and rechromatographed, and the components monitored and detected. The kind and extent of conjugations can be evaluated on comparison of the data from pre- and postsolvolytic procedures. Dr. Hirtz has stated in his introduction that the choice in metabolic research between the use of labeled molecules or "pure" analytical methods, between "hot" or "cold" assays, depends on the natures of the drug and its metabolites and gives the relative advantages of each.

However, after identification of the metabolites and their relative amounts from the different species, accurate and sensitive "cold" analytical methods should be devised for human biological fluids and tissues. The use of radiolabeled drugs in pharmacokinetics may not be applicable to man because of potential toxicities and the fact that such studies may be used in the future as a diagnostic tool in the routine establishment of human pharmacokinetic profiles. A typical program for the statistical evaluation of the analytical methodology would follow the following patterns.

Let us consider an analytical system where the drug or metabolite is to be extracted from an adjusted or modified biological fluid or tissue, where the extract undergoes certain technological manipulations (concentration, chromatographic separation, elution, reextraction, etc.) and the components are subsequently analyzed (chromatographically, instrumentally, by color development, etc.).

The first factor to consider is the validity and reliability of a statistically evaluated dose-response relation between the concentration of the drug (or metabolite) in the final solution and the analytical measurement. A calibration curve must be prepared based on graded amounts of the drug in the assayable final solvent system to determine the reliability, sensitivity, and statistical error of the analytical system. A desirable calibration curve of instrumental, chemical, or biological response plotted against concentration C should be linear and pass through the origin. Blank solutions without drug should be run simultaneously to obtain the constant background effects that have to be subtracted from the total response to obtain the corrected response.

In many instances, such as in gas chromatography, it is wise to employ an internal standard with a separated signal to correct for ambient instrumental variation. A specific amount of this standard with a known dose-response relation should be added to the final drug and blank solutions.

As soon as the metabolites are known, it is desirable to obtain simultaneous assays for them and the drug by using similar and noninterfering analytical responses. Admixtures of drugs and metabolites and varying concentration levels of each should be assayed from the final solution. These assays should be statistically evaluated to determine that there are no anomalous effects of the concentration level of the one component on the response level of the other, that the linearity of the drug-response curves is maintained.

The second factor to consider is the validity and reliability of statistically evaluated dose-response relation between the concentration of the drug (or metabolite) in the biological fluid or tissue and the analytical measurement.

The biological substance to be analyzed (blood plasma, urine, etc.) is to be "spiked" with graded amounts of drug or metabolites. The pertinent treatments are performed on that fluid or tissue and the pertinent extractions, separations, or manipulations are conducted to produce the final solution to be analyzed. The internal standard may then be added to this solution and the response obtained.

Analysis of the drug or metabolites in whole blood deserves special consideration. Whole blood should be spiked with graded concentrations of drug or metabolite and the plasma (and/or serum) obtained by centrifugation (and/or protein precipitation) after appropriate heparinization, citration, or oxalation. When the analytical procedure is followed through and the assay corrected for the volume of whole blood to that of plasma (or serum), it can be operationally assumed that there is no significant partition of the drug between the plasma and the erythrocytes, or that there is no irreversible binding to denatured proteins.

Also, drug or metabolite should be analyzed in plasma (or serum) prepared from aliquots of blood as a function of time. Reproducible analyses would indicate no significant time-dependent partition into the erythrocytes or no irreversible binding to proteins. This data must be obtained in preparation for those pharmacokinetic studies where the dynamic distributive, metabolic, and excretory processes are construed to be dependent on the concentrations of the free or unbound materials in the plasma water.

FOREWORD

It is important to determine the extent of protein binding of the drug and its metabolites with specific reference to the variation of the fraction bound as a function of the concentration of the compound. Typical methods are by ultracentrifugation, ultrafiltration, and equilibrium dialysis.

When the metabolites are known and isolated, protein binding studies of these metabolites should be also conducted. It is wise to do joint studies on drug and metabolites to gain assurance that the presence of one does not modify the binding properties of the other. Such effects will modify the pharmacokinetics and optimum dosage regimens on repetitive or chronic dosing. If it is anticipated by the clinical pharmacologist that any other drug is to be administered jointly for therapeutic reasons, this drug and its metabolites should also be studied for their effects on the protein binding of the others.

Drug may be incubated with liver slices or liver microsomes freshly isolated from the pertinent animal species and the preliminary metabolic profiles elucidated. The use of liver perfusion techniques with subsequent isolation and identification of the metabolites from the perfusate is also feasible.

These techniques do not give the same quantitative distribution of metabolites as do in vivo studies. However, they are qualitatively similar and permit systematic studies of: (a) effects of substrate or metabolite concentration on the rate of metabolite production to indicate whether or not enzymic factors are rate limiting in drug pharmacokinetics; (b) effects and need of cofactors in enzymic modification of drugs (such possible cofactors as uridine-5'-diphosphoglucuronic acid salts in glucuronidation); and (c) effects of enzyme inhibitors on enzymic rates (such as bromsulfophthalein as a potent inhibitor of glucuronidation with liver homogenate).

Cannulation of the common bile duct and subsequent drug administration to the animal will permit collection of the bile to be characterized for metabolites and to determine the fraction of the drug and its metabolites (frequently as conjugates) removed from the body by this route.

Blood levels of drug or metabolites, or amounts of drug or metabolites excreted within a particular time interval, can be monitored at specified times after each repetitive drug administration for chronic toxicity testing, and should be relatively invariant with time if the phenomenon of enzyme induction does not occur. Depression in the blood-level values of drug on elevations in levels of metabolites may be attributed either to the drug inducing the enzymes for its own metabolism or to the possible competition of an accumulating metabolite for a protein binding site.

Cessation of drug administration for a time interval and observation of elevated blood levels with subsequent decay on reintroduction of chronic testing would be indicative of reversibility in enzymic induction.

Observations of such phenomena are vital at an early stage. This information must be used in the study and planning of human dosage regimens since scheduled dosing to cure disease must take into account increased rates of removal of active drug from the compartments of the body on chronic administration.

Edward R. Garrett
Graduate Research Professor
College of Pharmacy
J. Hillis Miller Health Center
University of Florida
Gainesville, Florida 32601

July 31, 1970

INTRODUCTION

Any foreign substance that invades a living organism in which its presence is not nutritionally beneficial to that organism is expelled relatively rapidly. This excretion is frequently preceded by important in vivo modifications of its chemical structure. The activity of the administered material is attenuated, modified, or enhanced by these transformations into metabolites, and the distribution, localization, and excretion of the administered drug are considerably changed by these transformations.

The interest in drug metabolism spans the last century, commencing with quinine (1869), salicylic acid (1877), and morphine (1883). The number of such studies on the metabolism of drugs has greatly increased within the last 25 years and today all new drugs are subjected to exhaustive studies.

The study of the metabolism of drugs is a matter of pharmacology and biochemistry, but its elucidation is first a problem in chemical analysis, characterization, and identification. When a minute amount of a drug is introduced into a living organism by diffusion through an appropriate body membrane, it is transported by the circulatory system, diluted in the various fluids of the body, adsorbed reversibly or irreversibly in various tissues, transformed, and excreted. The transformations may be subtle, leading to derivatives similar to the administered drug, or they may be extensive, resulting in complete degradation of the molecule. The minute amounts of compounds closely related to the administered drug must be recovered from the tissues and organs, separated from the normal biological components, and liberated from structures that have adsorbed or complexed them. These structures must be defined, and quantitative methods devised to elucidate the stoichiometry and sequence of the pharmacokinetic and metabolic processes. These are the fundamental procedures necessary for the quantification of the time courses of the drug and its metabolites in the body.

Analytical metabolic chemistry may be specifically characterized by the following facts:

(1) It applies to the analysis of compounds in mixtures of great formular analogy: the unchanged drug and its metabolites.

(2) It applies to amounts that are generally small, especially when very active substances or small animals are used; or, specifically, when the drugs have high potency, or when metabolic reactions are followed in a single organ or part of an organ.

(3) It applies to those conditions when the drugs or the derived metabolites are in complex biological systems where the possibility of drug complexes or binding exist.

(4) It applies to those circumstances when repetitive routine assays are foreseen. Specific instances may be when biological variation is to be determined, when the metabolic and pharmacokinetic processes are to be studied as functions of time or dose, or when these processes are affected by other simultaneously administered drugs, etc.

These are the stringent requirements that analytical metabolic chemistry must fulfill.

At the end of World War II, Brodie and Udenfriend [136] proposed a general procedure based on the observation that basic compounds administered to organisms are almost always metabolized into more polar derivatives with smaller oil/water partition coefficients. Advantage may be taken of these solubility differences to extract the drug and its derived metabolites preferentially and quantitatively. The isolated and purified compounds may be analyzed by any valid method; a particular favorite for amine-containing drugs was methyl orange complex colorimetry. Verification of the nature of the unchanged drug or its metabolites was indicated by comparison of the partition coefficients in various solvents and at several pH values of the extracted compound produced in vivo with those of an authentic sample.

About the same time, Craig and coworkers [229] developed countercurrent distribution equipment and techniques that permitted the fractionation of mixtures of metabolites. Identification was indicated by comparison of the distribution curves of the compound produced in vivo with that of an authentic sample. The time consuming nature of this process has limited utility and the more rapid techniques of paper and thin-layer chromatography supplanted it after 1950.

The use of radioactively labeled molecules may permit the monitoring of the administered drug and its metabolite in all parts of the organism with a high sensitivity. First, a radioactive atom must be introduced into the drug molecule. It may be introduced randomly or specifically. In the former case,

INTRODUCTION

the elucidation of the metabolite pathways may be confused. In the latter case, new and complex synthetic procedures may have to be devised with possible multiple labelings that are at wide variance with the synthetic methods described for the unlabeled analog. The synthesized labeled drug must be carefully purified. The concept of radiochemical purity must be added to that of chemical purity. Since impurities in the preparation may also be labeled, it is necessary to eliminate them completely to avoid subsequent errors in interpretation. The labeled compound may be sensitive to radiochemical degradation, which introduces a unique stability problem before the drug may be used. The distinct advantage of using labeled compounds is that on administration to the animal the labeled drug and the derived labeled metabolites may be readily detected and the total amounts readily quantified. The distribution and excretion of the drug and its metabolites in the organism may be readily monitored.

However, this monitoring is a gross measurement and does not give any information on the nature of the molecule containing the isotope. Thus, extractive or separative procedures must be developed and the identity of the separated compounds determined. Isotopic dilution may be applied when the corresponding authentic sample and an isolation method are available. Ordinary analytical procedures may also be used. The radioactivity of the products is helpful as it indicates their presence.

The use of labeled molecules does preclude exhaustive isolation and analysis to elucidate the in vivo transformation processes.

The use of "cold" methods of analytical and physical chemistry makes it possible to administer the unmodified drug. The problem reduces solely to the characterization of the derived metabolites. It is possible to consider a priori the likely metabolites and to elaborate proper techniques for their isolation and determination. There are modern methods of analysis for amounts at the levels encountered in metabolism studies. Gas-liquid chromatography can reach the nanogram level with flame ionization detectors; the electron capture detector has a sensitivity 10 to 100 times greater. A general procedure for the determination of a large number of metabolites in urine or organ extracts by gas-liquid chromatography has been published recently [240]. Thin-layer chromatography may characterize and analyze micrograms of substances. Mass spectrometry has quantitative requirements comparable to those of gas-liquid chromatography. The coupling of these two techniques provides a most powerful analytical tool. Ultraviolet and infrared spectrophotometry with its special attachments can allow assays of 1 μg or less of substance.

Modern analytical techniques and methods are able to fractionate, isolate, characterize, and determine the metabolites in biological media. The only insoluble problem with "cold" methods is the identification and quantification of in vivo transformation products that are compounds normally present in the organism. Then the only recourse is the use of labeled compounds.

The choice between the two pathways for analytical metabolic research — labeled molecules or pure analysis, "hot" or "cold" methods — depends on the nature of the drug and its metabolites. When the labeled molecule is easily prepared, when there is no danger in administration of the radioactive material to a human being, it will be advantageous to use radio tracer techniques. The occurrence of metabolites and their distribution in organs or excreta can be readily monitored if an appropriate isolation technique can be formulated in pursuance of the labeled compounds.

When the labeled molecule is difficult to synthesize or prepare, when equipment is not available, when the radioactivity necessary to administer is above the safety level, or when separation of drug and metabolites are difficult, physical and chemical analytical methods are appropriate.

The literature presents interesting and well-referenced reviews on the metabolic processes and the enzyme systems which govern them. To date there is no critical review or examination of the often elaborate analytical methods which have contributed to the understanding of these metabolic processes. In two basic books, <u>Chemistry of Drug Metabolism</u> by Fishman [328] and <u>Detoxication Mechanisms</u> by Williams [1008], there is little information on the analytical methods used. They are dispersed in the pharmacological or biochemical journals and have not been compiled.

Thus, it should be useful to collect in a condensed form a large number of these scattered analyses and data to make them accessible to the analyst. This is the origin and the purpose of this book, which was preceded by a partial bibliography published in 1967 [424].

This book is intended primarily for the analyst. It is not a treatise on drug metabolism and will consider metabolic reactions only so far as they are necessary for the understanding of the analytical methods. It is not a specialized treatise on analytical chemistry: the background and theory of

INTRODUCTION

every method that will be mentioned are known to those "experienced in the art." This book is intended to be a compendium of references and, above all, a guide to easy access to the literature.

This book is devoted to analytical methods, to all physical and chemical techniques that enable one to separate, purify, identify, and determine. It considers these methods in their application to metabolism, the transformations of drugs in the organism. Methods that apply only to studies of absorption, distribution, blood levels, etc., are outside its scope. It is intended to be limited to drugs, which may be defined as "substances used or intended to be used to modify or study a physiological system or a pathological state."

However, some drugs, such as vitamins and hormones, are substances normally present in the organism. Since it is impossible to distinguish between the literature pertinent to the normal metabolism of these products when their origin is endogenous from that referring to their transformations after exogenous administration for a therapeutic purpose, these two categories of substances have been deliberately disregarded.

This book is divided into 20 chapters. Each considers a class of drugs on the basis of their chemical similarities and ignores the differences in their pharmacological utility. This partition may be arbitrary but it is helpful in the consideration of the analytical chemistry.

International common names (proposed or recommended) are used when available. An empirical formula index is available for easy access to the textual discussion.

The R_f values for chromatographic separations are only approximate and are given as indications of sensitivity of separation rather than as identifications or characterizations. When an author indicated R_f values both for the authentic sample and the corresponding metabolite, only the R_f for the authentic compound is recorded. When the literature mentions deviations of the R_f values, the mean values are generally given.

When a chromatographic development system is described, the reference to a pure solvent indicates that it, or the most concentrated solution of it commercially available, was used. Otherwise the concentration is given. The solvent proportions are in volumes. Unless otherwise stated, paper chromatography is always descending, thin-layer chromatography ascending. When an author gives several chromatographic systems, these are separated by semicolons. If the author applies several systems successively, they are separated by commas.

When we mention the concentration of an acid for an hydrolysis, this is the final concentration in the sample and not the concentration of the added acid.

All references are brought together at the end of the volume. Should the number of authors exceed two, only the first one is cited.

The assistance of my friend Edward R. Garrett, Graduate Research Professor at the College of Pharmacy, University of Florida, Gainesville, USA, who assumed the tedious and time consuming task of correcting the English translation of this book, is very gratefully acknowledged.

Jean L. Hirtz

CONTENTS

Foreword		iii
Introduction		ix
1.	PHENOLIC ACIDS AND DERIVATIVES	1
	I. Salicylic Acid and Salicylates	1
	II. Acetylsalicylic Acid	4
	III. Salicylamide	6
	IV. p-Aminosalicylic Acid	8
2.	AMINES	13
	I. Alkylamines	13
	II. Monoarylalkylamines	14
	III. Diarylalkylamines	23
	IV. Other Amines	27
3.	AMINOPHENOLS, CATECHOLAMINES	31
	I. Aminophenols	31
	II. Catecholamines	36
4.	PHENOTHIAZINES	53
	I. Phenothiazine	53
	II. Chlorpromazine	54
	III. Promazine	66
	IV. Other Phenothiazines	70
	V. Thioxanthenes	72
5.	DIBENZAZEPINES AND BENZODIAZEPINES	77
	I. Dibenzazepines	77
	II. Benzodiazepines	81
6.	CARBAMATES	87
	I. Mephenesine and Related Compounds	87
	II. Meprobamate and Related Drugs	89
	III. Chlorphenesine Carbamate	97

	IV.	Ethinamate	101
	V.	Other Carbamates	105
7.	ANILIDES		107
	I.	Acetanilide	107
	II.	Phenacetin	110
	III.	Other Analides	113
	IV.	Thiocarbanilides and Derivatives	116
8.	BARBITURATES		121
	I.	General Analytical Methods	121
	II.	Barbiturates with Alkyl Substituents	123
	III.	Barbiturates with Cyclic Substituents	134
	IV.	Thiobarbiturates	140
9.	DERIVATIVES OF UREA, GUANIDINE, AND OTHERS		147
	I.	Open Chain Ureides	147
	II.	Hydantoins	151
	III.	Guanidine Derivatives	154
	IV.	Miscellaneous	156
10.	SULFONAMIDES		159
	I.	Paraaminobenzenesulfonamides	160
	II.	Other Sulfonamides	177
11.	IMIDES		181
	I.	Thalidomide	181
	II.	Glutethimide and Derivatives	183
12.	HYDRAZIDES AND HYDRAZINES		191
	I.	Isoniazid	191
	II.	Other Hydrazides and Hydrazines	198
13.	HETEROCYCLES WITH ONE NITROGEN		203
	I.	Pyrrolidine Derivatives	203
	II.	Oxazolidine Derivatives	204
	III.	Thiazoline Derivatives	206
	IV.	Pyridine Derivatives	206
	V.	Piperidine Derivatives	212
	VI.	Indole Derivatives	216
	VII.	Quinoline Derivatives	219

14.	HETEROCYCLES WITH TWO NITROGENS		229
	I.	Pyrazole Derivatives	229
	II.	Other Derivatives	236
15.	HETEROCYCLES WITH MORE THAN TWO NITROGENS		245
16.	HETEROCYCLES NOT CONTAINING NITROGEN		257
17.	ALKALOIDS		261
	I.	Phenanthrene Derivative Alkaloids	261
	II.	Xanthine Derivative Alkaloids	272
	III.	Reserpine	275
	IV.	Other Alkaloids	276
18.	ANTIBIOTICS		285
	I.	Chloramphenicol	285
	II.	Tetracyclines	286
	III.	Penicillins	288
	IV.	Erythromycin	291
	V.	Streptomycin, Kanamycin	292
	VI.	Fusidic Acid	293
	VII.	Puromycin	294
	VIII.	Griseofulvin	295
	IX.	Pristinamycin	296
19.	GLYCOSIDES		297
	I.	Digitalis Heterosides	297
	II.	Other Glycosides	302
20.	MISCELLANEOUS DRUGS		309
	I.	Sulfur-Containing Drugs	309
	II.	Metal-Containing Organic Drugs	313
	III.	Halogenated Drugs	317
	IV.	Miscellaneous Compounds	318
	BIBLIOGRAPHY		323
	AUTHOR INDEX		365
	SUBJECT INDEX		383

Chapter I

PHENOLIC ACIDS AND DERIVATIVES

Most of the drugs considered in this chapter are widely used and have been known a long time. The study of their metabolism began more than half a century ago.

I. SALICYLIC ACID AND SALICYLATES

It had been noted early that salicylic acid was excreted partly unchanged and partly conjugated with glycine. This conjugate, salicyluric acid, was first isolated in 1933. However, the possible formation of a glucuronic acid conjugate remained in doubt for a longer period of time.

In 1942 Kapp and Coburn [486], extracting with ethyl ether and petroleum ether, separated salicylic, salicyluric, and gentisic acids from urine. These reasonably pure metabolites were identified by melting point, bromometric and potentiometric titrations, UV spectra, and elemental analysis. Salicylglucuronic acid could not be purified but a butanol extraction of urine, previously treated with ethyl ether, gave a fraction with equimolecular amounts of glucuronic and salicylic acids. On the basis of these determinations, Kapp and Coburn were able to establish a complete metabolic balance.

These techniques have also been used by Smith et al. [862] although the salicylic acid was colorimetrically assayed as the ferric salt. Since the acid hydrolysis of salicyluric acid is first-order, it was difficult to determine it quantitatively as salicylic acid.

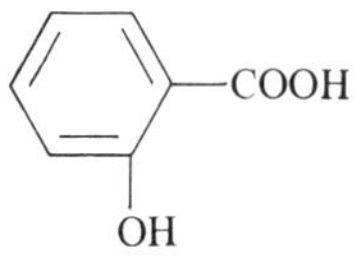

salicylic acid

A similar analytical procedure has been applied by Bray et al. [116]. These authors also [118] used paper chromatography (benzene : acetic acid : water, in various proportions — ascending). They detected the 2,3- and 2,5-dihydroxylated derivatives in rabbit urine, and salicyluric acid which had not been observed before [116].

In 1950, Schayer [802] studied ^{14}C-carboxyl labeled salicylate. Rat urine metabolites were separated by paper chromatography (butanol : pyridine : sodium chloride saturated aqueous solution 1 : 1 : 2 — ascending). Salicylic, salicyluric, and gentisic acids were detected by their R_f values and by isolation with a carrier. Two other metabolites were not definable.

Alpen et al. [22] administered the labeled product to the dog and man. Lyophilized urine was dissolved in 0.2M sodium citrate, pH 4.52, and fractionated by counter-current distribution with amyl alcohol as a secondary phase (24 transfers). Every fraction was assayed for radioactivity, salicylic acid (ferric salt), gentisic acid (hydroquinone), and glucuronic acid (naphtoresorcinol). Ultraviolet spectra and partition coefficients completed the identifications. The glucuronides fraction was analyzed on a paper "chromatopile" (0.5% of sodium acetate in 80% ethanol). Two compounds were detected, one of which had a free phenolic group.

Dumazert and El Ouachi [285] questioned whether salicylic acid metabolism proceeded through trihydroxylated derivatives. After acid hydrolysis of urine, they were able to show the occurrence of levulic acid by paper and cellulose powder column chromatography (ammonium hydroxide : ethanol). It was therefore likely that, before treatment, the urine contained 2,3,5-trihydroxybenzoic acid in a conjugated form.

Robinson and Williams [777] isolated the ester glucuronide from human urine. The glucuronide was precipitated as its lead salt and crystallized as its methyl acetyl derivative and identified by elemental analysis. Similarly, the ether glucuronide was obtained from rabbit urine.

Tsukamoto et al. [921] have described in detail the isolation of salicylic acid glucuronides. They were unable to get a crystalline product according to the procedure of Kamil et al. [480], which had been applied by Robinson and Williams [777]. The crude glucuronides, regenerated by hydrogen sulfide after separation as lead salts, were redissolved in ethanol and precipitated again by ethyl ether. They were redissolved in ethanol and precipitated with barium acetate. The precipitate was washed and redissolved in water. The aqueous solution was treated again with basic lead acetate and the glucuronide regenerated by hydrogen sulfide. A methylated and acetylated crystalline derivative, similar to the synthetic glucuronide, was obtained.

Tsukamoto et al. [922] also separated the two types of glucuronides, ester and ether, by paper chromatography (butanol : acetic acid : water 4 : 1 : 5; isopropanol : ethanol : water 4 : 1 : 2).

Schachter [796, 797] determined the ester glucuronides selectively through the preparation of their hydroxamate and the colorimetry of the ferric salt. The same reaction with hydroxylamine enabled the detection of the ester glucuronides after paper chromatography. Schachter showed the presence of an ester glucuronide of salicylic acid in human urine. This metabolite was hydrolyzed by 0.1N sodium hydroxide in 10 minutes at room temperature. β-Glucuronidase hydrolyzed it to an extent of 84% in 60 minutes at 38°C. In the system butanol : acetic acid 8 : 2, water saturated, the R_f of the metabolite on paper was 0.72. Since it is extractable with ethyl ether and the diglucuronide is insoluble in this solvent, the monoglucuronide structure was assigned.

Schachter also determined salicyluric acid after paper chromatography and the ether glucuronide by difference. Schachter and Manis [799] improved this method to determine successively the two types of salicylic acid glucuronides. After the ethyl ether extraction of salicylic hydroxamate described previously, the solution was deproteinized by perchloric acid. The ether glucuronide was hydrolyzed with 6N sulfuric acid and determined fluorometrically as sodium salicylate.

Schachter et al. [798] determined the specificity of this technique when applied to the study of salicylic glucuronide biosynthesis in vitro.

Gel filtration on Sephadex G-25 was also used to separate free salicylate from plasma protein bound salicylate. This technique, which has been applied by Potter and Guy [720], avoids solvent extraction.

MacIsaac and Williams [577], in a study on the metabolism of hydrazides and hydroxamic acids derived from salicylic acid, paper chromatographed salicylic acid metabolites (Table 1).

TABLE 1

Product	R_f in system:[a]		
	I	II	III
Salicylic acid	0.85	0.50	0.41
Salicyluric acid	0.45	0.08	0.06
Gentisic acid	0.66	0.25	0.14

[a] Solvent systems: I, ammonium hydroxide saturated propanol; II, methyl ethyl ketone : 2 N ammonium hydroxide 2 : 1; III, benzene : butanol : ammonium hydroxide 2 : 5 : 2.

II. ACETYLSALICYLIC ACID

The hydrolysis and excretion of acetylsalicylic acid as salicylic acid were observed as early as 1902. Subsequently (1946-1951), the presence of unchanged acetylsalicylic acid in blood or urine was argued.

Paul et al. [702] prepared ^{14}C-1-acetylsalicylic acid in 1951 and they briefly reported chromatographic evidence of some unchanged acetylsalicylic acid in blood.

Similarly, Mandel et al. [595] administered ^{14}C-acetyl- or ^{14}C-carboxyl-labeled acetylsalicylic acid. Acidified plasma was extracted with ethylene dichloride and the concentrated extract was chromatographed on paper (0.5% nitric acid — ascending). Acetylsalicylic and salicylic acids were detected and titrated (R_f 0.80 and 0.58).

The same chromatographic procedure was applied by Lange and Bell [536] who used fluorometric determinations after the administration of unlabeled acetylsalicylic acid.

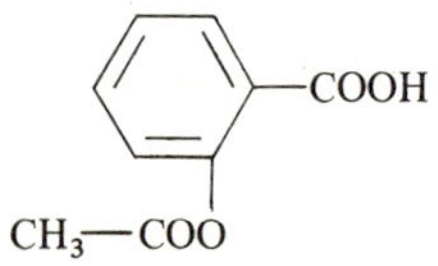

acetylsalicylic acid

Bedford et al. [71] determined total salicylic acid in urine after hydrolysis for 3 hours at 115°C of 4 ml of urine with 2 ml of 12N sulfuric acid. When 1 ml of 8N sulfuric acid was added to 2 ml of urine and heated 1 hour at 100°C, only salicylic and salicyluric acid glucuronides were hydrolyzed and salicyluric acid was not transformed to salicylic acid. Salicylic and salicyluric acids were extracted successively by carbon tetrachloride and chloroform and determined colorimetrically as ferric salts.

Cummings and King [233] used thin-layer chromatography on silica gel (benzene : ethyl ether : acetic acid : methanol 120 : 60 : 18 : 1) (Table 2). They demonstrated the presence of unchanged acetylsalicylic acid in urine and confirmed it by infrared spectrophotometry after preparative chromatography.

This thin-layer chromatographic procedure enabled Cummings et al. [235] to determine free salicyluric acid. After separation, it was eluted with methanol and assayed by spectrophotometry.

Hackenthal [390] studied salicylic acid excretion after injection of an aluminium acetylsalicylate suspension. Salicylic acid was determined colorimetrically and the sum of the salicylic and acetylsalicylic acids by UV spectrophotometry of the butyl ether extract.

The 3-methyl homologs of salicylic and acetylsalicylic acids have also been considered by Cummings and Martin [234]. Hydroxytoluic acid is

TABLE 2

Product	R_f (approximate)
Salicyluric acid	0.19
Gentisic acid	0.34
Acetylsalicylic acid	0.50
Salicylic acid	0.57

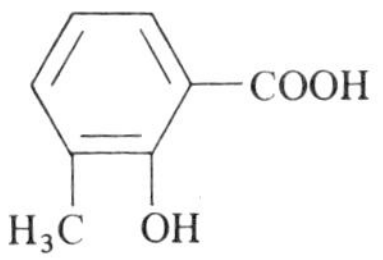

hydroxytoluic acid

determined by colorimetry of its ferric salt, as well as 3-methylacetylsalicylic acid after spontaneous hydrolysis. 3-methylacetylsalicylic acid can be extracted from urine at pH2 by chloroform. When the extract is washed with a 0.1% ferric nitrate aqueous solution, hydroxytoluic and 3-methylsalicyluric acids are eliminated. 3-Methylacetylsalicylic acid can also be separated by thin-layer chromatography on silica gel (benzene:ethyl ether:acetic acid: methanol 60:30:10:1). The same procedure was applied to separate 3-methylsalicyluric acid from hydroxytoluic acid after hydrochloric acid hydrolysis of the glucuronides.

III. SALICYLAMIDE

Mandel et al. [596] employed the ^{14}C-labeled compound obtained from radioactive salicylic acid. Urine concentrated by lyophilization was analyzed by paper chromatography (benzene:acetic acid:water 4:4:2) and showed two components: one is identified as salicylamide on the basis of its R_f and UV spectrum. The urine was fractionated by counter-current distribution (0.1M sodium citrate buffer pH 5.0:isoamyl alcohol — 24 transfers). The presence of the unchanged product was confirmed by its partition coefficient and by cocrystallization with salicylamide. A nonphenolic metabolite was also detected that could be hydrolyzed by β-glucuronidase.

Becher et al. [54] applied paper chromatography to acid hydrolyzed urine (butanol:ammonium hydroxide of various concentrations 1:1; butanol: acetic acid:water 4:1:5 with 10% of ethylene glycol — ascending). Salicylic acid, gentisic acid, and gentisamide were separated. Salicylamide and gentisamide glucuronides were analyzed by column chromatography. A salicylamide sulfate was also observed.

Foye et al. [336] also used paper chromatography to compare salicylamide metabolism by normal and stone-former subjects (Table 3).

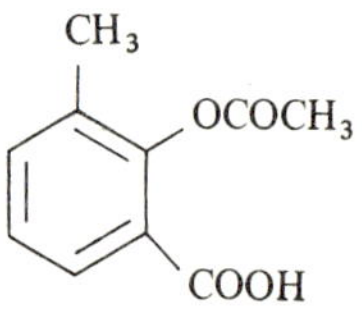

3-methylacetylsalicylic acid

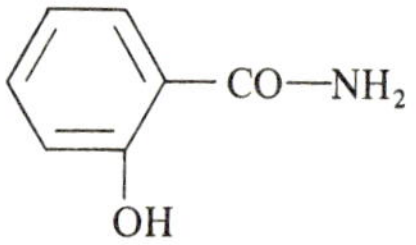

salicylamide

TABLE 3

Product	R_f in system:[a]	
	I	II
Salicylamide	0.61	0.96
Gentisamide	0.07	0.94
Salicylic acid	0.52	0.69
Gentisic acid	0.14	0.58
Salicyluric acid	0.12	0.40

[a] Solvent systems: I, 3N ammonium hydroxide saturated butanol; II, butanol : pyridine : sodium chloride saturated aqueous solution 1 : 1 : 2 (ascending chromatography).

Salicylamide glucuronide and sulfate, and salicylic acid glucuronide did not migrate. These glucuronides, as well as the sulfate, were synthesized. The separated components were eluted and determined by UV spectrophotometry.

Ethenzamide is the ethyl homolog of salicylamide and was investigated by Davison et al. [243]. Paper chromatography (benzene : acetic acid : water 4 : 4 : 2; t-butanol : acetic acid : water 78 : 6 : 16 — ascending) did not reveal any metabolite in the ethylene dichloride extract of urine. After acid hydrolysis, salicylamide appeared. It was present partly as sulfate, partly as glucuronide. The unchanged drug was not present in urine, but was determined in plasma by spectrophotometry at 288 mµ (salicylamide absorbs at 307 mµ and did not interfere). After hydrolysis, the presence of ethoxybenzoic acid was shown by paper chromatography of its dinitrophenylhydrazide

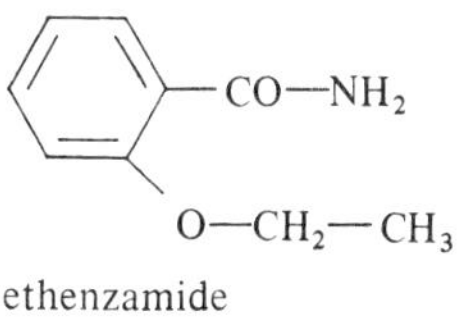

---

ethenzamide

and by gas-liquid chromatography (diethylene glycol succinate polyester column at 232° C). These experiments were completed by an in vitro metabolism study on ^{14}C-labeled material. The nature of the metabolites was confirmed by chromatography and isotopic dilution.

Cummings [232] demonstrated that o-carbamoylphenoxyacetic acid was a metabolite of etosalamide. This derivative was extracted from urine by ethyl ether at pH 2 and crystallized from its aqueous solution after acidification. It was analyzed by IR spectrophotometry, potentiometric titration, elemental analysis, and comparison with the authentic compound.

IV. p-AMINOSALICYLIC ACID

The first papers on p-aminosalicylic acid metabolism appeared in 1948. Venkataraman et al. [958] treated rabbit urine with normal, then basic lead acetate. The precipitate was regenerated by hydrogen sulfide and redissolved. The N-acetyl metabolite crystallized on cooling. It was identified by comparison of its melting point with the authentic sample. The unchanged drug was extracted from the mother liquors by ethyl ether.

Way et al. [980] fractionated urinary metabolites by counter-current distribution (isoamyl alcohol:1.6M acetate buffer pH 3.40 — 16 transfers). In every fraction, free and conjugated amines were determined by the Bratton and Marshall procedure. Five amino derivatives were found, two of which were conjugated. Unchanged drug, p-aminosalicyluric acid, and N-acetyl-p-aminosalicylic acid were characterized according to their partition coefficients.

Way et al. [978] also used paper chromatography (dioxane:water 5:1; butanol:ethanol:3 N ammonium hydroxide 4:1:5) on urine concentrated by lyophilization or on an ether extract of urine. The unchanged drug was

----

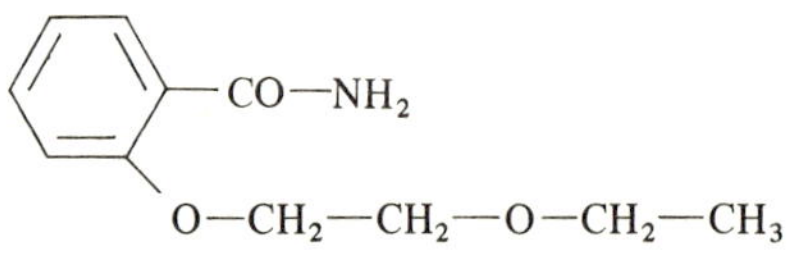

etosalamide

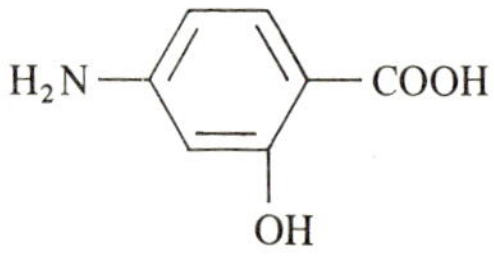

p-aminosalicylic acid

characterized by its R_f and partition coefficients. N-Acetyl-p-aminosalicylic acid was isolated by ethyl ether extraction, crystallized, and compared to the authentic sample on the basis of its melting point and UV spectrum. p-Aminosalicyluric acid was isolated by counter-current distribution, precipitated by bubbling gaseous hydrochloric acid from its ether solution, and identified by elemental analysis, melting point, and UV spectrophotometry. Small amounts of m-aminophenol and four other metabolites were revealed by paper chromatography.

Kawamata and Kashiwagi [493] used counter-current distribution (ethyl acetate: 0.5M citrate buffer pH 3.6 — 30 transfers) of urine. They demonstrated the presence of three free and two conjugated amino compounds. The urine was also chromatographed on paper [(I) butanol : sodium chloride saturated aqueous solution: ammonium hydroxide: pyridine 40 : 25 : 15 : 20; (II) water : phenol 1 : 4; (III) butanol: acetic acid: water 4 : 1 : 2]. The various metabolites were characterized by their partition coefficients, R_f, UV spectra, and chemical reactions.

Kawamata and Hiratani [492] demonstrated the presence of p-aminosalicylglutamine in rat urine by paper chromatography (systems II and III above).

Tsukamoto and Yamamoto [928] determined the major metabolite — N-acetyl-p-aminosalicylic acid — as well as small amounts of glucuronide by chemical procedures before and after hydrolysis. They also separated metabolites by paper chromatography (butanol: acetic acid: water 4 : 1 : 5).

Subsequently, these workers [929] isolated the ester glucuronide of p-aminosalicylic acid according to the Kamil et al. procedure [480]. The purified glucuronide was methylated and acetylated and then crystallized. It was further compared to the synthetic glucuronide in regard to melting point and IR spectrum. In the course of this isolation a substance insoluble in ethanol was separated. It had the same R_f on paper as the glucuronide obtained in administering m-aminophenol to the rabbit. The identification was confirmed by elemental analysis and hydrolysis. Moreover, paper chromatography (butanol: acetic acid: water 4 : 1 : 5) revealed p-aminosalicyluric acid, m-aminophenol sulfate, and the ether glucuronide of p-aminosalicylic acid.

The two types of p-aminosalicylic acid glucuronide were separated by paper chromatography (Tsukamoto et al. [922]; Table 4).

TABLE 4

Product	R_f in system:[a]	
	I	II
p-Aminosalicylic acid ester glucuronide	0.28	0.58
p-Aminosalicylic acid ether glucuronide	0.17	0.33
m-Aminophenol glucuronide	0.14	0.21

[a] Solvent systems: I, butanol:acetic acid:water 4:1:5; II, ethyl acetate: acetic acid:water 5:2:2.

The same authors [922] isolated the pure ester glucuronide in the free state. The glucuronides, precipitated as lead salts and regenerated by hydrogen sulfide, were chromatographed on a cellulose powder column (ethyl acetate:acetic acid:water 5:2:2). The ester glucuronide was obtained in an eluted fraction and crystallized. It could also be extracted from the crude glucuronides with ethyl acetate and crystallized from its aqueous solution. It was then compared with the synthetic glucuronide after methylation and acetylation as stated previously.

Nakao [660] has also studied in detail the metabolism of p-aminosalicylic acid. The urine was concentrated and desalted by adding ethanol and chromatographed on paper in two directions [(I) butanol : acetic acid : water 4:1:5; (II) methanol:benzene:butanol:water 2:1:1:1] (Table 5). Nine excretion products were identified on the basis of their reactions on paper, their behavior toward hydrolysis, and their UV spectrum after elution. The sulfate was characterized using the ^{35}S-labeled molecule.

Nakao and Yanagisawa [661] combined paper chromatography with acidic and enzymic hydrolysis and partial isolations through lead acetate precipitation in their study of the glucuronides.

Yoshinari [1035] studied the paper chromatography of p-aminosalicylic acid metabolites. The best solvent system was the mixture (butanol:water: acetic acid 5:4:1). A quantitative elution was obtained by 0.1N sodium hydroxide to determine the separated components.

TABLE 5

Product	R_f in system: I	R_f in system: II
N-Acetyl p-aminosalicylic acid	0.82	0.60
p-Aminosalicylic acid	0.79	0.50
p-Aminosalicyluric acid	0.46	0.38
N-Acetyl p-aminosalicylic acid ester glucuronide[a]	0.38	0.55
N-Acetyl p-aminosalicylic acid ester glucuronide[a]	0.33	0.35
p-Aminosalicylic acid ester glucuronide	0.28	0.35
Unidentified sulphate	0.28	0.32
Unidentified glucuronide	0.19	0.22
p-Aminosalicylic acid ether glucuronide	0.14	0.12

[a] As p-aminosalicylic acid has one hydroxyl and one carboxyl group, there are two possible ester glucuronides.

Kakemi et al. [476] have been especially interested in the determination of p-aminosalicyluric acid in urine in the presence of the other metabolites. p-Aminosalicylic acid and its acetyl derivative were decomposed by heating into m-aminophenol which was then colorimetrically determined with diazotized p-nitroaniline. Under these conditions, p-aminosalicyluric acid does not react.

Chapter II

AMINES

The aliphatic mono- and diaryl amines and miscellaneous amino compounds will be successively examined.

I. ALKYLAMINES

Butynamine is active as a hypotensor and its N-demethylation has been studied by MacMahon and Easton [585] in rats and man using the N-^{14}CH$_3$-labeled drug. In addition to the measurement of the excreted ^{14}CO$_2$, thin-layer chromatography on silica gel (butanol:acetic acid:water 3:1:1) was applied to separate butynamine and N-desmethylbutynamine.

The biotransformation of the two isomers (+) and (-) of ethambutol has been examined by Peets and Buyske [704] who used derivatives ^{14}C labeled on the two carbons of ethylenediamine.

Crude dog urine was fractionated by counter-current distribution (butanol:acetic acid:water 4:1:5; butanol:ammonium hydroxide:water 4:1:5 — 220 transfers). Two metabolites were revealed in addition to the unchanged drug. Since the presence of urinary salts modified the displacement of the compounds during the distribution, the identification cannot be certain. Paper chromatographic assay was done for confirmation.

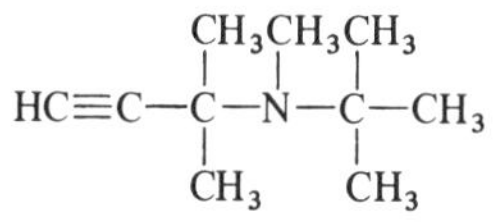

$$HC\equiv C-\underset{\underset{CH_3}{|}}{\overset{\overset{CH_3}{|}}{C}}-\underset{}{\overset{CH_3}{|}}{N}-\underset{\underset{CH_3}{|}}{\overset{\overset{CH_3}{|}}{C}}-CH_3$$

butynamine

$$OH-CH_2-\underset{\underset{C_2H_5}{|}}{CH}-NH-CH_2-CH_2-NH-\underset{\underset{C_2H_5}{|}}{CH}-CH_2-OH$$

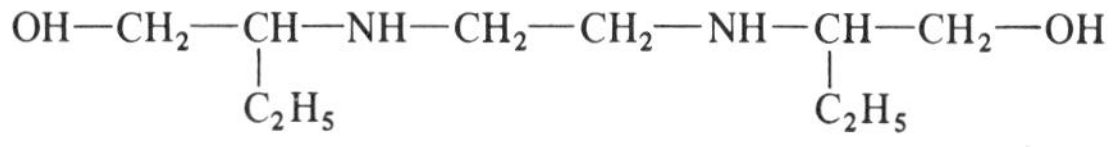

ethambutol

The first metabolite was identical with synthesized 2,2'-ethylenediiminodibutyric acid. The metabolite was paper chromatographically identical with the synthetic compound in five solvent systems. The hydroxamic derivatives of the synthetic reference and the metabolite had the same R_f values. The second metabolite appeared to be an aldehyde and gave the first on oxidation. Hydrogen peroxide treatment of ethambutol also yielded two products that corresponded to these two metabolites.

This study was completed with enzymic assays. Ethambutol was incubated with various amounts of alcohol dehydrogenase in the presence of DPN at pH 8.0. The DPNH production was followed by measuring the UV absorption at 340 mµ. Also, rat livers were homogenized in a 0.25M sucrose solution, then centrifuged for 15 minutes at 9,000g. The supernatant was centrifuged for 60 minutes at 105,000g. A sediment of microsomes and a supernatant containing the soluble enzymes were thus obtained. When ethambutol was incubated with either of these fractions in the presence of TPN, an aldehydic compound was formed. Its behavior was consistent with the second metabolite in counter-current distribution and paper chromatographic experiments. For these in vitro assays ethambutol was tritiated in the gas phase and the specific radioactivity was higher than that of the ^{14}C-labeled derivative.

Peets et al. [705] considered the metabolism of ethambutol in man using the ^{14}C-labeled drug. The urinary metabolites were separated by counter-current distribution of both crude and lyophilized urine (butanol:ammonium hydroxide:water 4:1:5 — 200-220 transfers). They were identified by paper chromatography (butanol:acetic acid:water 4:1:5; butanol:ammonium hydroxide:water 4:1:5; isopropanol:ammonium hydroxide:water 4:1:5). The two metabolites found were the same as those excreted by the dog.

II. MONOARYLALKYLAMINES

The majority of studies under this category relate to amphetamine. Richter [761] monitored urinary excretion in 1938 by colorimetry with picric acid.

Jacobsen and Gad [451] in 1940 applied this technique, after previously separating amphetamine by steam distillation. Harris et al. [405] adsorbed the interfering substances on magnesium oxide. Beyer and Skinner [84] formed an assayable colored complex of amphetamine with p-nitrobenzene diazonium chloride. McNally et al. [589], after purification and distillation steps, determined amphetamine in urine, blood, and tissues by this method. Keller and Ellenbogen [500] used the methyl orange method of Brodie et al. [139].

Axelrod undertook the first large scale studies on amphetamine and methamphetamine metabolism [36] in addition to studies on ephedrine and norephedrine [35]. The methods were the same in both cases. The bases were extracted by benzene from the alkalinized urine or plasma and analyzed by methyl orange colorimetry. Hydroxyamphetamine was extracted at pH 9-10 by ethyl ether, hydroxyephedrine (or hydroxynorephedrine) by isoamyl alcohol. The first compound was determined colorimetrically by nitroso-β-napthol; the two others by 4-aminoantipyrine and potassium ferricyanide. Axelrod isolated and identified the various products, as well as their p-hydroxylated metabolites, by counter-current distribution. The nature of the excretion products was elucidated unequivocally by paper chromatography for the amphetamine (butanol:acetic acid:water 5:4:1), or elemental analysis and UV, and IR spectrophotometry for the ephedrine.

With the same analytical procedures Axelrod [37] undertook the study of the in vitro metabolism of amphetamine. He was also able to demonstrate phenylacetone formation by obtaining its dinitrophenylhydrazone.

Asatoor et al. [32] investigated the biotransformations of amphetamine in man and the rat by paper chromatographies combined with methyl orange determinations of the unchanged drug. The p-hydroxyamphetamine was extracted with ethyl ether from the alkaline medium and subsequently assayed colorimetrically with nitroso-β-naphtol. The dinitrophenyl derivative of amphetamine, prepared according to the procedure of Asatoor and Kerr [33], was chromatographed on paraffin oil-impregnated paper by Asatoor's procedure [30]. Amphetamine and p-hydroxyamphetamine were also separated

---

amphetamine

methamphetamine

directly by bidimensional paper chromatography (I, isopropyl alcohol : ammonium hydroxide : water 8 : 1 : 1 — II, butanol : acetic acid : water 12 : 3 : 5). The phenolic acids were separated by the bidimensional chromatographic procedure of Armstrong et al. [29]. Phenylacetone was precipitated as the dinitrophenylhydrazone after El Masry et al. [307] and chromatographed on paper in the reversed phase system of Asatoor [31].

Young and Gordon [1037] followed the fate of ^{14}C-labeled amphetamine in the rat brain as a function of time. Paper chromatography in various solvent systems showed that the radioactivity that appears in brain between the 30th and the 60th minute after injection was essentially due to unchanged amphetamine. The absence of p-hydroxyamphetamine was demonstrated by the nonradioactivity of the fraction extracted by ethyl ether at pH 9.

Ellison et al. [305, 306] examined the comparative metabolism of amphetamine in man, rat, dog, and monkey using the ^{14}C label in the α-position. The unchanged product and its metabolites were separated by paper chromatography (88% formic acid : isoamyl alcohol : tertioamyl alcohol : water 2 : 5 : 5 : 10). Three metabolites were identified by their R_f in various solvent systems and their IR spectra. They were free p-hydroxyamphetamine and p-hydroxyamphetamine conjugated to glucuronic acid and hippuric acid in small amounts. The glucuronide was quantitatively hydrolyzed by β-glucuronidase in 24 hours at 37°C and pH 5.

Similar studies were conducted in man by Baltes et al. [48].

In 1966 Dring et al. [278] performed a detailed comparison of amphetamine metabolism in man, rat, rabbit, and dog. They used a labeled compound (α-methyl-[β-^{14}C]-phenethylamine), and separated the metabolites by paper chromatography. Details of the procedure were not given. The identifications were achieved by reverse isotopic dilution. The conjugates were hydrolyzed in heated urine with the same volume of 10N hydrochloric acid for 2 hours at 100°C. They detected and determined phenylacetone, 1-phenylpropan-2-ol, and benzoic acid, as well as amphetamine and its p-hydroxylated derivative. These compounds accounted for 56-82% of the administered dose in the various animal species.

Eberhardt and Debackere [289] extracted amphetamine or methamphetamine, as well as their metabolites, from urine with the mixture methyl acetate : ethyl ether at pH 9-10. They used thin-layer chromatography on silica gel (dimethylformamide : ethyl acetate 1 : 9).

Amphetamine can be detected in horse urine, saliva, or blood by thin-layer chromatography on silica gel GF 254 with the mixture butanol:acetic acid:water as developing solvent. Debackere and Massart-Lëen [246] devised this procedure and also studied the metabolism of both enantiomers of amphetamine in vitro in accordance with Axelrod's procedure [36].

Another metabolic reaction of amphetamine was observed in the rat by Goldstein and Anagnoste [375]. (+) p-Hydroxyamphetamine underwent a β-hydroxylation into (+) p-hydroxynorephedrine. After the administration of ^{3}H-labeled amphetamine to the rat, the organs were homogenized in 0.4N perchloric acid. The amines were retained on a Dowex 50 X-8 column and eluted by 2N hydrochloric acid after reference substances had been introduced. p-Hydroxynorephedrine was identified by paper chromatography (toluene:ethyl acetate:methanol:water 9:1:5:5).

The gas-liquid chromatography of amphetamine, like any amine, presents various difficulties: peak tailing and partly irreversible adsorption on the support. Derivatization is recommended. Horning et al. [432] and Brooks and Horning [140] have studied such methods.

Beckett and Rowland [62] steam distilled amphetamine from alkalinized urine into an hydrochloric acid solution. Amphetamine hydrochloride was separated from the ammonium chloride by extractions with chloroform. The chloroform had been purified by refluxing for 2 hours with 1% of amphetamine base and stabilized by 2% of ethanol to avoid the carbonyl compounds, which could react with amphetamine. The amphetamine was liberated by triethylamine and changed into its ene-amino derivative in acetone containing 10% of water. Gas-liquid chromatography on a column of Celite 545 impregnated with 10% of polyethylene glycol 6000 and 5% of potassium hydroxide permitted determination of 1 to 6 µg of amphetamine per ml of urine with a 100 ± 5% yield. An internal standard was used.

This method was improved and simplified by the same authors [63]. The slightly acid urine was treated with ethyl ether and then made strongly alkaline, and the amphetamine was extracted into ethyl ether. The concentrated solution was determined directly by gas-liquid chromatography. In situ treatment with hexamethyldisilazane with an alkalinized support gave symmetrical peaks and did not necessitate ene-amino derivativization. No interference was observed. The standard curve was linear between 0.1 and 10 µg/ml of urine. The yield was 100 ± 5% and the method was sensitive up to 0.1 µg/ml.

Amphetamine may be easily identified and separated from methamphetamine by making its ene-amino derivative. The retention time was increased by 10% whereas that of methamphetamine remained unchanged. The direct separation of an amphetamine : methamphetamine mixture was not very satisfactory. β-Phenylethylamine was also separated from amphetamine and methamphetamine as the ene-amine; the retention time of the derivatives was 60% higher than the respective base.

These methods have been extensively applied by Beckett et al. to the study of amphetamine metabolism and its excretion as a function of urinary pH. They showed, incidentally, that nicotine in smoking subjects gives a gas chromatographic peak distinct from amphetamine, although both compounds would be determined by the methyl orange colorimetric method [64].

A similar method was published by Beckett and Wilkinson [65] for ephedrine and related compounds. The reaction with acetone enabled them to separate ephedrine and pseudoephedrine which had the same retention time (t_R) as the bases (Table 6).

TABLE 6

Product	t_R in minutes	
	Base	Ene-amine
Ephedrine	8.2	4.0
Pseudoephedrine	8.2	3.6
Norephedrine	11.3	4.0
Norpseudoephedrine	10.5	4.2
Methylephedrine	6.8	6.8

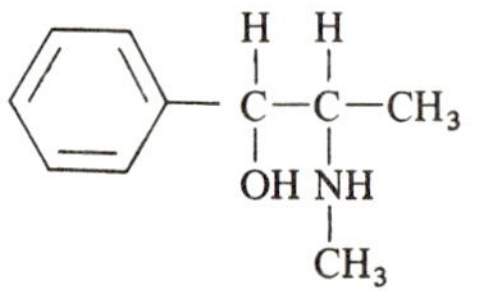

ephedrine

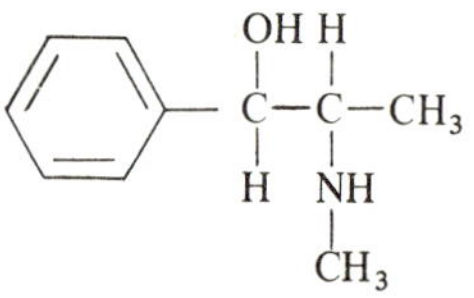

pseudoephedrine

Thus, it was possible to determine 5 to 100 µg of base per ml of urine.

Thompson and Smith [902] were able to obtain a satisfactory amphetamine peak by injecting the acidified urine directly into the chromatograph. The analytical column contained Chromosorb W impregnated with 10% of 85% potassium hydroxide and 10% of Apiezon L. The injection was made in a small fore-column containing Chromosorb W impregnated with 20% of 85% potassium hydroxide which is cleaned out after each analysis with one injection of 4 µl of concentrated ammonium hydroxide.

Hall et al. [393] extracted amphetamine or methamphetamine with carbon disulfide from urine alkalinized with ammonium hydroxide. The amine is thus converted into its isothiocyanate. This derivative was chromatographed in the gas phase at 118°C on a column packed with Chromosorb G, acid washed, dimethylchlorosilane treated, and impregnated with 2% of silicone SE-30.

Cartoni and de Stefano [195] extracted amphetamine and methamphetamine from rat urine with ethyl ether at alkaline pH values. The two compounds were separated and determined by gas-liquid chromatography (capillary column coated with Carbowax) at 120°C. The limit of detection was 1 µg with less than 5% error. On methamphetamine administration, 5.5% was renally excreted unchanged and 1% was excreted as amphetamine.

Gordis [379] has succeeded in separating the two optical isomers of amphetamine by gas-liquid chromatography. He used trifluoroacetyl-L-prolyl chloride as a resolving agent. The benzene solution of racemic amphetamine was heated 1 minute with this reagent, then neutralized by tributylamine. The isomers were separated on a column of Gas-Chrom P impregnated with 3% of silicone SE-30 at 175°C. It was also possible to cause derivative formation by mixing the reagent and the amphetamine solution in the syringe just before injection. No racemization occurred after

---

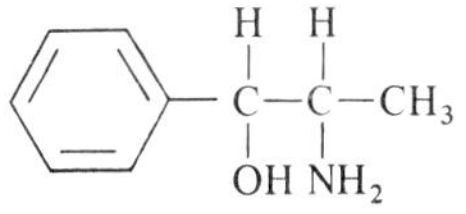

norephedrine

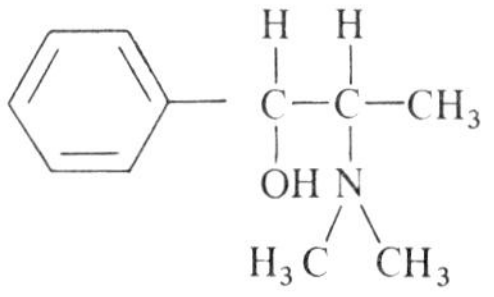

methylephedrine

administration of either isomer of amphetamine to man and to the rat.

Eberhardt et al. [290] investigated <u>phentermine</u> and its metabolites in a suicide case. They extracted the urine with ethyl ether and thin-layer chromatographed on silica gel (dimethylformamide : ethyl acetate 1 : 9).

Dubnick et al. [280] examined the distribution of <u>chlorphentermine</u> in the mouse by the methyl orange method, by distribution between 0.1M phosphate buffer pH 7.4 and benzene and by paper chromatographies. They demonstrated the unchanged product to be concentrated in brain.

The metabolism of <u>mephentermine</u> has been studied by Walkenstein et al. [965] by means of paper chromatography of the labeled drug. A somewhat poor separation is achieved in the solvent system (butanol : acetic acid : water 4 : 1 : 5)

		R_f
○	Mephentermine	0.76
○	p-Hydroxymephentermine	0.78
○	Desmethylmephentermine	0.79
○	p-Hydroxydesmethylmephentermine	0.68

The components are best characterized by the colors developed with bromocresol green, diazotized sulfanilic acid, and diazotized p-nitroaniline.

The <u>fenfluramine</u> metabolism investigation by Duhault and Fenard [283] also involved paper chromatography. The metabolites were extracted with chloroform from alkalinized urine or plasma. The developing solvent was

---

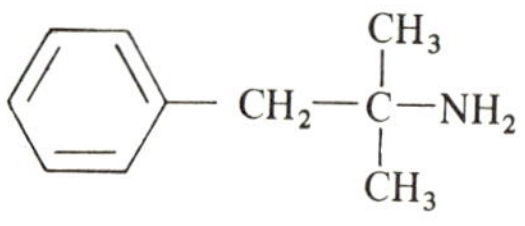

phentermine

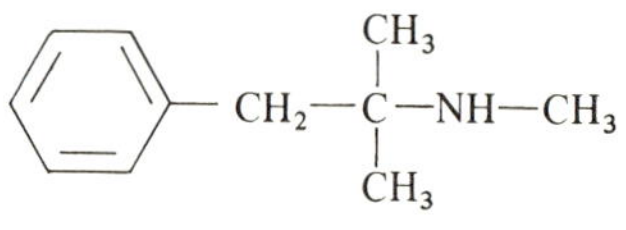

mephentermine

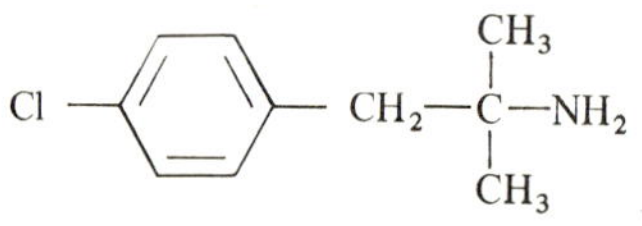

chlorphentermine

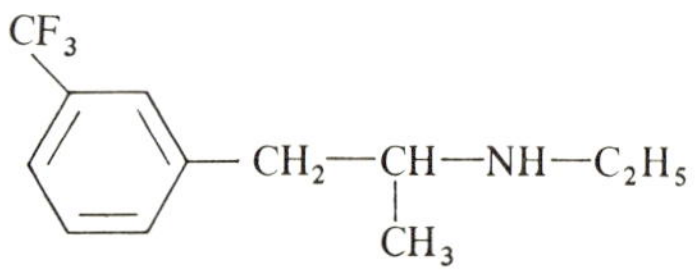

fenfluramine

the mixture toluene : butanol : acetic acid : water 2 : 2 : 1 : 1. Various reagents were used for the detections to distinguish between components of close R_f (Table 7) and included bromocresol green, Dragendorff, ninhydrin, sodium nitroprusside + acetaldehyde (primary amine reagent), and sodium nitroprusside + acetone (secondary amine reagent).

The metabolism of tranylcypromine has been compared by Alleva [19] to amphetamine by means of paper chromatography. The two products, ^{14}C labeled on the α-carbon of the benzene ring, were administered to rats. The urine was chromatographed directly. Six radioactivity peaks were recorded in the system (isoamyl alcohol : tertiary amyl alcohol : water : formic acid 5 : 5 : 10 : 2). The major peak (R_f 0.88) corresponded to hippuric acid, with trace amounts of benzoic acid, and was developed in butanol saturated with 1.5N ammonium hydroxide-ammonium carbonate buffer. Hippuric acid was confirmed by ether extraction, paper chromatography, and formation of its anilide. In this method the glucuronides of the hydroxylated derivatives do not migrate and were not studied. This investigation was previously aimed at obtaining evidence on the opening of the cyclopropane ring.

The urinary excretion of various anorectic phenylalkylamines has been investigated by Opitz and Weischer [694]. These amines were converted into

TABLE 7

Product	R_f
Fenfluramine	0.84
m-Trifluoromethylphenylisopropylamine	0.79
N-Ethylphenylisopropylamine	0.80
Amphetamine	0.71
p-Hydroxyamphetamine	0.64

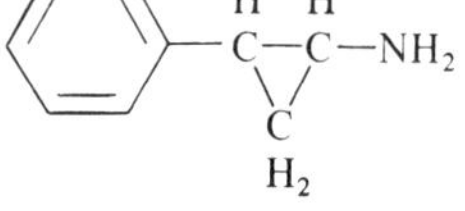

tranylcypromine

their dinitrophenyl derivatives, extracted into cyclohexane and separated from the dinitrophenyl derivatives of the normal urinary components by thin-layer chromatography on silica gel (xylene and several consecutive developments). The separated products were then eluted by methanol and determined spectrophotometrically. p-Hydroxyphentermine was separated in the same way from the unchanged drug by the mixture benzene:chloroform 5:1.

The paper or thin-layer chromatography of sympathomimetic amines, as of catecholamines, may be perturbed by the multiple spot formation phenomenon. This question has been the subject of definite studies and will be discussed in Chapter III, "Aminophenols, Catecholamines."

Brodie et al. [134] in 1948 undertook the study of the simple metabolism of procaine. They extracted the unchanged procaine with ethylene dichloride, reextracted it into acid solution, and determined it colorimetrically by the diazo coupling reaction. The metabolites were diethylaminoethanol and p-aminobenzoic acid and were identified by comparing their partition coefficients in ethylene dichloride:water at various pH values with those of the authentic compounds. The first diethylaminoethanol was determined by the methyl orange procedure. p-Aminobenzoic acid was colorimetrically determined after hydrolysis by the diazo coupling reaction. In vitro experiments demonstrated an enzyme in plasma which rapidly split procaine.

The diazotation method was later used by Morvillo and Garattini [646]. They separated the colored derivatives obtained with p-aminobenzoic acid, p-aminohippuric acid, and procaine by extracting the aqueous solution with ethyl ether at pH 2 and then at pH 10. The acetylated conjugates were hydrolyzed by 2N hydrochloric acid for 1 hour at 100°C.

Procaine metabolites were analyzed by paper electrophoresis according to Makisumi et al. [592].

Procaine ^{14}C labeled on the carboxyl was administered to the rat by Shlyakhman et al. [841]. The urine was subjected to paper chromatography

$$H_2N-\langle\rangle-COO-CH_2-CH_2-N\langle{}^{C_2H_5}_{C_2H_5}$$

procaine

(butanol : water : formic acid 75 : 20 : 5) and revealed four metabolites. Eight radioactive components were found when labeled glycine and unlabeled procaine were administered. Four of these corresponded to those stated previously.

Kalser et al. [478] were interested in the biliary excretion of procaine and its metabolites and employed perfusion of an isolated liver with a ^{14}C-labeled drug. Five components were separated by chromatography (conditions not given) from the bile. They included procaine, and p-aminobenzoic and p-aminohippuric acids.

III. DIARYLALKYLAMINES

Hespe et al. [421] used orphenadrine, monotritiated on CH, in their metabolism and excretion studies in the rat. Urine, bile, and tissues were alkalinized and extracted for 4 to 8 hours with ethyl ether, ethylene dichloride, or heptane. The extracts were chromatographed on paper (benzene : water : butanol : acetic acid 9 : 9 : 1 : 1) or on silica gel thin layer (butanol added with 2% of 25% ammonium hydroxide) (Table 8).

TABLE 8

Product	R_f on: Paper	R_f on: Thin layer
Orphenadrine	0.85-0.95	0.50-0.55
N-Desmethylorphenadrine	0.85-0.95	0.30-0.35
N,N-Didesmethylorphenadrine	0.85-0.95	0.38-0.43
2-Methylbenzhydrol	0.30-0.50	0.65-0.75

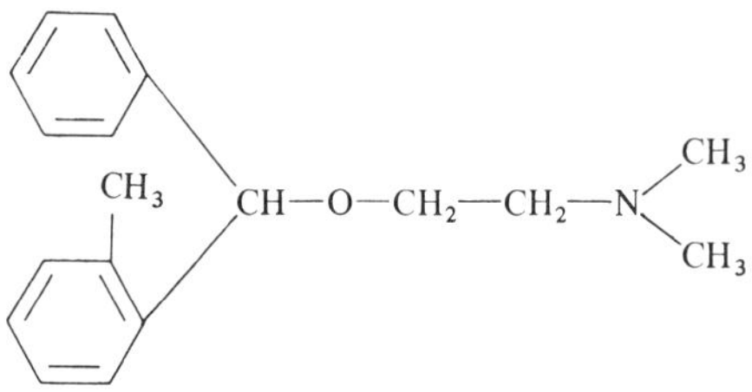

orphenadrine

The chromatograms were submitted to radioactivity measurements of the labeled drug. Methyl orange colorimetry was applied to the unlabeled compound. However, this technique does not distinguish orphenadrine from its demethylated metabolites, and a more detailed analytical scheme was elaborated. The reaction of salicylaldehyde with N,N-didesmethylorphenadrine determined the sum of the orphenadrine and N-desmethylorphenadrine present. Since acetic anhydride reacted with both N-demethylated metabolites only orphenadrine would be titrated. It is also possible to determine small quantities of the metabolites in blood by fluorimetry of their complexes with Tinopal GS.

The in vivo transformations of methadone have been more extensively investigated. Way et al. [981] first determined methadone in blood, feces, and urine by the Brodie et al. [139] methyl orange method. Interfering substances were eliminated by washing the ethylene dichloride solution twice with a 0.5M phosphate buffer, pH 7. Counter-current distribution was used to verify that the isolated product actually was pure methadone.

This procedure has been criticized by Rickards et al. [762] as lacking specificity. They preferred digestion of the tissues by heating with concentrated alkali rather than with trichloroacetic acid, extraction of the bases with ethyl ether, and the subsequent determination of methadone by dinitrophenyl derivatization and accompanying colorimetry.

Way et al. [979] used counter-current distribution to show that the drug was excreted unchanged in urine, but unidentified metabolites were eliminated in the feces.

The metabolism of methadone has also been studied by Vidic [960]. The extraction of urine at pH 9.1 with an organic solvent and paper

---

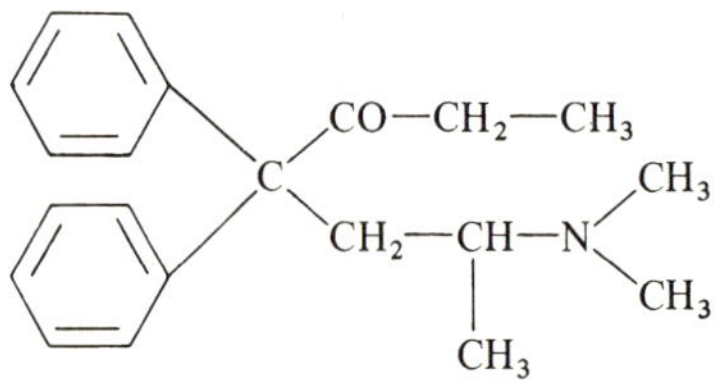

methadone

chromatography (butanol : formic acid : water 12 : 1 : 7) revealed a metabolite with a different R_f than that of the original compound, which reacted differently with diazotized sulfanilic acid. This color reaction was not observed when the urinary extract was diazotized or methylated. Moreover, primary amine reagents gave positive reactions on the chromatograms. The author concluded an N-demethylation without the necessity of characterizing the metabolites.

Axelrod [38] reported the N-demethylation of methadone by hepatic enzymes. The metabolite reacted with ninhydrine with formaldehyde evolution during incubation.

Pohland et al. [714] synthesized the cyclic compound resulting from the dehydration of the hemicetal form of N-desmethylmethadone. They demonstrated that the metabolite obtained either enzymically, in vitro, or in the dog, in vivo, was identical with the synthetic compound on paper chromatography.

The quaternary ammonium compound N-methylmethadone, another methadone metabolite, was discovered by Schaumann [801] after incubation of the drug with liver slices. It was analyzed by methyl orange after extraction with benzene at pH 5. Crystallization of a salt form agreed with an authentic sample. N-methylmethadone might correspond to one of the unidentified metabolites reported by Vidic [960].

McMahon et al. [584] studied <u>acetylmethadol.</u> They synthesized derivatives ^{14}C labeled in the 1-position and on N-methyl. They also synthesized the corresponding monodemethylated products. The metabolites extracted from urine or bile at pH 10 with chloroform were chromatographed on silica

---

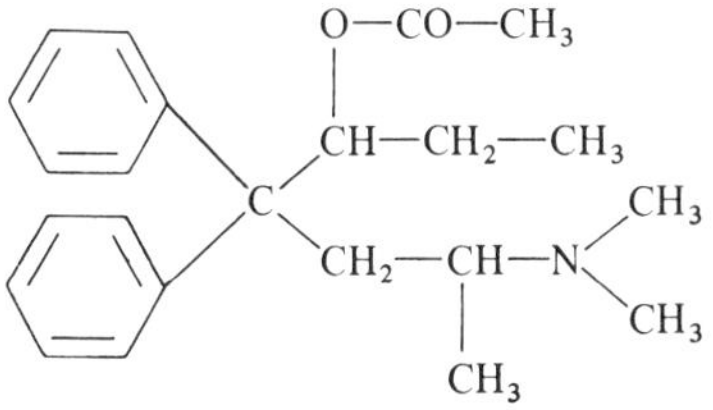

acetylmethadol

gel thin layer in various solvent systems. They were identified by adding a colorimetrically detectable nonradioactive tracer. The detected area was monitored by radioactivity. These studies were completed by in vitro assaying of liver preparations from rats. The rats were also treated with phenobarbital (50 mg/kg intraperitoneally for 4 days) to increase the enzymic activity.

The fate of propoxyphene in the organism has been examined by Lee et al. [547]. The excretion was investigated in the rat with the drug ^{14}C labeled on the N-methyl and in man with the unlabeled drug by methyl orange colorimetric determination. The latter determination was not specific, as countercurrent distribution revealed two compounds that reacted with methyl orange. Incubation assays with rat liver slices produced radioactive CO_2 evolution and implied that demethylation was the major metabolic pathway. This hypothesis was confirmed by increases in the amount of substance reacting with fluorodinitrobenzene during the incubation.

Extraction at pH 9 with carbon tetrachloride and the formation of a dinitrophenyl derivative permitted isolation of the metabolite from human urine. Infrared spectrum and elemental analysis indicated desmethylpropoxyphene.

This biological reaction was confirmed by Amundson et al. [23] with thin-layer chromatography. The drug and its metabolite were separated on silica gel by the system ammonium chloride saturated methanol : methanol 1 : 2. No differential determination was performed.

The biotransformation of Dibenamine was the subject of investigations by Axelrod et al. [40]. They demonstrated that the produced metabolite was dibenzylamine and not, as previously supposed, the corresponding alcohol (N,N-dibenzyl-β-hydroxyethylamine). Dibenamine was extracted at pH 3

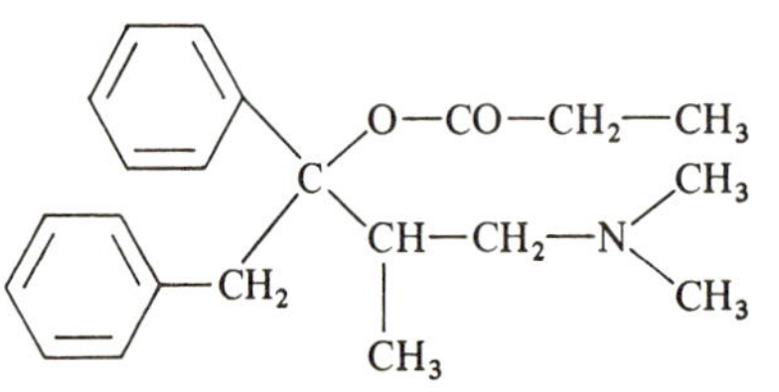

---

propoxyphene

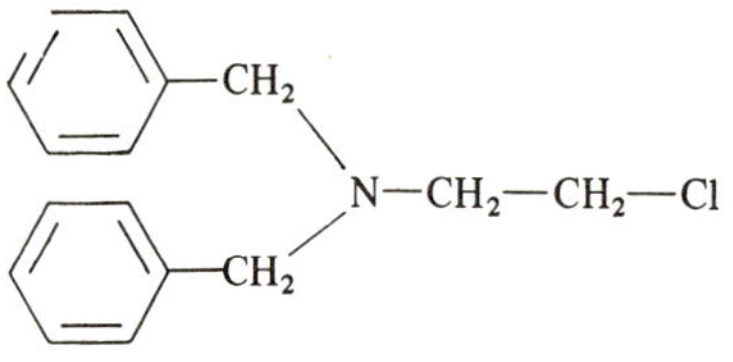

Dibenamine

OTHER AMINES

with heptane and titrated with methyl orange. Dibenzylamine was then extracted at pH 7 and also determined with methyl orange. This metabolite was detected and isolated by counter-current distribution. It was characterized by its elemental composition, melting point, and UV spectrum and by making its benzenesulfonyl derivative.

The same authors have extended their work to the closely related phenoxybenzamine [124]. It behaved similarly to give N-phenoxyisopropyl-N-benzylamine. Similar procedures were applied.

Smith and Grostic [854] briefly reported the isolation of four metabolites of methyl-N-(o-aminophenyl)-N-(3-dimethylaminopropyl)-anthranilate by thin-layer chromatography. These metabolites were identified by NMR, mass spectrometry, and by comparison to authentic compounds.

IV. OTHER AMINES

The in vivo transformations of amitriptyline have been investigated by Hucker [434, 438]. No information on the methods used to isolate and

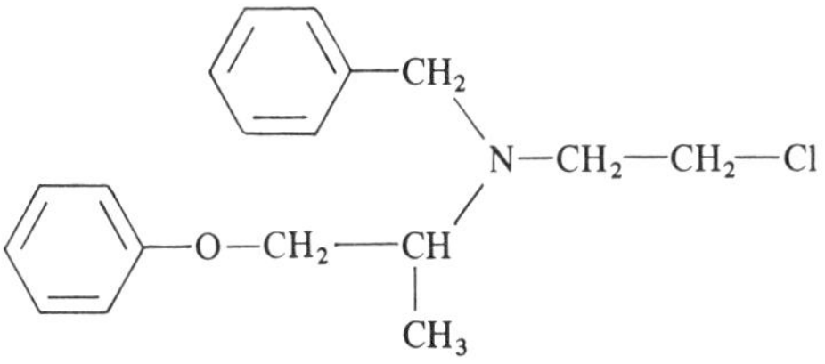

phenoxybenzamine

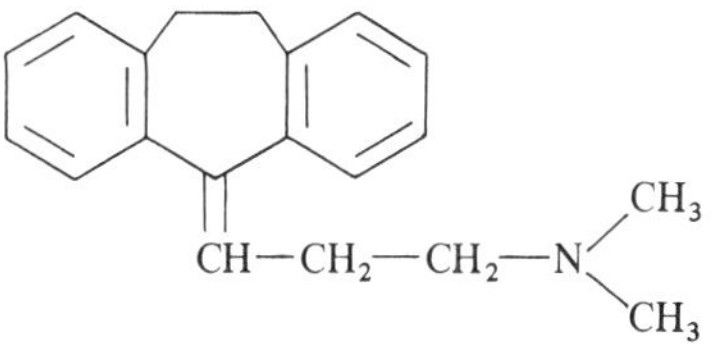

amitriptyline

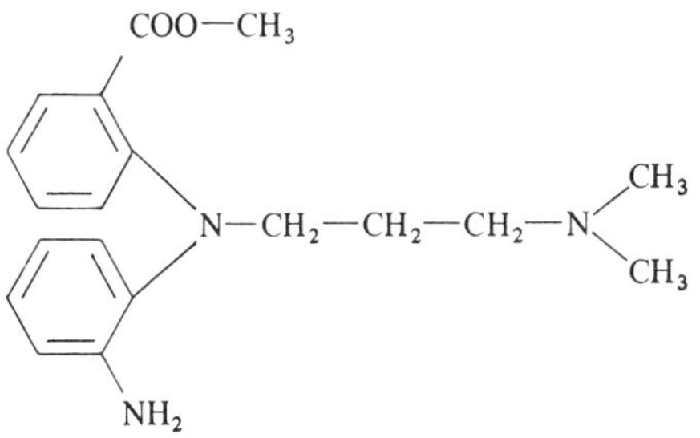

methyl-N-(o-aminophenyl)-
N-(3-dimethylaminopropyl)-anthranilate

identify desmethylamitriptyline, the 10- and 10,11-hydroxy derivatives, and their glucuronides were given.

Cassano et al. [196] administered ^{14}C-labeled amitriptyline and extracted urine or bile at pH 9-10 with chloroform. The unchanged drug was separated from its mono- and didemethylated derivatives by thin-layer chromatography on fluorescent silica gel (benzene:dioxane:ammonium hydroxide 60:35:5).

McMahon et al. [587] used <u>nortriptyline</u>, ^{14}C labeled on the N-methyl. The study included radiochemistry: $^{14}CO_2$ excretion, radioactivity of the various tissues, total biliary and urinary excretion, and investigations of the urinary metabolites. These are not directly extractable but after hydrolysis with Glusulase (β-glucuronidase + sulfatase) they are soluble in isoamyl alcohol at pH 11. Preliminary thin-layer chromatography on silica gel (ethanol:ethyl acetate 1:1) revealed a major metabolite of low R_f and practically no nortriptyline or N-acetylnortriptyline.

The metabolites were acetylated and mildly hydrolyzed to cleave the O-acetyl derivatives that might have been produced. In this way only the N-acetyl derivatives are separated by chromatography on an alumina column. One metabolite was obtained on elution with the solvent mixture benzene:ethyl acetate 4:6. A second was obtained with the mixture ethyl acetate:ethanol 20:1.

The UV and IR spectrophotometry of these two metabolites, as well as the influence of alkalinization on their UV spectra, the preparation of the O- and N-diacetyl derivatives, and their mild hydrolysis provide data which permit the conclusion that the two products are identical. They appear to be the cis and trans isomers of 10-hydroxynortriptyline.

Amundson and Manthey [24] studied the nortriptyline metabolism in man, in contrast to McMahon's study in the rat. Thin-layer chromatography on

---

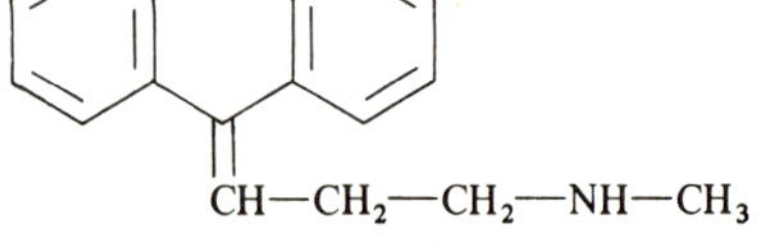

nortriptyline

OTHER AMINES

silica gel (isopropanol:water 88:12 one-third saturated with sodium chloride) separated four metabolites in an ethyl ether extract from alkalinized urine. The most important appeared to be 10-hydroxynortriptyline. The others have not been identified. A UV spectrophotometric determination of nortriptyline together with its metabolites was reported.

Drabner et al. [276] extracted urine with chloroform and employed thin-layer and paper chromatography. The thin layer was silica gel containing 13% of plaster and the developing solvent was methanol:chloroform 1:2. An acid system, butanol:formic acid, or an alkaline system, 5N ammonium hydroxide saturated butanol, was used in paper chromatography. The thin-layer chromatography revealed two metabolites, whereas the paper chromatography only revealed one. The same procedure applied to opipramol enabled the detection of six metabolites on thin layer, two on paper with the acid solvent system, and three with the alkaline system.

Amitriptyline and nortriptyline are structurally similar to imipramine, the metabolism of which will be discussed in Chapter V, "Dibenzazepines."

Amantadine is a somewhat peculiar amine. Its metabolism has been investigated by Bleidner et al. [96] in four different animal species and in man. Gas-liquid chromatography was used for urine, blood, and tissue extracts. Only unchanged drug was found. Its identity was established by conversion into its acetylated derivative, which corresponded to the authentic compound.

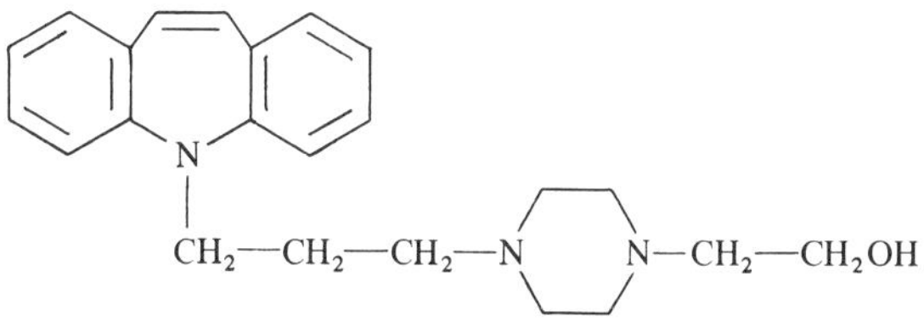

opipramol

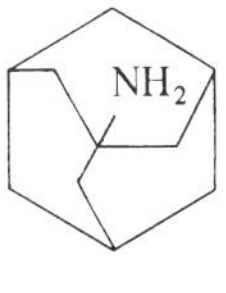

amantadine

Decamethonium is also not appreciably modified by the body. Lüthi and Waser [569] administered the $^{14}CH_3$-labeled molecule to cats and tried to separate possible metabolites by paper chromatography in various solvent systems. Their results were confirmed by Christensen [201], who also used the labeled drug and paper chromatography (ethanol : water : ammonium hydroxide 75 : 25 : 2; butanol : ethanol : acetic acid : water 8 : 2 : 1 : 3).

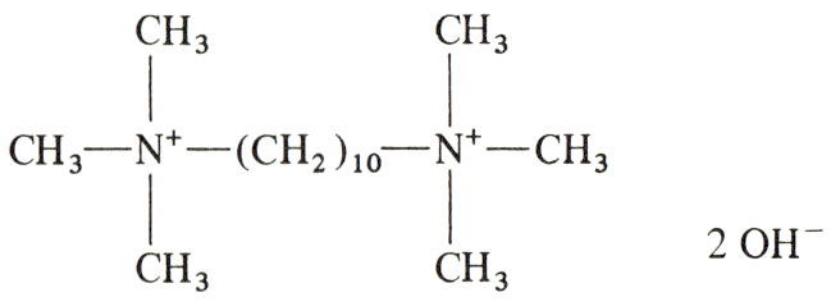

$$CH_3-\underset{\underset{CH_3}{|}}{\overset{\overset{CH_3}{|}}{N^+}}-(CH_2)_{10}-\underset{\underset{CH_3}{|}}{\overset{\overset{CH_3}{|}}{N^+}}-CH_3 \quad 2\,OH^-$$

decamethonium

Chapter III

AMINOPHENOLS, CATECHOLAMINES

The aminophenolic catecholamines are a family of substances of considerable biological importance and have been the subject of extensive studies on their metabolism. Primary emphasis will be placed on compounds used in therapeutics rather than on in vivo transformations of catecholamines of endogenous origin.

I. AMINOPHENOLS

Paracetamol is the major metabolite of phenacetin (see page 110). Gwilt et al. [389] varied the analytical procedures of Brodie and Axelrod [127]. Blood, triturated with sodium sulfate, was extracted with ethyl ether. Paracetamol was recovered in acid solution and reacted with α-naphtol. The colored derivative was extracted in butanol and determined at 635 mμ. Glucuronide and sulfate complexes did not interfere.

Welch et al. [995] compared paracetamol metabolism in the cat and other animals. The conjugated metabolites were hydrolyzed by β-glucuronidase or sulfatase (24 hours at 37°C, pH 5). The liberated paracetamol was determined by methyl orange colorimetry after identification by thin-layer chromatography on alumina (benzene : toluene : water : acetic acid 2 : 2 : 1 : 2; benzene : ethanol : water : acetic acid 20 : 20 : 20 : 1; chloroform : methanol : water : acetic acid 20 : 10 : 20 : 1).

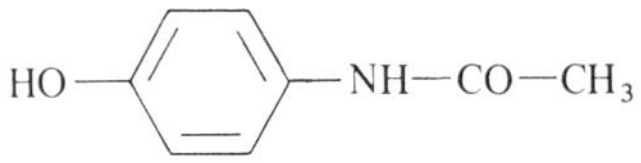

HO—⟨C₆H₄⟩—NH—CO—CH₃

paracetamol

p-Methylaminophenol gave an ether glucuronide, isolated according to the lead acetate procedure of Kamil et al. [480] and recrystallized from ethanol by Shibasaki and Nakamura [838]. After methylation and acetylation its formula was established precisely by elemental analysis, and by comparison of its melting point and IR spectrum with those of the synthetic glucuronide.

The biological transformations of mescaline in man and animals mainly involve desamination to the 3,4,5-trimethoxyphenylacetic acid (Slotta and Müller, [850]). Blood and urine levels of mescaline were determined by the technique of Woods et al. [1022]. The procedure, specific for mescaline, consisted of chloroform extraction of the amine and, after dehydration of the extract on a sucrose column, colorimetric determination with bromocresol purple.

Spector [868] unequivocally identified 3,4,5-trimethoxyphenylacetic acid in the urine of dogs dosed with $2'-^{14}C$-labeled mescaline. The fractionation of the acidified urine on a Dowex-1 column (4N ammonium formate, 5N formic acid, 24N formic acid) gave two radioactivity peaks. The first one (butanol:acetic acid:water 4:1:4) had an R_f of 0.73 on paper chromatography, identical to the one of mescaline. The second peak, R_f 0.91, corresponded to a compound that, purified by sublimation, was identical to 3,4,5-trimethoxyphenylacetic acid in melting point, IR spectrum, and elemental analysis.

Charalampous et al. [199] studied the metabolism of $2'-^{14}C$-labeled mescaline in man and detected the metabolite by paper chromatography in various solvent systems (Table 9).

The nature of the metabolite was confirmed for urine and blood by melting point and IR and UV spectra.

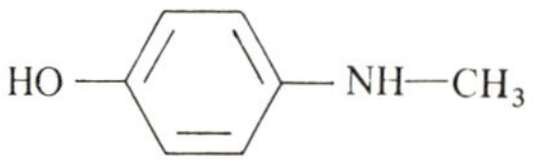

---

p-methylaminophenol

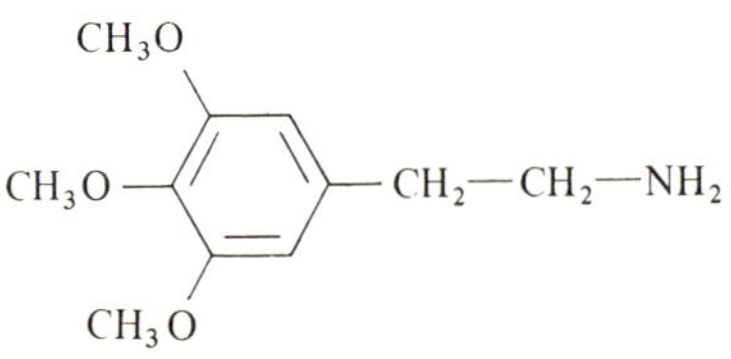

mescaline

TABLE 9

Solvent systems	R_f 3,4,5-Trimethoxy-phenylacetic acid	R_f Mescaline hydrochloride
Ethyl acetate : pyridine : water, 2 : 1 : 1	0.78	0.82
Chloroform : pyridine : water, 40 : 51 : 7	0.73	0.56
Butanol : acetic acid : water, 4 : 1 : 1	0.87	0.65
Chloroform : acetic acid : water, 4 : 1 : 1	0.80	0.90
Butanol : acetic acid : water, 4 : 1 : 5	0.89	0.64
Chloroform : acetic acid : water, 2 : 1 : 1	0.92	0.71
Isopropanol : ammonium hydroxide : water, 8 : 1 : 1	0.54	0.82
Isopropanol : formic acid : water, 8 : 1 : 1	0.86	0.75

The cerebral metabolism of mescaline was examined in the cat by Neff et al. [666] with the 2'-^{14}C-labeled drug. Various parts of the brain were homogenized with twice their volume of ethanol. The carrier solution (mescaline + 3,4,5-trimethoxyphenylacetic acid) was added and paper chromatography performed with three solvent systems:

(I) Butanol : acetic acid : water, 4 : 1 : 5

(II) Benzene : acetic acid : water, 2 : 1 : 1

(III) 2-Propanol : ammonium hydroxide : water, 20 : 1 : 1

Although a number of metabolites were considered, only 3,4,5-trimethoxyphenylacetic acid was detected in brain, cerebrospinal fluid, blood, and urine (Table 10).

Seiler et al. [825] devised methods for the direct fluorimetric determination of compounds separated by thin-layer chromatography and applied them to mescaline [826] and its metabolism in mouse brain [824]. Fluorescence was engendered by treatment with formaldehyde in the presence of ammonium hydroxide and subsequent heating with hydrochloric acid.

TABLE 10

Products	R_f in system:		
	I	II	III
3,4,5-Trimethoxyphenethylamine	0.74	0.36	0.92
3,4-Dihydroxy-5-methoxyphenethylamine	0.40		
3,5-Dimethoxy-4-hydroxyphenethylamine	0.55		
3,5-Dihydroxy-4-methoxyphenethylamine	0.50		
3,4,5-Trihydroxyphenethylamine	0.28		
3-Hydroxy-4,5-dimethoxyphenethylamine	0.62		
3,4,5-Trimethoxyphenylethanol	0.85		0.95
3,4,5-Trimethoxyphenylacetic acid	0.91	0.88	0.62
3,4,5-Trimethoxymandelic acid	0.88		0.64

Friedhoff and Hollister [344] compared the metabolisms of mescaline and 3,4-dimethoxyphenethylamine in man. The chloroform extract of urine was analyzed by gas-liquid chromatography on a column of Diatoport S coated with 3.8% of silicone SE-30 at 210-230°C. The identifications depended on comparative retention times with reference compounds added to the samples. Mescaline and trimethoxyphenylacetic acid were thus separated.

Friedhoff and Goldstein [343] hydrolyzed the urine of labeled mescaline-treated rats with Glusulase (β-glucuronidase + sulfatase) at pH 2. The ethyl acetate extract was chromatographed on paper (chloroform : acetic acid : water 2:1:1; isopropanol : ammonium hydroxide : water 8:1:1). An inhomogeneous radioactive peak with the same R_f as 3,4,5-trimethoxyphenyl-acetic acid was obtained on treatment with the latter chromatographic system. The corresponding area was eluted, acetylated, and extracted with methylene chloride. Paper chromatography revealed a substance with the R_f of the highly active metabolite 3,4,5-trimethoxyphenylethanol.

Harley-Mason et al. [403] isolated small amounts of a conjugated metabolite of mescaline in 1958 from human urine. The synthesis of a similar 3,4-dihydroxy-5-methoxyphenylacetic acid was described by the authors [402].

Ratcliffe and Smith [745] demonstrated similarly the presence in human urine of 3-hydroxy-4,5-dimethoxyphenethylamine. Extraction with ethylene dichloride:amyl alcohol and paper chromatography was followed by acid hydrolysis to show traces of 4-hydroxy-3,5-dimethoxyphenethylamine also. These two products were authenticated by synthesis of the reference compounds.

Daly et al. [241] performed in vitro experiments with 2'-^{14}C-labeled mescaline. The various demethylation products were separated by paper chromatography (butanol:acetic acid:water 4:1:5) and identified by comparison.

Charalampous et al. [200] published in 1966 a detailed study of mescaline metabolism on administration of the 2'-^{14}C-labeled drug to man. Bidimensional paper chromatography (I, butanol:acetic acid:water 4:1:1; II, butanol:isopropanol:ammonium hydroxide:water 3:1:1:1) revealed 12 radioactive metabolites in urine. The urine was fractionated on a Dowex 50 W-X4 (H$^+$) column eluted with water and then with 50% ethanol containing 5% ammonium hydroxide.

The mescaline was extracted from the acidified ethanol eluate with ethyl acetate and identified by its melting point and that of its α-naphtylthiourea derivative. The residual neutral fraction contained N-acetylmescaline, characterized by isotopic dilution and R$_f$ comparison on paper.

The aqueous eluate was extracted by ethyl acetate at neutral, then at acid pH values. The acid fraction contained 3,4,5-trimethoxyphenylacetic acid and another demethylated phenylacetic acid, separable by paper chromatography. The unextracted metabolite appears to be N-acetyl-3,4-dimethoxy-5-hydroxyphenethylamine, as suggested by a study of N-acetylmescaline metabolism. The same metabolites were found in the cerebrospinal fluid.

Gold and Stormann [370] roughly examined hexobendine metabolism. Trimethoxybenzoic acid was recognized by extraction of urine at pH 2-3 with

---

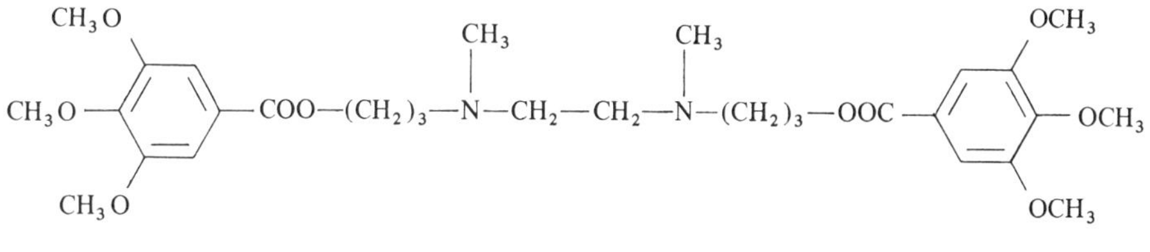

hexobendine

ethyl acetate and thin layer chromatography on silica gel in one dimension (chloroform : ethyl acetate : acetic acid 4 : 5 : 1) and in two dimensions (I, ethyl acetate : methanol : 3 N ammonium carbonate : dimethylformamide 12 : 2 : 1 : 1; II, carbon tetrachloride : methanol : acetic acid 77.5 : 20 : 2.5).

II. CATECHOLAMINES

The catecholic aminoacids (dopa and derivatives) will be considered before the true catecholamines (adrenaline and derivatives).

A. Dopa Derivatives

The metabolism of dopa which is not used therapeutically proceeds by decarboxylation to dopamine. Methyldopa has been proposed as a hypotensive drug and its in vivo transformations investigated in parallel with those of dopa.

Porter and Titus [716] performed a complete study of methyldopa metabolism in the rat. The decarboxylation was easily demonstrated by using two methyldopa compounds, ^{14}C-labeled in positions 1 and 2. After administration of the ^{14}COOH-labeled compound, the radioactivity was found in the expired carbon dioxide. No radioactivity was detected in the respiratory gases when the 2-labeled compound was administered.

Paper chromatography of urine (butanol : acetic acid : water 4 : 1 : 1) after administration of 2-labeled methyldopa revealed six radioactive bands, of which four were important. Injection of 1-labeled methyldopa produced only two bands corresponding to nondecarboxylated products.

Butyl acetate extracted from acidified urine the radioactivity corresponding to the two bands of higher R_f. Two components separated on

---

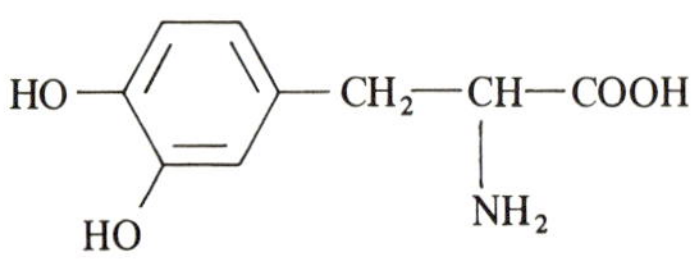

dopa

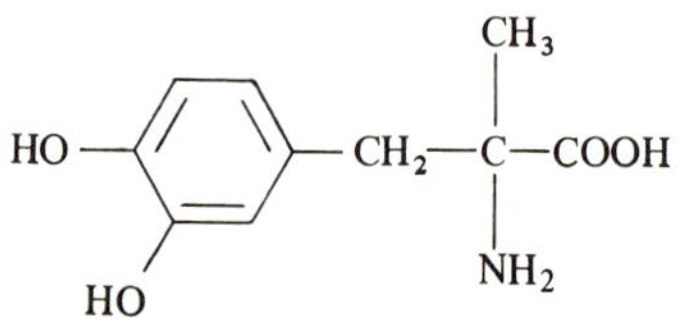

methyldopa

counter-current distribution; one behaved as 3-methoxy-4-hydroxyphenylacetone, the other behaved as 3,4-dihydroxyphenylacetone. These two ketones were identified by their semicarbazones and the R_f and counter-current distribution values of these were similar for reference products.

A third band had the same R_f as 3-methoxymethyldopa. The corresponding area was eluted from a series of chromatograms and the extract was acetylated. The acetylated product had exactly the same counter-current distribution as the reference amino acid acetylated in the same way.

Different fractions were isolated by chromatography on a sulfonic resin at pH 6.5 after water washings of the column. One fraction showed the R_f of methyldopa on paper, and its UV spectrum confirmed the identification. When the column was washed with 1N hydrochloric acid a fraction was eluted with an R_f comparable to 3-methoxy-α-methyldopamine and confirmated by the UV spectrum. A small fraction seemed to be α-methyldopamine.

The band with the lowest R_f on paper chromatograms was eluted from a great number of sheets with 0.1N hydrochloric acid and hydrolyzed with β-glucuronidase. Three radioactive bands appeared after incubation and one corresponded to 3-methoxy-α-methyldopamine. This band was attributed to a mixture of glucuronides of α-methyldopamine and its m-methoxy derivative.

This detailed study is summarized in Table 11.

TABLE 11

Metabolite	R_f	% of total radioactivity
Glucuronides mixture	0.20	4
Methyldopa	0.40	56
α-Methyldopamine	0.51	1
3-Methoxymethyldopa	0.58	16
3-Methoxy-α-methyldopamine	0.68	3
Mixture of:		
○ 3,4-Dihydroxyphenylacetone	0.78 ⎫	11
○ 3-Methoxy-4-hydroxyphenylacetone	0.82 ⎬ 0.78-0.84	7
		98

Buhs et al. [152] used the same labeled compounds as Porter and Titus in their studies of methyldopa metabolism in man. The urine was fractionated on ion-exchange resins by chromatographies directly or after acidification and autoclaving:

- CG-50: carboxylic acid cation-exchange resin
- CG-120: sulfonated polystyrene cation-exchange resin
- AG-1 X 8: quaternary amine anion-exchange resin

Methyldopa, 3-methoxymethyldopa, and a basic fraction were retained on normal chromatography of the crude urine on CG-120 resin (H^+). Methyldopa mono-O-sulfate and the neutral conjugated fraction were in the effluent. 1N hydrochloric acid elution separated methyldopa and 3-methoxymethyldopa, whereas the basic fraction remained on the column. Chromatography of urine on CG-50 (Na^+) retained this fraction and permitted isolation. The column was eluted after water washings with 1N hydrochloric acid. When the effluent of the CG-120 column was hydrolyzed by autoclaving at pH 1, it contained only methyldopa, which was also separable on CG-120 (H^+).

The chromatography of the hydrolyzed urine on CG-120 (H^+) gave all the methyldopa and methoxy-3-methyldopa. The chromatography on CG-50 (Na^+) gave the entire basic fraction.

These several metabolites were identified by comparing their chromatographic behavior with that of the labeled authentic compounds. After isolation they were analyzed by UV, IR, and NMR spectrometry.

Methyldopa mono-O-sulfate was isolated by a special chromatographic procedure on AG-1 X 8 (acetate) to separate it from the unconjugated amino acid, the basic fraction, and the impurities. After elution, its UV spectrum showed the reversible shift of the absorption maximum from 273.5 to 292 mμ on alkalinization, a characteristic of the monosubstituted catechols. The sulfate ion and methyldopa were easily identified after hydrolysis. Elemental analysis indicated only one sulfate group.

The basic fraction was mostly α-methyldopamine, free and conjugated. After the first chromatography on CG-50, a second was performed on alumina at pH 7, followed by an acid elution. The extract was acetylated and fractionated on a cellulose powder column by partition between the mixture isooctane:butanol 95:5 (mobile phase) and 1M acetic acid (stationary phase). The pure acetyl derivative was identified by NMR. The other components of this fraction remain unknown.

The neutral fraction contained an undefined metabolite. A sulfur determination by isotopic dilution showed it as a monosulfate.

This accounted for the metabolism of 90% of the administered product. Unchanged methyldopa represented 24% of the urinary excretion products.

The analytical procedure of Buhs was used by Prescott et al. [721] to compare clinical effects and metabolism of methyldopa in man. A relationship was established between increases in 3-methoxymethyldopa levels and the blood pressure decreases.

Sjoerdsma et al. [849] studied the mechanism of action of methyldopa and its hydrazine. They determined methyldopa and methyldopamine fluorometrically after potassium ferricyanide oxidation. Methyldopa was separated from methyldopamine by column chromatography on Dowex-50 X4 at pH 6.5. Methyldopamine was eluted by 2N hydrochloric acid. These two products were also identified by paper chromatography (butanol:acetic acid:water 12:3:5 — ascending; 1N hydrochloric acid saturated butanol; isopropanol:ammonium hydroxide:water 20:1:2 — ascending).

Young and Edwards [1036] studied the metabolism of the (+) and (-) isomers of methyldopa in the rat without radio labeling. The metabolites were separated by paper chromatography in one or two dimensions. Identifications were achieved by comparison with known substances. R_f measurements were supplemented by the use of various detection reagents (ninhydrin, isatin, Ehrlich reagent, diacetyl reagent, and diazotized p-nitroaniline). The glucuronides were hydrolyzed by β-glucuronidase at 37°C, pH 4.5. The experiments were performed on urine and bile. The distinctive R_f are given in Table 12.

Methyldopa was the only compound determined after chromatography. The ninhydrin-stained spot, after fixation with copper nitrate, was eluted in 70% methanol and determined colorimetrically.

Duhm et al. [284] synthesized the D(+) and L(-) isomers of methyldopa, β-^{14}C labeled. After administration, rat urine was analyzed by a combination of electrophoresis and paper chromatography. The metabolites were separated by electrophoresis with a potential gradient of 40 V/cm for 75 minutes in the buffer pyridine:acetic acid:water 10:1:89 (pH 6.2); then by chromatography in the rectangular direction [(I) butanol:acetic acid:water 4:1:1 or (II) secbutanol:pyridine:water:acetic acid 12:6:4:1]. Ten radioactive components were detected after L-isomer administration. Six

TABLE 12

Products	R_f in system:[a]			
	I	II	III	IV
Methyldopa	0.55	0.60	0.73	0.54
3-Methoxymethyldopa	0.61	0.70	0.76	0.88
Methyldopamine	0.53	0.80	0.81	0.62
3-Methoxymethyldopamine	0.67	0.80	0.85	0.86
Dopa	0.47	0.55	0.47	0.34
Dopamine	0.54	0.76	0.75	0.55

[a] Solvent systems: I, butanol : acetic acid : water 4 : 1 : 2; II, butanol : pyridine : water 1 : 1 : 1; III, methanol : pyridine : water 20 : 1 : 5; IV, phenol : water 500 g : 125 ml.

of them were identified on comparison with reference compounds. Only four metabolites were found after D-isomer administration (Table 13).

B. True Catecholamines

The metabolism of adrenaline and noradrenaline has been studied extensively since 1956, partly with compounds of endogenous origin and partly with adrenaline or related products administered to animals or man. It is obviously difficult to include all the details and those that are most interesting from an analytical point of view have been selected.

The main metabolites of <u>adrenaline</u> and their structure elucidation will be examined first.

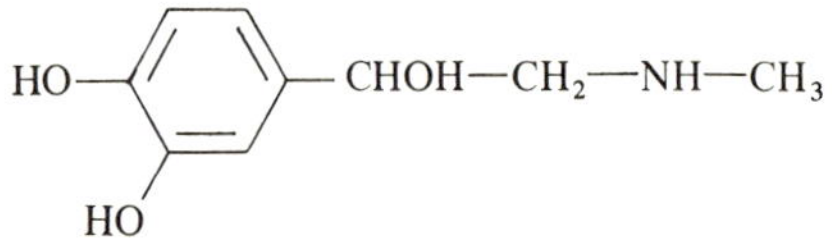

HO—⟨benzene ring⟩—CHOH—CH$_2$—NH—CH$_3$
HO

adrenaline

TABLE 13

Metabolite	R_f in system: I	II	Electrophoretic migration (mm)
Methyldopa	0.30-0.32	0.29	-4
3-Methoxymethyldopa	0.42	0.34	-6
Glucuronides of α-methyldopamine and 3-methoxy-α-methyldopamine[a]	0.14	0.11	-4
3-Methoxy-4-hydroxyphenylacetone +	0.45	0.60	+65
3,4-Dihydroxyphenylacetone[b]	0.45	0.70	+65
α-Methyldopamine	0.52	0.59	-75
3-Methoxy-α-methyldopamine	0.58-0.63	0.58-0.63	-80

[a] After β-glucuronidase treatment, the amines were characterized by chromatography and electrophoresis.

[b] The combination of these two components can be split by acid hydrolysis and the two ketones characterized by chromatography and electrophoresis.

1. <u>3-Methoxy-4-hydroxymandelic Acid.</u>

This metabolite was isolated, initially without identification, by Armstrong et al. [29] in 1956. It was extracted from urine at pH 1-2 with ethyl acetate and submitted to a bidimensional paper chromatography (I, isopropanol : ammonium hydroxide : water 8:1:1; II, benzene : propionic acid : water 2:2: 1). These authors [28] later identified it by paper chromatographic comparison with a synthetic compound after separation by counter-current distribution. At the same time they proposed its determination by two-dimensional paper chromatography and comparison with dilutions of a reference sample.

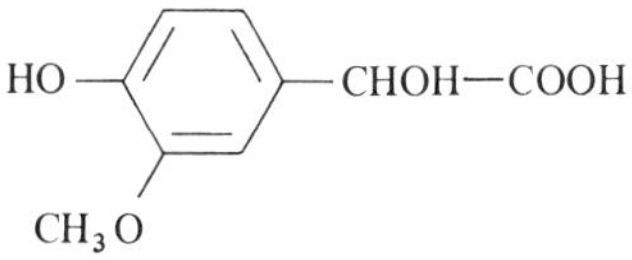

3-methoxy-4-hydroxymandelic acid

2. Metadrenaline.

Shortly thereafter, a second metabolite was isolated and characterized in the rat by Axelrod [39]. Metadrenaline was extracted from urine at pH 9 by the mixture ethyl ether : isoamyl alcohol, reextracted into hydrochloric acid, and then into butonal at pH 9. Bidimensional paper chromatography (I, isopropyl alcohol : 5% ammonium hydroxide 8 : 2; II, butanol : acetic acid : water 8 : 2 : 2) enable separation and identification by comparison with a reference compound. The fluorometric determination of metadrenaline in rat urine was achieved (Axelrod et al. [41]) after extraction in ethylene dichloride containing 2% of isoamyl alcohol at pH 10 and reextraction in 0.1N hydrochloric acid. The glucuronides were hydrolyzed by β-glucuronidase (2000 U/ml of urine, 3 hours at 37° C).

Metadrenaline appeared to be the major metabolite in man also. Labrosse and Axelrod [534] isolated it from the urine of patients administered with β-^{3}H-labeled adrenaline. Metadrenaline was extracted from urine at pH 10 by amyl alcohol and identified by paper chromatography (the same systems as [39]).

3. 3-Methoxy-4-hydroxyphenylglycol.

This metabolite was found by Axelrod et al. [42] as a sulfate in the urine of rats receiving adrenaline or noradrenaline. Its R_f values (same systems as [39]) after sulfatase hydrolysis were similar to the authentic compound as synthesized by Benigni and Verbiscar [76].

4. Other Metabolites.

Secondary metabolites that have been detected are 3,4-dihydroxymandelic acid, 3,4-dihydroxyphenylglycol, and the adrenaline conjugates.

In general studies of adrenaline metabolism, Kirshner et al. [507] used 2-^{14}C-labeled adrenaline and a fractionation procedure that combined

------

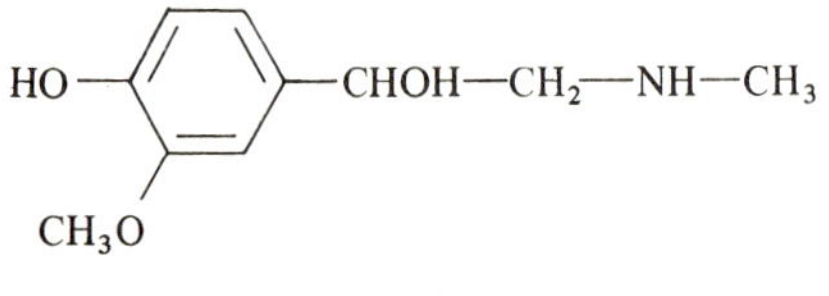

metadrenaline

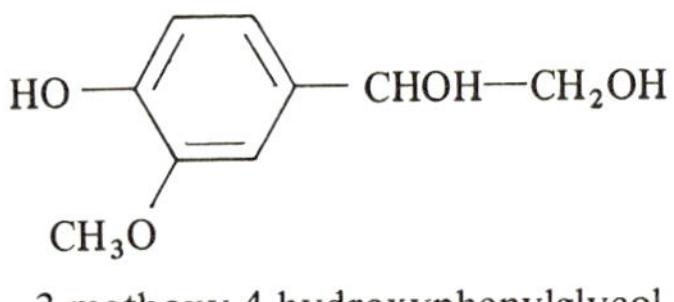

3-methoxy-4-hydroxyphenylglycol

chromatographies on paper and ion-exchange resins. Urine was percolated through an Amberlite IRC-50 column and the effluent was chromatographed on paper (1N hydrochloric acid saturated butanol). Three radioactive peaks were recorded: the first one corresponded to adrenaline and the second to metadrenaline; the third one was not identified. The effluent was then adsorbed on Dowex-1 (acetate). The column was eluted with 0.3M ammonium acetate at pH 4.8 and one radioactive fraction was obtained. Two other radioactive components were eluted when the concentrations of the ammonium acetate solution were increased to 1M, then 3M. The 0.3M eluate was a metadrenaline conjugate, hydrolyzed by sulfatase but not by β-glucuronidase. After hydrochloric acid hydrolysis, the base was characterized by paper chromatography. The 1M eluate was identified by paper chromatography with 3-methoxy-4-hydroxymandelic acid and the 3M eluate with 3,4-dihydroxymandelic acid.

Axelrod et al. [45] pursued the study of ^{3}H-labeled adrenaline in the cat. The drug was separated on an alumina column and eluted by 0.1M acetic acid. Metadrenaline was extracted at pH 10 by the mixture toluene:isoamyl alcohol 3:2. The glucuronides were hydrolyzed by β-glucuronidase for 3 hours at 37°C. The mandelic acids were extracted by isoamyl alcohol from the acid medium.

Whitby and Axelrod [1001] extended this study to ^{3}H-labeled noradrenaline in the cat. They examined many organs, blood, and urine by the same analytical methods.

Kopin and Axelrod [520] investigated the metabolism of ^{3}H-labeled adrenaline and ^{14}C-labeled metadrenaline in the rat. Urine was percolated through an alumina column that retained the free catechols, whereas the conjugates, metadrenaline and 3-methoxy-4-hydroxymandelic acid, were found in the effluent.

Free catechols were desorbed by 0.2N hydrochloric acid. The eluate, saturated with sodium chloride, was made strongly acid and shaken with ethyl acetate to extract the nonamine catechols: 3,4-dihydroxyphenylglycol and 3,4-dihydroxymandelic acid. These two compounds could not be separated directly by paper chromatography and it was necessary to change them into their methoxylated derivatives with catechol-O-methyltransferase. The unextracted adrenaline was determined by difference.

The effluent of the alumina column was divided in two portions. The first was hydrolyzed by 6N hydrochloric acid and liberated the catechols from their conjugates. They were then separated as above and determined. The second portion was chromatographed on a Dowex CG-50 column. The effluent contained 3-methoxy-4-hydroxymandelic acid, which was extracted by ethyl acetate at pH 1. The column was eluted by 3N ammonium hydroxide to separate metadrenaline.

Total metadrenaline and 3-methoxy-4-hydroxyphenylglycol were obtained from hydrolyzed urine with Glusulase (β-glucuronidase + sulfatase) 24 hours at 37°C. The fractionation reported above was applied. The entire separation is summarized in Fig. 1 (page 45).

The urinary metabolites of adrenaline and noradrenaline were investigated in the chicken by Rennick et al. [751]. Intravenous infusion was used to study overall body metabolism. Direct infusion into the renal venous portal circulation gained evidence on the urinary metabolites produced only by the kidney parenchyma.

The urinary metabolites were fractionated according to the procedure of Kopin et al. [520] and identified by paper chromatography; the solvents used were:

- Butanol : propionic acid : water, 47 : 22 : 31
- Butanol : acetic acid : water, 8 : 2 : 2, for the free amine and nonamine catechols
- Isopropanol : ammonium hydroxide : water, 8 : 1 : 1, for free metadrenaline and normetadrenaline
- 1N ammonium hydroxide saturated butanol for 3-methoxy-4-hydroxymandelic acid and 3-methoxy-4-hydroxyphenylglycol.

A fractionation procedure different from that of Kopin et al. was applied to the study of the human metabolism of 1-^{14}C-labeled normetadrenaline by Goodall et al. [378]. Urine was first chromatographed on Amberlite IRC-50 which retains basic compounds. The effluent was poured on to a Dowex 1-X2 (acetate) column to fix the acid substances. The neutral compounds were found in the last effluent. The bases were eluted by 0.5N acetic acid and separated by paper chromatography. Metadrenaline and one unidentified compound were found. The acids were eluted by a solvent of uniformly varying composition (water → 6M ammonium acetate buffer pH 4.85). Eight radioactive fractions were separated:

CATECHOLAMINES

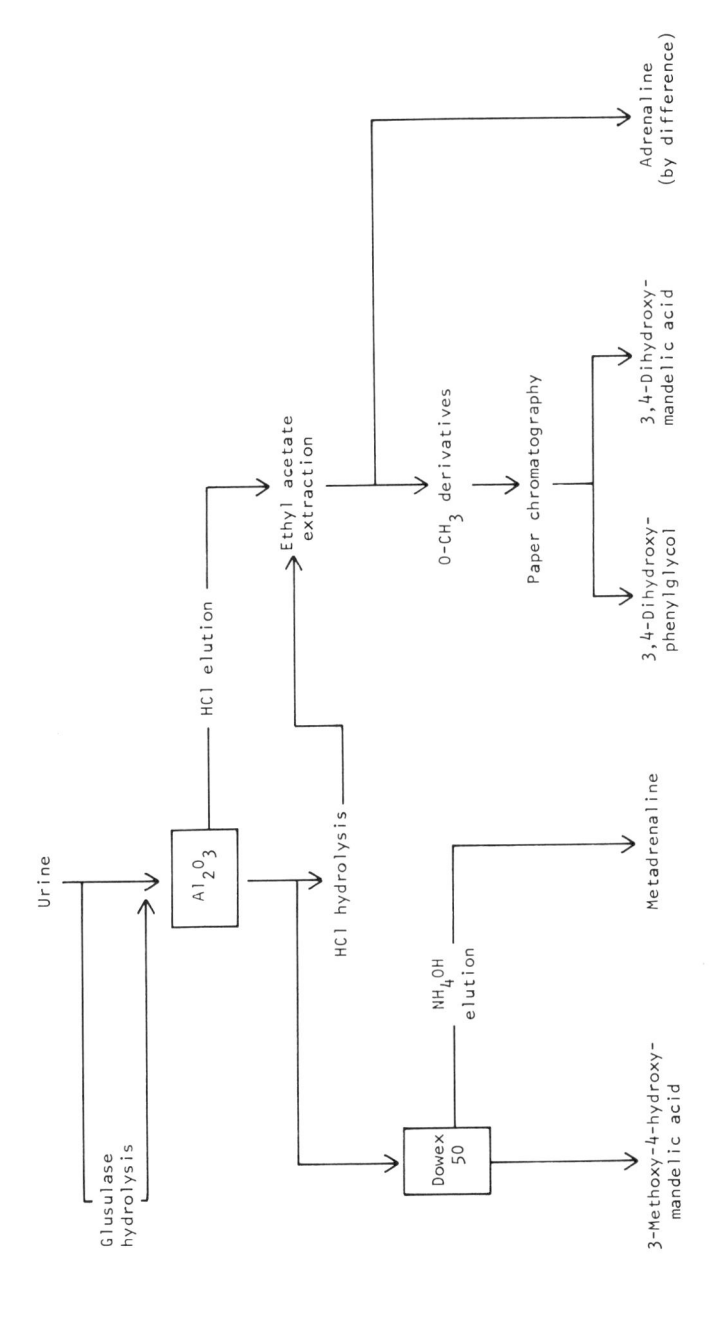

Fig. 1. Separation of adrenaline and metadrenaline metabolites; after Kopin and Axelrod [520].

1. Normetadrenaline 4-O-sulfate
2. Undefined normetadrenaline glucuronide
3. Unidentified metabolite
4. 3-Methoxy-4-hydroxymandelic acid
5. Dihydroxymandelic acid
6. Vanillic acid
7. 3-Methoxy-4-hydroxyphenylglycol sulfate
8. Dihydroxyphenylglycol sulfate.

The authors do not report how they defined these various metabolites.

Goodall and Alton [377] applied the same separation procedures to the urinary excretion of adrenaline metabolites in man at 2, 5, and 10 minutes after intravenous injection. The drug was ^{14}C-labeled in the 2 position. The authors delineated the way they followed the identifications of the acid metabolites eluted from the Dowex-1 column. Unlabeled reference compounds were chromatographed under the same conditions. The fractions corresponding to each radioactivity peak were chromatographed also on paper in the presence of a reference sample.

Kopin [519] demonstrated with adrenaline that it is possible to evaluate the relative importance of the various metabolic reactions by administering the labeled drug and one of its labeled metabolites simultaneously. He gave patients 7-^{3}H-labeled adrenaline and ^{14}C-methoxyl-labeled metadrenaline. The various metabolites were isolated by an unreported method and the amounts of ^{3}H and ^{14}C were determined for each metabolite. The importance of the different metabolic steps was deduced from the calculations and minimized the number of necessary experiments.

Isoprenaline metabolism was investigated by Hertting [420] with the 7-^{3}H-labeled drug, obtained by reducing isopropylnoradrenolone with tritium.

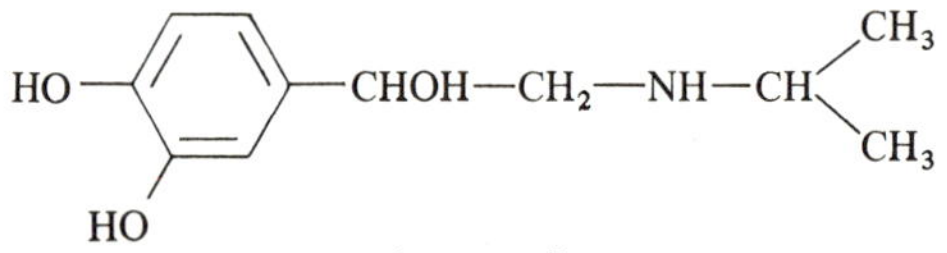

isoprenaline

The crude urine was chromatographed on paper (butanol : acetic acid : water 4 : 1 : 1; isopropanol : ammonium hydroxide : water 8 : 1 : 1; butanol : ethanol : water 4 : 1 : 1). 3-Methoxyisoprenaline was detected. Its identity was confirmed by chromatography in parallel with the reference product. Enzymic and acid hydrolysis assays showed that isoprenaline and its metabolite were conjugated in part to glucuronic acid.

In experiments on guinea pigs dosed with <u>metaraminol</u>, Maitre and Staehelin [591] observed a discrepancy between fluorometric and biological noradrenaline determinations. The measured hypotensive activity was higher than the one that corresponded to the fluorometric determination. Catecholamines were extracted by 10% trichloroacetic acid, adsorbed on alumina, and eluted by sulfuric acid according to the procedure of Euler and Orwen [314]. Paper chromatography (phenol : 0.1N hydrochloric acid 85 g : 15 ml) separated a substance with the same R_f as α-methylnoradrenaline from noradrenaline. This eluted metabolite had blood pressure activity. Confirmation was effected by a series of comparisons with reference α-methylnoradrenaline on paper chromatography and by paper electrophoresis. It thus appears that metaraminol and α-methyldopa might have a common metabolite.

Catecholamines, as do many other amines, exhibit the phenomenon of double or multiple spots on paper or thin-layer chromatography. This artifact has been reported frequently, particularly with pure substances. Shepherd and West [834] noted it for adrenaline in 1952 and it was extensively studied in 1959 by West [1000]. Adrenaline and the related compounds were extracted in general from tissues after a trichloroacetic acid treatment. The paper chromatography of such a trichloroacetic extract (butanol : acetic acid : water 4 : 1 : 5) gave, in addition to the adrenaline spot (R_f 0.32), a second spot of R_f 0.63. When the extract was first percolated through an Amberlite IRA-400 (chloride) column, only one spot (R_f 0.32) appeared.

------

metaraminol

When the product R_f 0.63 was chromatographed again, two spots were observed. The first (R_f 0.32) corresponded to adrenaline and the second (R_f 0.90) to trichloroacetic acid. This phenomenon has been observed in other solvent systems and with other catecholamines and noncatecholamines.

Beckett et al. detailed the study of this duplication on both paper [55] and thin layer [59]. They summarized the precautions to be taken in chromatographic metabolism studies. The simplest is to extract the free base in a solvent after alkalinization of the aqueous solution of its salt. When the tissues have been first treated by trichloroacetic acid, the removal of this acid is often incompletely achieved by washing. Such traces are sufficient to alter the chromatographic behavior of the bases. Alkalinization with ammonium hydroxide is to be avoided, since ammonium hydroxide promotes the passage of trichloroacetic acid into the organic phase.

Bidimensional chromatography may give useful information: the spots corresponding to an amine salt in the one dimension should duplicate in the second. The use of several different solvent systems is strongly recommended, as well as the use of reference compounds, to detect the presence of artifacts.

Broadley and Roberts [121] studied 22 sympathomimetic amines by paper chromatography in phenol with 15% of 0.1N hydrochloric acid. Adrenaline, noradrenaline, N-ethylnoradrenaline, N-butylnoradrenaline, N-isobutylnoradrenaline, as well as isoprenaline, carbadrine, oxedrine, and metadrenaline, all gave at least four spots in the presence of 10N hydrochloric acid. The numbers of these multiple spots varied with the structure of the amine and did not appear to be simply the separate migration of anion and cation. One of these spots, for instance, corresponded to a dibenzocycloheptatriene-like compound.

These studies show that it is necessary to be cautious before concluding the existence of a metabolite from chromatography of catecholamines. As a further example, Roberts [770] observed a "metabolite" of adrenaline that resembled isoprenaline by paper chromatography. He demonstrated later [771] that pure adrenaline, in the presence of the hydrochloric acid used for extraction, also gave several spots. One of these spots had the same R_f as isoprenaline.

Since the catecholamines give asymmetrical peaks in gas chromatography, as do most polar amines, the preparation of convenient derivatives has been investigated.

Gas chromatography of acetylated derivatives by Brochmann-Hansen and Svendsen [122] was at first disappointing. However, Brooks and Horning [140] achieved better results by complete acetylation in a pyridine medium. The separation required a column of silanized Gas-Chrom P coated with 7% of silicone F 60 and 1% of phenylmethylsiloxane-ethyleneglycol succinate copolymer*. An argon ionization detector was used. Adrenaline was separated from noradrenaline, and metadrenaline from normetadrenaline by operating at two temperatures. The relative retention values (anthracene = 1.00) are given in Table 14.

It was also possible to separate the acetylated methyl hippurate and methyl-3-methoxy-4-hydroxymandelate on a column coated with 7% of silicone F 60.

The acetylated derivatives of adrenaline, noradrenaline, metadrenaline, and normetadrenaline were separated by Horning et al. [432] on a column coated with 0.6% of methylsiloxane polymer †and 0.2% cyclohexane-dimethanol succinate polyester** with a programed temperature rise between 165° and 225°C. The propionyl derivatives gave no advantage over the acetyl derivatives. However, the pentafluoropropionyl derivatives were well separated and the sensitivity was improved with the use of an electron capture detector.

Capella and Horning [192] developed a convenient gas-liquid chromatographic method for catecholamines which improved on the incomplete and poorly reproducible acetylation procedures. The hydroxyl groups were converted into their trimethylsilyl ethers and the primary amino groups into eneamines or Schiff bases. The mixture of amines was first treated with hexamethyldisilazane at room temperature. Acetone or cyclobutanone was added after 30 minutes and the mixture chromatographed after 12 hours. Silanization of the glass vessels inhibited degradation of noradrenaline or octopamine. The only trimethylsilyl derivativization did not permit separation of adrenaline-noradrenaline or metadrenaline-normetadrenaline pairs (Sen and McGeer [827]). Determinations with the flame ionization detector were possible with an accuracy of ± 0.5% to ± 1.5% on a mixture of reference

---

* EGSS-Z (Applied Science Laboratories).
†JXR (Applied Science Laboratories).
**CHDMS (Applied Science Laboratories).

TABLE 14

Column temperature	Relative retention	
	198°C[a]	216°C[b]
Noradrenaline	19.50	12.05
Adrenaline	17.30	11.20
3,4-Dihydroxynorephedrine	14.10	9.18
Normetadrenaline	11.00	7.43
Metadrenaline	9.81	6.80
Dopamine	7.53	5.44
Octopamine	5.91	4.39
Synephrine	5.35	4.17
Phenylephrine	4.71	3.36
Homovanillylamine	4.44	3.43
Mescaline	4.34	
Vanillylamine	3.45	2.75
Homoveratrylamine	2.35	2.00
Tyramine	2.20	1.77
β-(4-Methoxyphenyl)-ethylamine	1.20	
Ephedrine	1.12	1.05
β-Hydroxyphenylethylamine	1.08	
Norephedrine	0.96	0.84
Desoxyephedrine (methylamphetamine)	0.49	
β-Phenylethylamine	0.39	
Amphetamine	0.37	
Hordenine	0.36	

[a] The retention time for anthracene was about 5 minutes.

[b] The retention time for anthracene was about 2.8 minutes.

standards. Retention times of catecholamines and related compounds were obtained on a column coated with 10% of silicone F 60 and with the temperature increasing at 1.5°C per minute. The retention times are expressed (Table 15) in methylene units, calibrated with respect to the paraffin series with even numbers of carbon atoms from tetradecane (14.0 MU) to eicosane (20 MU).

TABLE 15

| | Retention Times (Methylene Units) Ketone used ||
	Acetone	Cyclobutanone
β-Phenylethylamine	13.2	14.5
Ephedrine	13.9	13.9
Norephedrine	14.5	16.3
β-hydroxy-β-phenylethylamine	14.9	16.6
β-(4-Methoxyphenyl)-ethylamine	15.6	17.3
Tyramine	16.5	18.2
Phenylephrine	16.7	16.7
β-(3,4-Dimethoxyphenyl)-ethylamine	17.4	19.1
β-(3-Methoxy-4-hydroxyphenyl)-ethylamine	18.0	19.7
Octopamine	18.2	20.0
Metadrenaline	18.2	18.2
Dopamine	18.6	20.3
Adrenaline	18.8	18.8
Normetadrenaline	19.5	21.1
Noradrenaline	20.0	21.6
5-Hydroxytryptamine	21.8	—

A similar method was described by Kawai et al. [491]. Catecholamines were treated by hexamethyldisilazane, condensed with pentan-2-one, and separated on a column of acid-washed and silanized Gas-Chrom P coated with 3.5% of silicone SE.30.

Gas-liquid chromatography has not yet been applied to problems of catecholamine metabolism because the extraction of these substances from biological materials is difficult.

Chapter IV

PHENOTHIAZINES

Phenothiazine has been widely used in veterinary practices since 1938 and has been the subject of detailed metabolic studies because of colored urines and photosensitized keratitis.

Although the metabolism of phenothiazine appeared simple, about 30 metabolites of chlorpromazine have been numbered, largely by Fishman and Goldenberg. The biotransformations of promazine, other phenothiazines, and the chemically related thioxanthenes also will be reviewed.

I. PHENOTHIAZINE

The urine of phenothiazine-treated animals turned red and suggested that the drug was readily oxidized in vivo. A detailed study of phenothiazine metabolism in various animals and in man was undertaken by de Eds et al. [247] as early as 1938. The urine became recolored, presumably by oxidation, after the chloroform extraction of the red derivative of the drug. The red dye also lost its color on reduction and confirmed a dye/leuco-derivative system. A mixture of phenothiazine and its metabolite was precipitated from acidified urine and fractionated by recrystallization from ethanol. The metabolite was first erroneously identified as thionol on the premise that its polarographic behavior was similar to this compound, although the authors were unable to prepare the expected acetyl derivative. Ultimately, the phenothiazone structure was assigned [248].

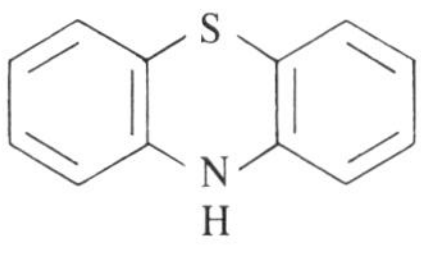

phenothiazine

Lipson [557] observed that chloroform eluted phenothiazone but not thionol in the column chromatography of phenothiazine metabolites on alumina. A crystalline metabolite was extracted from sheep urine by Collier [214] and was chloroform eluted by the Lipson [557] method. Its identity with phenothiazone sulfate was confirmed by the coincidence of spectral behavior with the synthetic compound in various media [215]. An almost complete balance of phenothiazine excretion was given by Collier.

Benham [75] demonstrated by spectrophotometry that the major metabolite in the rabbit is thionol. Hydrochloric acid hydrolysis liberated large amounts from a conjugate with glucuronic acid. Although the conjugate has not been isolated, the glucuronic acid was determined by a variant of the napthoresorcinol method.

Clare [207] directly extracted a crystalline metabolite with ethyl acetate from calf's blood and aqueous humor. Its melting point and spectral properties were the same as phenothiazone sulfoxide. Sulfoxide as well as phenothiazone was identified in the alimentary canal by similar methods.

II. CHLORPROMAZINE

The sulfoxide was first characterized by Salzman, Moran, and Brodie in 1955 [792] in dog and human urine as a major metabolite of chlorpromazine. It was extracted by ethyl ether from alkalinized urine, separated from chlorpromazine by counter-current distribution, and isolated as the picrate. Its elemental analysis and UV spectrum were synonymous with those of authentic phenothiazine sulfoxide.

Salzman and Brodie [791] published in 1956 a method for determining chlorpromazine in the presence of its sulfoxide. Chlorpromazine was

---

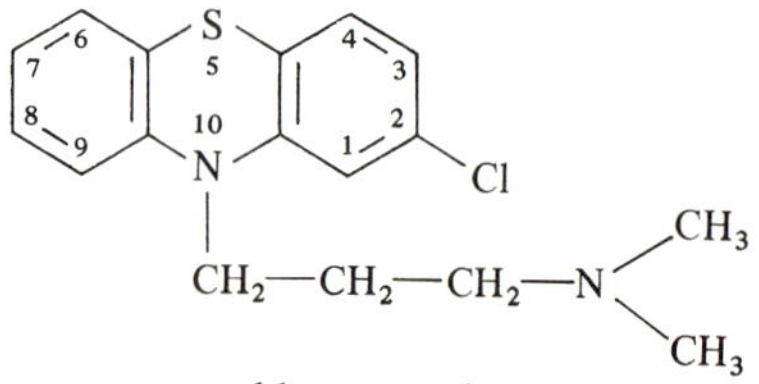

chlorpromazine

extracted from the alkalinized biological medium by heptane containing 1.5% of isoamyl alcohol. The sulfoxide was extracted simultaneously and further extracted by washing the heptane with a pH 5.6 buffer. Both products were then analyzed by UV spectrophotometry, after correcting for foreign substances carried through the procedure with a blank. Paper chromatography proved that phenothiazine extracted in this manner was not contaminated by any other phenothiazine.

Henriksen et al. [415] extracted chlorpromazine in ethyl ether from alkalinized urine and, after reaction with concentrated sulfuric acid, performed a colorimetric determination. No information was given on the method's specificity, but only 10% of the injected drug was recovered in urine.

Gillette and Kamm [358] applied the Salzman and Brodie [791] analytical procedure to demonstrate that sulfoxidation is an enzymic reaction which proceeds at a hepatic microsomal level. Guinea pig liver was fractionated by ultracentrifugation into microsomes sedimented at 78,000g and soluble enzymes. Neither isolated fraction had any activity on chlorpromazine, but their mixture was active. Chlorpromazine sulfoxide was identified by counter-current distribution, UV spectrophotometry, and paper chromatography (Table 16).

Flanagan et al. [330] published a method for the determination of chlorpromazine and its sulfoxide in biological media by ethyl ether extraction and UV spectrophotometry.

Porter and Beresford [717] determined chlorpromazine and its sulfoxide by polarography. The sulfoxide was directly reducible, whereas both the sulfoxide and chlorpromazine were determined after bromination.

TABLE 16

Solvent systems	R_f Chlorpromazine	Chlorpromazine sulfoxide
Butanol:acetic acid:water 50:12:50	0.83	0.79
Water saturated isoamyl alcohol	0.76	0.45
95% acetone	0.93	0.74

The intricacy of chlorpromazine metabolism was delineated by Fishman and Goldenberg [323], who examined in detail the ethylene dichloride extract of urine of chlorpromazine treated patients. Three paper chromatography systems were used:

(I)   Benzene:acetic acid:water, 2:2:1
(II) Ethylene dichloride:benzene:88% formic acid:water, 3:1:4:2
(III) Butanol:1M ammonium acetate, pH 7, 2:1

The separated components were detected by various reagents, or eluted in methanol for UV spectrophotometry or for a second chromatographic elucidation.

Chromatography of the extracted urine in system I separated seven components detectable by the Dragendorff reagent and persulfate. When the third component (R_f 0.25) was rechromatographed in systems II and III, three other metabolites were isolated. Six components exhibited the sulfoxide UV spectrum. One component reacted with ninhydrin as a primary amine and another with nitroprusside/acetaldehyde as a secondary amine. These two products had the same R_f values as authentic samples of mono- and didesmethylchlorpromazine sulfoxide.

Lin et al. [554] had previously shown that ethylene dichloride-extracted alkalinized urine has several polar metabolites separable on a cationic-exchange resin. The metabolites from human urine were separated into two groups by chromatography on Dowex 50 (H^+) followed by elution at increasing pH ranging from water to 1.4% ammonium hydroxide. Each group contained several metabolites separable by paper chromatography (isoamyl alcohol: water:ethanol:formic acid 100:100:15:10). The β-glucuronidase hydrolysis of urine resulted in an increase of free glucuronic acid and the disappearance of three water soluble metabolites.

Ross et al. [785] confirmed the N-demethylation by detecting $^{14}CO_2$ in the expired air of rats dosed with chlorpromazine ^{14}C-labeled on the N-methyl.

Goldenberg and Fishman compared chlorpromazine metabolism [371] in various animal species. Urine was successively extracted with ethylene dichloride at pH 11.5, at pH 9, and again at pH 9 after β-glucuronidase hydrolysis. The three fractions were chromatographed on paper in the previously described systems I and II [323], as well as in systems IV (butanol: ethanol:water 15:2:5) and V (58% ammonium hydroxide:methanol 1:1). The latter system used paper impregnated with a 10% acetone solution of peanut oil. Quantitative evaluations were effected by visual comparison, reflectometry, or colorimetry after elution. The mono- and didemethylated derivatives were more precisely identified after chromatographic separation by comparing their acetyl derivatives to authentic samples in the chromatographic systems.

The same authors found a new metabolite, chlorpromazine-N-oxide [327] in human urine. The ethylene dichloride extract of urine was chromatographed in system I. The component 5 (R_f 0.64) was eluted and chromatographed again or submitted to paper electrophoresis. It had the properties of a reference N-oxide. Color reactions, UV spectrum, and paper chromatography of its oxidation products confirmed its identity.

The N-oxide directly injected into a gas-liquid chromatograph decomposed almost quantitatively into 10-alkyl-2-chlorophenothiazine. This decomposition product had a different retention time than chlorpromazine according to Craig et al. [228].

Forrest et al. [333] isolated a metabolite, from human urine, that had a UV spectrum between chlorpromazine and its sulfoxide but was not the sulfone. It was reducible to chlorpromazine or oxidizable to the sulfoxide. A compound resembling this metabolite was produced by UV irradiation at 50-60° C of an aqueous chlorpromazine solution. Electron spin resonance spectrometry suggested a free radical structure, in agreement with the oxidation-reduction properties of the product and the [S—OH]· function was assigned by the authors.

Beckett et al. [56] did not confirm these findings. The UV irradiation of a chlorpromazine solution did not give any product with such a UV spectrum. Only that assignable to the sulfoxide was observed.

The first really quantitative investigation is the 1962 work of Emmerson and Miya [308] who administered ^{35}S-labeled chlorpromazine to rats. Total radioactivity of urine and feces was monitored and the urine was extracted by the method of Fishman and Goldenberg [323] before and after β-glucuronidase hydrolysis. The metabolites were paper chromatographed by the procedure of Eisdorfer and Ellenbogen [297], who had studied the paper chromatography of sulfoxide, N-desmethyl, or N-desmethyl sulfoxide derivatives in detail. The system ethylene dichloride : benzene : formic acid : water 3 : 1 : 4 : 2 (system II of Fishman and Goldenberg) appeared the most useful. It was recommended that two developments be run, one for 12 and the other for 24 hours, to achieve a good separation of the components of the mixture. Emmerson and Miya chromatographed in the dark and under nitrogen atmosphere. They monitored the radioactivity of every separated component and visualized them with various reagents and under UV light.

Eight metabolites and unchanged chlorpromazine were found in rat urine. The total comprised 44% of the administered dose whereas 37% was found in the feces. The three main metabolites were identified by specific reagents: one was a primary amine (ninhydrin) and another a secondary amine (nitroprusside acetaldehyde). These three metabolites appeared to correspond to the sulfoxides of chlorpromazine, mono-, and didesmethylchlorpromazine. An additional 10 to 15% of the administered radioactivity was extractable from urine after 18 hours incubation with or without β-glucuronidase.

Later Fishman and Goldenberg directed their efforts toward the identification of the phenolic metabolites of chlorpromazine [325]. Human or dog urine, before or after β-glucuronidase hydrolysis (18 hours at 37°C, pH 4.5) was extracted with ethylene dichloride at pH 9. The extracts were chromatographed on paper in various solvent systems that included the five already mentioned.

After 3 to 4 days of prolonged chromatography in system II, three components were detected with persulfate. A second chromatography in the second dimension with system IV resulted in four spots. Three were lavender and the fourth was purple-blue on persulfate development. One reacted as a primary amine, the second as a secondary amine, and the two others as tertiary amines. The fourth component was isolated by paper chromatography and reduced by zinc and acetic acid to give one of the three phenolic metabolites; it must therefore have been its sulfoxide. This phenolic metabolite, when treated by diazomethane, gave an O-methyl derivative similar to 7-methoxychlorpromazine. The two other phenolic metabolites were primary or secondary amines that, on reaction with methyl iodide, produced 7-hydroxychlorpromazine. These four metabolites were both free and conjugated with glucuronic acid in urine, as shown by comparison of assays on hydrolyzed and unhydrolyzed urine.

Beckett et al. [56] in 1963 separated and purified chlorpromazine metabolites on a preparative scale using 15 to 18 l of human urine. Nevertheless several excretion products have not been identified with certainty. The amounts were small and reference compounds were unavailable.

The urine was continuously extracted with butanol. The solid matter that precipitated from the extract on concentration was purified by redissolving it in dimethylformamide and reprecipitating at pH 4.5. This fraction gave five spots on paper chromatography (isoamyl alcohol : water : ethanol : 88% formic acid 10 : 10 : 1.5 : 1 according to [554]). These five metabolites were separated by chromatography on 16 sheets of paper and eluted. Purification of the major component was effected by paper electrophoresis and it was now homogeneous on paper chromatography. Elemental analysis, potentiometric molecular weight, IR spectrum, and various spot tests identified it as N-didesmethylchlorpromazine glucuronide. It was hydrolyzed with β-glucuronidase (24 hours at 37°C, pH 4.5). The aglycones were extracted in ethyl ether and chromatographed on paper. The three found were also obtained on acid hydrolysis of the whole urine (3% of hydrochloric acid, 5 minutes in a boiling water bath). Spot tests and UV or IR spectra could only provide indicative information on the identity of two of the other metabolites which were present in much smaller quantities. No analyses were possible on the last two of the five.

The filtrate of the butanol extractions, after separation of the precipitate, was concentrated and extracted with 0.1N hydrochloric acid. Four nonpolar metabolites were recovered and identified by paper chromatography as chlorpromazine and the sulfoxides of chlorpromazine and its two demethylated derivatives. The butanol solution was concentrated further and chromatographed on an alumina column. Chloroform elution yielded two fractions, each homogeneous on paper chromatography but not precisely identified.

The sum of the isolated metabolites in the 24 hour urines represented 7% of the ingested chlorpromazine dose.

Posner [718] reported in 1959 that the chlorpromazine-administered patients had urine that, after β-glucuronidase hydrolysis and paper chromatography of the butanol extract, showed three components which differed from those observed without hydrolysis.

He published a method in 1963 for the determination of the urinary metabolites of chlorpromazine [719]. Urine (0.5 ml) adjusted to pH 9, was extracted with ethyl ether to give the unconjugated basic metabolites (I). β-Glucuronidase was added to the aqueous fraction adjusted to pH 5.0 and the mixture was incubated for 18 hours at 37°C. The aglycones of the conjugated metabolites were extracted by chloroform after alkalinization (II).

The polar metabolites, conjugated and unconjugated, were extracted with butanol from the aqueous solution at pH 3.0 (III). The three extracts were chromatographed on Whatman 3MM paper, buffered by dipping in a 0.25M citrate solution pH 5.7, and dried.

A mixture of dioxane:water 3:1 was distilled over ferrous sulfate and the azeotrope 87-88° C collected. The developing solvent was azeotrope: citric acid:ammonium hydroxide 400 ml:1.6 g:0.6 ml.

The development was ascending after one day's saturation of the cage. Detection was accomplished by spraying a mixture of concentrated sulfuric acid:water:95% ethanol 1:1:8 and the evaluation was quantified by densitometry.

Seventeen metabolites were separated under these precise conditions. The total represented about 16% of the ingested dose (Table 17).

The identifications in this table were obtained through a series of color reactions and comparison with authentic samples. The color reactions on paper with 2,6-dichloroquinonechloroimide and sodium metaperiodate indicated the hydroxyl position on the phenothiazine ring (see promazine). Sulfuric acid (50%) developed a purple color with monooxygenated derivatives and a blue color with dioxygenated derivatives. The known compounds exhibited on paper chromatography and electrophoresis (500 V, 6.5 hours, 0.05M phosphate buffer pH 7.0) the characteristics given in Table 18.

The authors claimed a routine method applicable to small urine samples although some secondary metabolites of chlorpromazine were neglected.

Johnson et al. [461] fractionated the metabolites by extracting urine first at pH 11.5 with ethylene dichloride to remove relatively nonpolar unconjugated metabolites. A subsequent extraction at pH 2-3 with tetrahydrofuran after the addition of sodium chloride removed the polar conjugated metabolites, 7-hydroxychlorpromazine sulfates and glucuronides.

Price et al. [737] described a procedure for the nearly quantitative isolation of 7-hydroxychlorpromazine in human urine. Urine (250 ml) was concentrated under vacuum at room temperature and 100 ml of methanol were added to the resulting syrup. The insoluble portion was discarded and the extract dried. This solution was redissolved in 0.2M phosphate buffer, pH 6.8, and the metabolite extracted into methylene chloride, reextracted into 0.1N hydrochloric acid and, again, into methylene chloride at pH 10. The residue had the same UV spectrum and the same R_f on paper as the authentic compound.

TABLE 17

Number		R_f	Color	Probable identity
	1	Front	Purple-pink	Chlorpromazine + its N-oxide + phenol
	2	0.88-0.90	Purple	Chlorpromazine monophenol
	3	0.56-0.66	Purple	Monodesmethylchlorpromazine monophenol
I	4	0.40-0.56	Purple	Didesmethylchlorpromazine monophenol
	5	0.38	Pink	Chlorpromazine sulfoxide
	6	0.29-0.39	Pink	Monodesmethylchlorpromazine sulfoxide
	7	0.17-0.32	Pink	Didesmethylchlorpromazine sulfoxide
	8	0.21	Pink	?
	9	Front	Purple	Neutral phenol related to chlorpromazine
	10	0.83-0.93	Purple	Chlorpromazine monophenol
	11	0.41-0.62	Purple	Monodesmethylchlorpromazine monophenol
II	12	0.33-0.58	Purple	Didesmethylchlorpromazine monophenol
	13	0.19-0.40	Blue	Similar to 17 (?)
	14	0.16	Purple	?
	15	0.96-1.00	Purple	?
III	16	0.73-0.77	Purple	?
	17	0.33-0.57	Blue	Diphenolic compound

Goldenberg and Fishman [372] sharply criticized the paper of Price et al. and demonstrated that the product isolated by their method was not unique. After paper chromatography according to the technique of Price et al. (butanol : ethanol : water 5 : 2 : 2), the spot was eluted and submitted to bidimensional thin-layer chromatography (I, chloroform : acetone : diethylamine 2 : 7 : 1; II, ethyl acetate : acetone : methanol : diethylamine 68 : 2 : 20 : 15). At least 14 metabolites were detected and most of them identified. Five of them were in the 7-hydroxychlorpromazine series.

Thin-layer chromatography was described by Cochin and Daly [211] for various phenothiazines and their metabolites after extraction at pH 9 with ethylene dichloride containing 10% of isoamyl alcohol. The following systems were applied:

TABLE 18

Products	R_f	Electrophoretic migration (cm)
2-Chlorophenothiazine	0.98	0.0
2-Chloro-10-(3'-hydroxypropyl) phenothiazine	0.98	0.0
Chlorpromazine	0.92	-13.0
Chlorpromazine-N-oxide	0.92	- 3.0
Monodesmethylchlorpromazine	0.67	—
Didesmethylchlorpromazine	0.61	—
Chlorpromazine-N-oxide sulfoxide	0.67	- 4.5
Chlorpromazine sulfoxide	0.52	-20.0
Monodesmethylchlorpromazine sulfoxide	0.41	-20.0
Didesmethylchlorpromazine sulfoxide	0.31	-20.0

- Benzene : dioxane : ammonium hydroxide, 60 : 35 : 5
- Ethanol : acetic acid : water, 50 : 30 : 20
- Methanol : butanol, 60 : 40
- Benzene : dioxane : ammonium hydroxide, 10 : 80 : 10

For bidimensional separations, the second of the above-listed systems was followed by benzene : dioxane : ammonium hydroxide 50 : 45 : 5. Cochin and Daly detected only chlorpromazine and desmethylchlorpromazine sulfoxides.

Eberhardt et al. [292] employed the system pyridine : petroleum ether : methanol 10 : 45 : 1 to separate on thin layer the metabolites extracted with chloroform after acidification of the medium to pH 1 by hydrochloric acid.

Bidimensional thin-layer chromatography was applied by Goldenberg and Fishman in their investigations on promazine metabolism [373] and was used for various phenothiazines including chlorpromazine [326]. The procedure is described in Section III, on promazine. The unconjugated fraction revealed 14 components with 24 found after hydrolysis. Most of these derivatives were identified on the basis of their chromatographic behavior and the colors they gave with various reagents as compared to authentic samples. The action of oxidizing or reducing reagents completed the identifications.

In 1962 VandenHeuvel et al. [953] briefly reported the application of gas-liquid chromatography to the study of chlorpromazine metabolism, whereas Martin et al. [611] in 1963 separated chlorpromazine from its mono- and didemethylated derivatives after acetylation. A column of 5% of silicone SE-30 on a highly silanized support at a temperature of 270°C with a flame ionization detector was used.

These authors [279] were also able, under the same conditions, to separate and determine chlorpromazine, its mono- and didemethylated derivatives, and the three corresponding sulfoxides from human urine with a 70-80% yield. The demethylated derivatives were acetylated and the sulfoxides reduced before injection.

Gudzinowicz recently reported [388] that the electron capture detector is 10 to 50 times more sensitive than flame ionization for chlorpromazine and its derivatives in the range 0.05 to 1.50 µg/µl.

Only in 1965 was the first large-scale application of gas-liquid chromatography to chlorpromazine metabolites investigation published. Johnson et al. [462] used the microcoulometric determination equipment of Dohrmann as a detector. The chlorine or sulfur was titrated after combustion and provided selective detection, since urine does not contain any component that may interfere, whereas the flame ionization detector makes it necessary to check that every supposed metabolite does not correspond to a normal urinary component. This detection system is highly sensitive (0.1 µg) and does not require calibration; its response is absolute. Although the electron capture detector is even more sensitive (0.1 ng), it does record various urinary components which are not metabolites.

Urine was extracted at pH 13 with ethylene dichloride and chromatographed either directly or after zinc reduction or acetylation. The chromatographic column was filled with glass beads coated with 0.15% of XE-60, 0.1% of Carbowax 20M and 2% of diatomaceous earth according to the procedure of Johnson et al. [464]. The system was completely glass from the injection block to the detector. The column temperature was programed from 130° to 230°C (10°/minute).

The retention times for reference compounds (in minutes) are given in Table 19.

The demethylated derivatives and 7-hydroxychlorpromazine must be acetylated to avoid tailing. Chlorpromazine sulfoxide must be reduced.

TABLE 19

Product	Retention time (minutes)
Promazine	0.9
Phenothiazine	1.0
Chlorpromazine-N-oxide	1.5
Chlorpromazine	2.0
Monodesmethylchlorpromazine	3.0
2-Chlorophenothiazine	4.0
Didesmethylchlorpromazine	4.1
Chlorpromazine-N-oxide sulfoxide	4.8
Chlorpromazine sulfoxide	5.0
Acetylmonodesmethylchlorpromazine	6.0
Monodesmethylchlorpromazine sulfoxide	6.2
Acetyldidesmethylchlorpromazine	6.5
Didesmethylchlorpromazine sulfoxide	6.7
2-Chlorophenothiazine sulfoxide	7.0
Acetylmonodesmethylchlorpromazine sulfoxide	7.0
Acetyldidesmethylchlorpromazine sulfoxide	8.0
Acetyl-7-hydroxychlorpromazine	8.1

Under these conditions nine metabolites in addition to chlorpromazine were identified in human urine (no assay has been performed on the conjugated fraction):

- Chlorpromazine-N-oxide;
- Chlorpromazine sulfoxide;
- 2-Chlorophenothiazine;
- Monodesmethylchlorpromazine;
- Didesmethylchlorpromazine;
- Monodesmethylchlorpromazine sulfoxide;
- Didesmethylchlorpromazine sulfoxide;
- 7-Hydroxychlorpromazine;
- 2-Chlorophenothiazine sulfoxide.

Their identification was achieved by comparison with authentic compounds and by modifying their retention times by acetylation or reduction. It was confirmed by thin-layer chromatography, which proved that there was no transformation of the metabolite in the injection block or on the column. Quantification was easy and calibration experiments demonstrated a yield higher than 80% from urine. The entire analysis took 30 to 45 minutes.

This gas-liquid chromatography procedure with microcoulometric detection concomitant with thin-layer chromatography permitted Rodriguez and Johnson [779] to find a new metabolite in the chloroform extract of urine at pH 3 [463]. After extraction with ethylene dichloride according to the method of Johnson et al. [461], the urine at pH 2 was extracted with chloroform. 2-Chlorophenothiazine was identified by gas-liquid chromatography under the conditions previously mentioned. Since it is not extractable in chloroform, it must have been a degradation product of the metabolite. Three other peaks were recorded. Thin-layer chromatography on silica gel in five different solvent systems gave four spots, one of which corresponded to authentic 2-chlorophenothiazine-N-propionic acid. This metabolite was isolated by chromatography on a Dowex 1 X-8 (Cl$^-$) column by means of a solvent with a variable formic acid concentration. It was precisely identified by its IR spectrum.

Burchfield and Wheeler [154] again emphasized the advantages of the microcoulometric detector in metabolic studies on products containing Cl, S, P, or even N. They described the procedures used and mentioned two detector types.* Type C-100 detects 0.1 to 0.01 µg of Cl; type C-200 as little as 0.001 µg of Cl and approaches the sensitivity of the flame ionization detector. Johnson and Burchfield [460] reviewed the gas-liquid chromatography of chlorpromazine and its metabolites.

In addition to their wide use of gas-liquid chromatography, Rodriguez and Johnson briefly reported in 1966 [780] a procedure for the separation and determination of 15 chlorpromazine metabolites by thin-layer chromatography. The metabolites were determined colorimetrically after elution. This procedure was more rapid than gas-liquid chromatography as it permitted the analysis of several samples simultaneously.

Miscellaneous studies on chlorpromazine metabolism included Robinson's [776] investigation on in vitro metabolism by thin-layer chromatography.

---

* Dohrmann Instrument Co.

Ragland and Kinross-Wright [742] performed a total determination of phenothiazines metabolites by spectrofluorimetry after hydrogen peroxide oxidation in acetic medium (sensitivity: 0.5 μg). The gross method of Bolt et al. [99] permitted the determination of all conjugated and nonconjugated metabolites in urine. Forrest et al. [332] concentrated the metabolites in quickly frozen urine in plastic bottles. They were thawed slowly at 5-8°C and the melted fractions collected at intervals. In the first 10.6% of the original volume, 61.8% of the total nonconjugated metabolites and 51.1% of the conjugated metabolites were found.

III. PROMAZINE

The study of promazine metabolism commenced in 1958, after that of phenothiazine and chlorpromazine. Walkenstein and Seifter [969] administered ^{35}S-labeled promazine to dogs and rats and examined urinary excretion and blood levels. Nonmetabolized promazine was determined, after addition of the unlabeled compound, by extraction and crystallization as the maleate until constant specific activity was obtained.

Labeled promazine was used to follow distribution in different organs. Paper chromatography (butanol:1M ammonium acetate, pH 7.0 2:1) revealed at least six metabolites, three of which were identified:

	R_f
Promazine sulfoxide	0.73
Monodesmethylpromazine	0.74
Monodesmethylpromazine sulfoxide	0.43

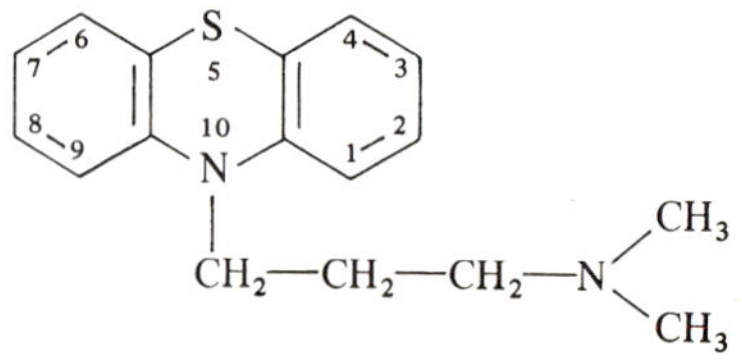

promazine

The distinction between the two first components was established by using an unlabeled carrier. The sum of promazine and the three metabolites represented 10% of the administered dose.

Posner et al. [719] in their investigations on chlorpromazine studied the properties of hydroxypromazines (for experimental conditions, see Section II, on chlorpromazine, p. 59). The results are summarized in Table 20.

Beckett and Curry [60] were able to distinguish between the various hydroxypromazines by means of their absorption spectra when they were dissolved in a mixture of concentrated sulfuric acid : 70% ethanol 1 : 1 The results are shown in Table 21.

Four glucuronides were isolated from the urine of promazine-dosed patients by means of continuous electrophoresis and paper, column, and thin-layer chromatography. Three of them were sufficient to give spectra in sulfuric acid which were similar to, but not the same as, 3-hydroxypromazine. Study of corresponding aglycones led to similar conclusions. The authors postulated dihydroxylation with one OH in position 3 [61].

A detailed analysis of promazine biotransformations was published by Goldenberg, Fishman et al. [373], who detected 26 to 30 metabolites in human and animal urine. Thin-layer chromatography on silica gel was used, complemented in some instances by the paper chromatographic methods they developed for chlorpromazine.

The unconjugated metabolites were extracted from urine at pH 9 with ethylene dichloride. The remainder of the urine was divided into three parts. One was hydrolyzed with β-glucuronidase and the second with sulfatase; the third one was kept as a reference. All three were extracted in the same manner as the crude urine. The extracts were chromatographed on silica gel thin layer (activated 1 hour at 110° C) with the following systems:

- Acetone : isopropanol : 25% ammonium hydroxide, 9 : 7 : 4
- Acetone : isopropanol : 1% ammonium hydroxide, 9 : 7 : 4
- Acetone : diethylamine, 9 : 1
- Ethanol : dioxane : benzene : ammonium hydroxide, 5 : 40 : 50 : 5
- Ethyl acetate : acetone : methanol : diethylamine, 13.6 : 0.4 : 4 : 3
- Chloroform : acetone : diethylamine, 8.6 : 0.4 : 1

In addition, bidimensional thin-layer chromatographies were run in the systems:

- for the unconjugated metabolites:

TABLE 20

Product	Paper chromatography (R_f)	Paper electrophoresis (cm)	Color after spraying with: 2,6-Dichloroquinonechloroimide	Color after spraying with: Sodium metaperiodate
Promazine	0.80	−20.0		
1-Hydroxypromazine	0.77	− 5.5	Light blue	Blue-purple
2-Hydroxypromazine	0.69	−12.0	Green	Light green
3-Hydroxypromazine	0.69	−13.8	Purple	Purple
4-Hydroxypromazine	0.68	−13.8	Dark blue	Pink

TABLE 21

Metabolite	Wavelength of the absorption maximum (mμ) [a]					
1-Hydroxypromazine		(270)	281	343	477	513
2-Hydroxypromazine	219		278	343	(440)	490	558
3-Hydroxypromazine			278	372			568
4-Hydroxypromazine		(262)	271	292	486	500

[a] Figures in parentheses correspond to shoulders.

Primary dimension: chloroform : acetone : diethylamine, 8.7 : 0.3 : 1
Secondary dimension: chloroform : acetone : diethylamine, 2 : 7 : 1
- for the conjugated metabolites (after hydrolysis):
Primary dimension: ethyl acetate : isopropanol, 65 : 35
Secondary dimension: ethyl acetate : acetone : methanol : diethylamine, 13.6 : 0.4 : 4 : 3

The metabolites were detected with persulfate. Primary amines were detected with ninhydrin and secondary amines with nitroprusside-acetaldehyde. Quantification was achieved by comparison with dilutions of reference-compound solutions chromatographed simultaneously. The current reactions of oxidation, reduction, and O-methylation were applied, as well as UV spectrophotometry after elution. Most of the detected metabolites were identified by using authentic compounds as references. Of the ingested promazine 33% were recovered in human urine and 10-13% from the dog or rabbit.

Fishman and Goldenberg slightly modified their procedure [326]. The following system was employed for the conjugated fraction after hydrolysis:

- Primary dimension: acetone : isopropanol : 1% ammonium hydroxide, 9 : 7 : 4
- Secondary dimension: ethyl acetate : acetone : methanol : diethylamine, 68 : 2 : 20 : 15

By means of these sophisticated chromatographic techniques, Fishman and Goldenberg demonstrated metabolites not only of the various demethylated, hydroxylated, and oxidized chlorpromazine derivatives but also of the chain scission products: phenothiazine, phenothiazine sulfoxide, and 3-hydroxyphenothiazine.

IV. OTHER PHENOTHIAZINES

Pecazine studies have been published in the German literature. Kleinsorge et al. [508] extracted urine with ethyl ether and paper chromatographed the extract (butanol:isoamyl alcohol:acetic acid:water 6:18:4:5). Two metabolites were detected in addition to the unchanged drug.

Hoffmann et al. [426] showed that the sulfoxide is the major metabolite under the same conditions. However, Block [97], using a ^{14}C-labeled pecazine, observed several metabolites and reported that Hoffmann's results were erroneous. When a peroxide containing ethyl ether was used, pecazine was changed in part into its sulfoxide during the extraction.

Symchowicz et al. [886] used ^{35}S-labeled perphenazine by paper chromatography (tertioamyl alcohol:acetic acid 95:5) and separated the unchanged drug from its sulfoxide in urine and organ extracts. Other metabolites have not been identified.

Fluphenazine enanthate was studied by Ebert and Hess [294], who prepared the ^{14}C-labeled compound and studied the mechanism by which this drug exerts a prolonged pharmacological action. Fluphenazine base and its sulfoxide were isolated from tissues, blood, and urine, as was the unchanged drug. The investigation of other metabolites have not been reported. Paper chromatography (benzene:acetic acid:water 2:2:1) was used and in some circumstances was complemented by thin-layer chromatography on silica gel (benzene:dioxane:15N ammonium hydroxide 60:35:5). The metabolites were extracted with ethylene dichloride, which appeared to be the best extraction solvent. The conjugated metabolites were hydrolyzed with

---

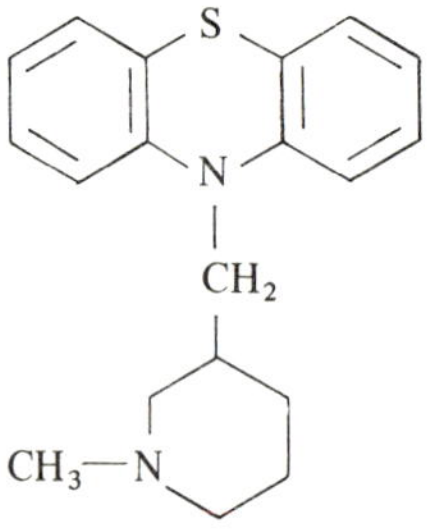

pecazine

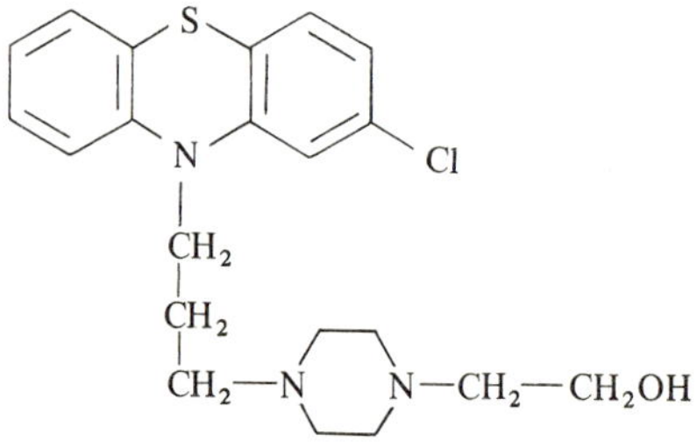

perphenazine

β-glucuronidase (1000 U/2.5 ml) or phenosulfatase (1 U/ml) for 18 hours at 37°C, pH 5, or by treatment with 1N hydrochloric acid for 24 hours under reflux.

Homofenazine is related to fluphenazine and was studied by Thiemer and Sadtler [899]. Its major metabolite was the sulfoxide, which was separated from the unchanged drug by thin-layer chromatography on silica gel (methanol : acetic acid 99.5 : 0.5). The two products can be determined in the presence of each other by UV spectrophotometry after ethylene dichloride extraction at pH 9.0.

The fate of thioridazine was studied by Zehnder et al. [1042] in the rat. These authors first prepared the drug labeled by ^{35}S on the phenothiazine ring or by ^{14}C on the N-methyl, and a series of nonradioactive metabolites. Oxidation of thioridazine gave two monosulfoxides, the disulfoxide, and the disulfone.

The analysis of metabolites in urine and bile was performed by reversed isotopic dilution. Known amounts of each of the three sulfoxides were added to the biological fluids, which were lyophilized and extracted with methanol.

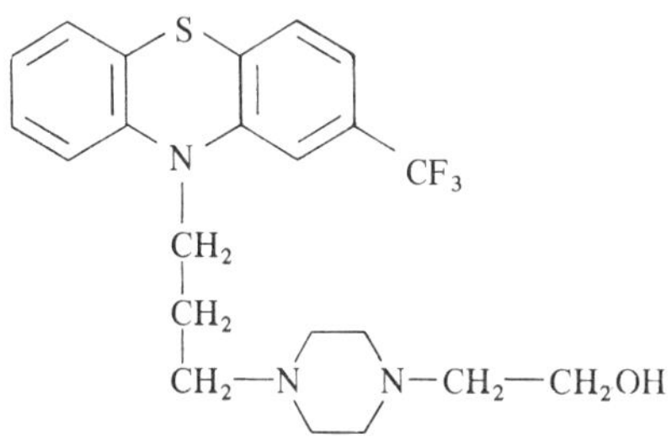

---

fluphenazine

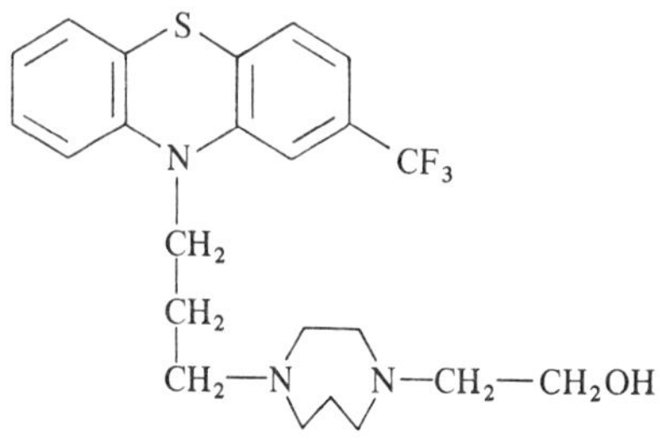

homofenazine

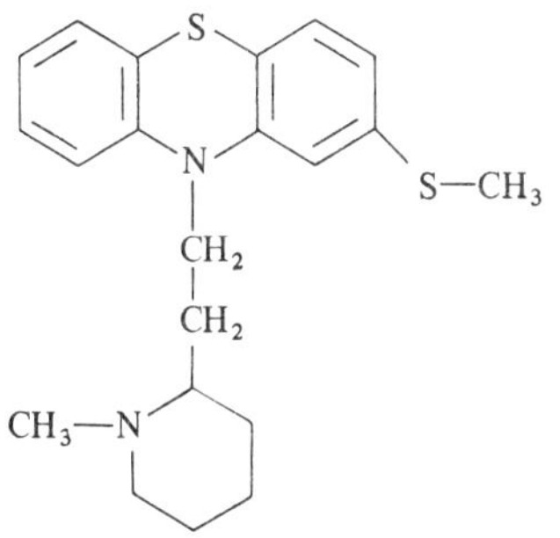

thioridazine

The methanol extract, after evaporation and redissolution in chloroform, was chromatographed on an alumina column. The mixture of the two sulfoxides was obtained by elution with benzene: chloroform 3:1 and then the disulfoxide was eluted with chloroform alone. The two monosulfoxides were separated by counter-current distribution (cyclohexane: 60% methanol). The three isolated metabolites were crystallized as their fumarates until constant activity. Another fraction of urine or bile to which the disulfone had been added was extracted directly with methanol and chloroform and crystallized until constant activity.

The total of the metabolites derived from thioridazine or desmethylthioridazine were oxidized with hydrogen peroxide into the corresponding disulfones and their total amount was determined by radioactive measurements after extraction.

The metabolites thus separated and determined represented, with respect to the total excreted radioactivity, 20% in bile and 50% in urine. The remainder was the glucuronides of the hydroxylated derivatives and represented the main fraction of the administered dose (about 80%).

Philipps and Miya [710], after administration of ^{35}S-labeled prochlorperazine, found three unidentified metabolites in urine, in addition to the unchanged drug, by means of paper chromatography (isoamyl alcohol: tertioamyl alcohol: formic acid: water 5:5:4:8).

V. THIOXANTHENES

The drugs derived from thioxanthene are chemically related to the phenothiazines and their metabolism is analogous.

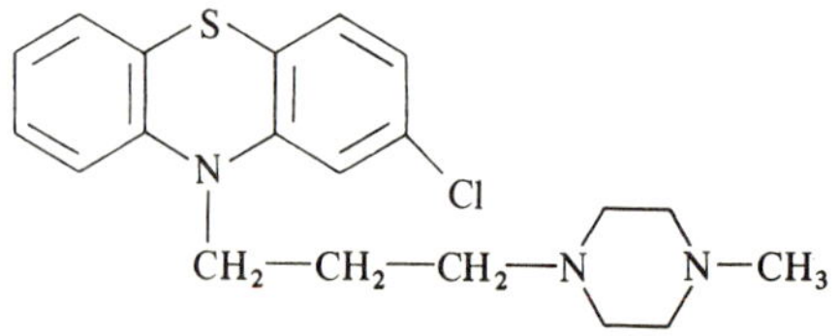

prochlorperazine

Lucanthone was studied after injection to various animals and man by Gönnert [376]. Paper chromatography permitted the isolation of several metabolites. The principal was a chromopeptide, the chromophore was lucanthone sulfoxide, and the peptide chain contained at least four amino acids. The sulfoxide was obtained on oxidation of lucanthone in vitro. The analytical details were not published.

Strufe [880] separated various groups of metabolites from human urine. Urine concentrated tenfold was extracted with ethyl ether at pH 9.6. The extract was shaken with diluted sulfuric acid, alkalinized, and extracted again with ethyl acetate to yield the basic metabolites. The ethyl ether solution was treated with sodium hydroxide and extracted at pH 2.2 to yield the acid fraction. The neutral fraction remained in the ether after these two treatments. Each fraction was evaluated for in vitro schistosomicidal activity. Chromatography on an alumina column isolated a metabolite as active as lucanthone. Hydrolysis in hydrochloric acid gave a lucanthone sulfoxide identified by its IR spectrum, and amino acids characterized by paper chromatography: thus the metabolite was a chromopeptide. The sulfoxide was the actual active metabolite.

Rosi et al. [784] found a new active metabolite of lucanthone that resulted from hydroxylation of the methyl. Its structure was established by IR and NMR spectrometry but no details have been given on its isolation or analysis.

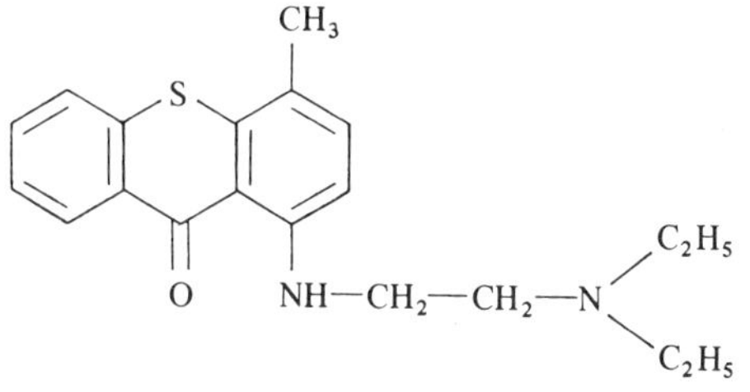

lucanthone

Chlorprothixene was excreted in rat or human urine as its sulfoxide. This was identified by Allgén et al. [21] by a UV spectrophotometric study of its reaction with concentrated sulfuric acid and by paper chromatography as compared to a reference compound. It was also found in feces with the unchanged drug but was alone in bile.

Methixene exhibited several metabolites in rabbit and human urine. Lehner et al. [550] extracted them into ethylene dichloride from the alkalinized urine and applied thin-layer chromatography on silica gel to compare with reference compounds in four solvent systems (Table 22).

Chromatograms were sprayed with Dragendorff reagent and diazotized p-nitroaniline to detect the hydroxylated derivatives. Beside the administered drug, the two stereoisomeric sulfoxides (α and β) of methixene and desmethylmethixene were identified in this manner.

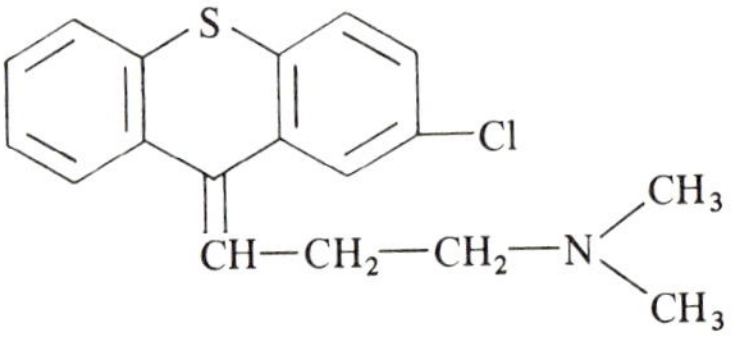

chlorprothixene

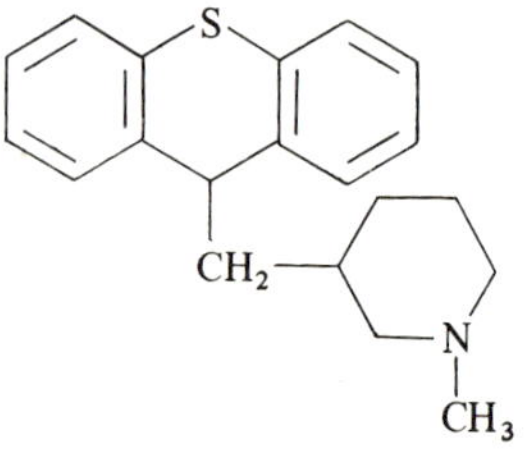

methixene

TABLE 22

Product[a]	R_f in system:[b]			
	a	b	c	d
Methixene base	1.00	1.00	1.00	1.00
Methixene sulfoxide A	0.42	0.66	0.76	0.31
Methixene sulfoxide B	0.53	0.79	0.76	0.59
Methixene sulfone	0.57	0.69	0.67	0.53
Methixene N-oxide	1.10	0.45	0.00	0.11
2-Hydroxymethixene	1.00	0.83	0.08	—
N-Desmethylmethixene	1.30	0.53	0.51	0.44
N-Desmethylmethixene sulfoxide A	0.83	0.29	0.23	0.14
N-Desmethylmethixene sulfoxide B	1.00	0.41	0.25	0.23
N-Desmethylmethixene sulfone	0.77	0.35	0.61	0.11

[a] A and B indicates the two stereoisomers, α and β.

[b] Solvent systems: (a) butanol:acetic acid:water 4:1:1; (b) 2N ammonium hydroxide saturated butanol; (c) chloroform:cyclohexane:diethylamine 5:4:1; (d) heptane:chloroform:95% ethanol 1:1:1.

Chapter V

DIBENZAZEPINES AND BENZODIAZEPINES

The metabolic investigations of these drugs are relatively recent. It is of interest that several of the metabolites that have been discovered (desmethyl-imipramine, oxazepam) are now themselves used as drugs.

I. DIBENZAZEPINES

Schindler [804] isolated the first metabolite of imipramine in urine by paper chromatography (toluene : butanol : 2N hydrochloric acid 8 : 2 : 10; ammonium hydroxide : water : ethanol 2 : 8 : 5). The oxidation product of imipramine by molecular oxygen in the presence of Fe^{II} and ascorbic acid appeared similar to this metabolite and was identified as 2-hydroxyimipramine by UV and IR spectrophotometry and by a series of chemical reactions.

Herrmann and Pulver [418] detected five metabolites of imipramine and succeeded in identifying three of them. Urine was extracted at pH 9-10 with ethylene dichloride and the extract chromatographed on paper (2.5% ammonium hydroxide : ethanol 200 : 15). Diazotized p-nitroaniline revealed four spots and the 2-desmethyl and 2-hydroxyl derivatives of imipramine, and imipramine itself, are characterized by comparison with authentic compounds

---

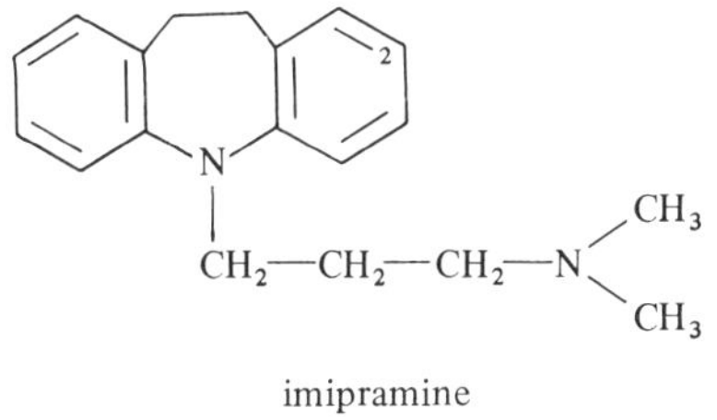

imipramine

similarly chromatographed. The metabolites which remained in aqueous solution were adsorbed on carbon and eluted by aqueous ethanol. The eluate was chromatographed on paper (butanol : ethanol : water 15 : 2 : 5). The three spots that appeared did not correspond to any of the above-mentioned derivatives.

The identification of these metabolites was completed by the preparative isolation of unchanged imipramine and its hydroxylated derivative from patients' urine. Urine (175 l) was repetitively extracted with ethylene dichloride. These extracts were washed by a 1N sodium carbonate solution, concentrated, and then chromatographed on alumina. The chromatography fractions containing iminodibenzyl derivatives were pooled and partitioned among the system water : methanol : pentane. The pentane fraction contained imipramine isolatable as the hydrochloride. 2-Hydroxyimipramine crystallized from the methanol solution after various treatments.

The water-soluble metabolite was isolated from the urine of rabbits that received high doses of imipramine per os. It was adsorbed on carbon and eluted by 40% and then 80% ethanol. The addition of an equal volume of ethyl ether to the ethanolic solution permitted its recrystallization. The glucuronide was hydrolyzed either with β - glucuronidase or 4N sulfuric acid to yield 2-hydroxyimipramine.

Three other metabolites were found with improvement of the chromatographic procedures [419]. When the paper is dipped in a 10% peanut oil-acetone solution, the developing solvent ammonium hydroxide : methanol 1 : 1 is used for polar metabolites, whereas the 4 : 1 ratio is used for the bases resulting from the hydrolysis of water-soluble metabolites. Didesmethylimipramine, 2-hydroxydesmethylimipramine, and its glucuronide were thus identified and compared to specially synthetized reference compounds.

Herrmann [416] described a quantitative paper chromatographic method for determining imipramine and five of its metabolites in urine and organs. Urine at pH 9-9.5 was extracted with ethylene dichloride. The organs were ground in the mixture 1N sodium bicarbonate solution : sodium hydroxide 1 : 1 and extracted similarly. The separation was performed on a paper dipped in a 10% peanut oil-acetone solution. Two solvent systems have to be used on two different paper sheets to achieve a complete separation:

- Methanol : ammonium hydroxide, 1 : 1
- Methanol : ammonium hydroxide : chloroform, 36 : 39 : 15

The separated products were eluted and determined by spectrophotometry after staining.

This method was used by Kuhn [529] to study the relationship between the urinary excretion of imipramine and its metabolites, and the clinical changes in patients.

The demethylation was confirmed by Bernhard and Beer [79], who found, after administration of imipramine, ^{14}C labeled on a methyl group, an important part of the radioactivity of the expired carbon dioxide.

Subsequent work in this area was conducted by Gillette, Brodie et al. Their first paper [357] reported the isolation of desmethylimipramine in rat brain. Pharmacological assays had shown that the antireserpine activity of imipramine was not related to its concentration in the brain and it was thus assumed that an active metabolite was present. Desmethylimipramine was extracted into heptane containing 1.5% of amyl alcohol and separated from imipramine by reextraction into a 0.2M phosphate buffer, pH 5.9. It was identified by the previously discussed paper chromatographic method [419] and by gas-liquid chromatography on a column coated with 0.5% of silicone SE-30.

This work was later extended to metabolic differences among various animal species [261]. The same paper chromatographic procedure was used [419] and imipramine and desmethylimipramine were fluorometrically determined after separation by extraction from heptane with a phosphate buffer. The R_f values are given for the paper chromatography in Table 23; concomitant enzymic assays were performed.

Gillette and his collaborators conducted a clinical study on desmethylimipramine and its determination in plasma was developed by Yates et al. [1030]. Unfortunately, the various metabolites were not distinguishable from the unchanged drug by this method.

Fishman and Goldenberg [324] extracted imipramine-N-oxide from urine at pH 10.7 with ethylene dichloride and applied paper chromatographies in eight different solvent systems to the extract with a synthetic N-oxide as a reference. In all cases a metabolite comparable to the reference N-oxide was found.

Cramer and Scott [230] have discovered new imipramine metabolites in the urine and feces of patients. After ethylene dichloride extraction, the aqueous phase was incubated for 4 hours with β-glucuronidase. The

TABLE 23

Product	R_f
Desmethylimipramine	0.37
Didesmethylimipramine	0.50
2-Hydroxyimipramine	0.73
2-Hydroxydesmethylimipramine	0.81

metabolites were separated by thin-layer chromatography with various solvents and detected under UV light, or with the mixture 50% sulfuric acid: ethanol 4:1, or with 0.1N hydrochloric acid, 10 minutes at 100°C. In this way 10-hydroxyimipramine and 10-hydroxydesmethylimipramine were found in plasma and urine, but not in the feces. The presence of free iminobenzyl and 10-hydroxyiminobenzyl can be considered likely.

Bickel et al. have published several papers on imipramine metabolism. They sometimes operated in vivo [87], although mainly in vitro [85, 86], with their analytical techniques described in detail in [86]. The paper chromatograph procedure of Herrmann [416], in addition to thin-layer chromatography on silica gel, was used. This method enabled imipramine to be separated from all its conjugated metabolites with one solvent system (chloroform:propanol:ammonium hydroxide 100:100:2). The quantitative evaluation was based upon the relationship between the square root of the spot area and the logarithm of the amount of substance.

The metabolism of opipramol has been studied by Herrmann [417]. Several metabolites were detected by thin-layer chromatography on silica gel

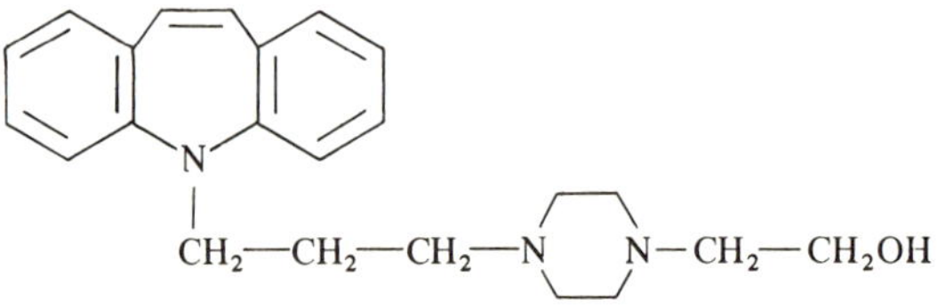

opipramol

(methanol : chloroform : ammonium hydroxide 75 : 25 : 1.5). Drabner et al. [276] included opipramol in their studies on nortryptiline (see p. 28).

Melitracene metabolism has been investigated by Eberholst and Huus [293]. Only the monodemethylated metabolite was detected in rat organs by thin-layer chromatography on silica gel (chloroform : acetone : diethylamine 50 : 50 : 1). The determinations were based on the areas of the spots.

II. BENZODIAZEPINES

The investigations on chlordiazepoxide, diazepam, and, lastly, diazepam derivatives will be reviewed.

The most important research on chlordiazepoxide originated from the US laboratories of Hoffmann-La Roche. Koechlin et al. [515] prepared chlordiazepoxide ^{14}C-labeled in the 2 position and they studied its metabolism in man, dog, and rat in 1965. These experiments followed preliminary researches by Koechlin and d'Arconte [512], who developed in 1963 a sensitive and specific spectrofluorometric method for determining chlordiazepoxide in plasma. The drug is hydrolyzed under well-controlled conditions to give the corresponding lactam. This lactam itself is a metabolite that may be detected after ingestion of high doses of the drug. No unchanged chlordiazepoxide was found in urine.

^{14}C-2-Labeled chlordiazepoxide appears to be very sensitive to daylight and also to the β-emission of ^{14}C incorporated into the molecule. It was necessary [871] to study this isomerization reaction in order to avoid it before using the labeled molecule in metabolic studies. The isomer may be

---

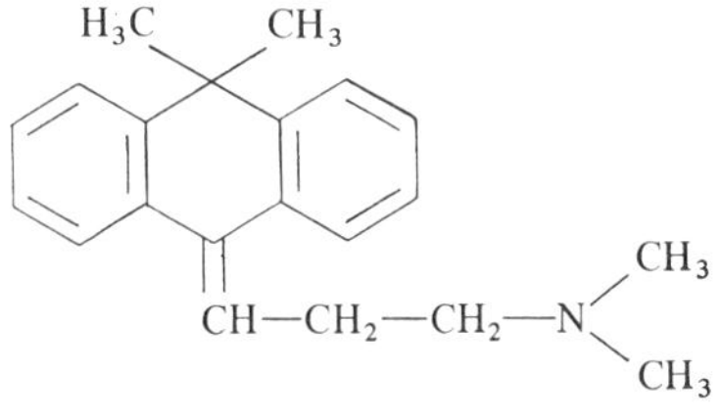

melitracene

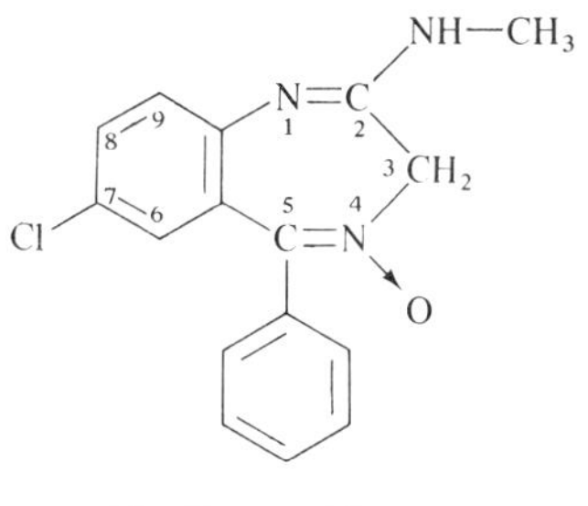

chlordiazepoxide

quantitatively retransformed into chlordiazepoxide by sublimation under vacuum at 200°C and it is possible to identify it by paper chromatography. Thus in all experiments labeled chlordiazepoxide was sublimed just prior to administration and monitored by chromatography. All operations were run in the dark.

The main investigation has used paper and silica gel thin-layer chromatography. In man, plasma and urine extraction with ethyl acetate enabled the lactam and unchanged chlordiazepoxide to be detected. In urine, the identity of the lactam was confirmed by reverse isotopic dilution. Five metabolites were detected when urine was extracted with butanol. Three different solvent systems were necessary to distinguish them by paper chromatography. Two of them showed R_f values corresponding to the lactam and the corresponding amino acid (open lactam). The third was labile and progressively went toward a more stable form, which was isolated by chromatography on a cellulose powder column. It has not been possible to establish its structure, but its acid and alkaline hydrolysis suggested a derivative of the open lactam. The two other metabolites were not identified.

In the rat, the main metabolic fraction was extracted from urine at pH 7 with ethyl acetate. Chromatography on a silica column with a series of solvents of increasing polarity separated three components: the unchanged drug and two others not definitely identified; however, one appears, on the basis of its chemical properties, to be a result of an attack on the N_1-C_2 bond.

Pribilla [735] systematically studied the physicochemical characteristics of chlordiazepoxide and five of its possible metabolites. He reported a thin-layer chromatography method to separate and determine them, but it does not seem to be applicable to the analysis of the urine of chlordiazepoxide-dosed patients.

Another chlordiazepoxide metabolite has been discovered by Schwartz and Postma [819] in man, dog, and rat: it is 7-chloro-2-amino-5-phenyl-3H-1,4-benzodiazepine-4-oxide, which results from an oxidative N-demethylation. This metabolite was detected in the course of in vitro experiments on liver slices, homogenates, and enzyme preparations. It was separated from the drug and the corresponding lactam by ethyl ether extraction and chromatography on fluorescent silica gel (ethyl acetate : ethanol 9 : 1; heptane : chloroform : ethanol 1 : 1 : 1) and identified by comparison with an authentic sample. It was then found in the blood by the same method.

The American investigators of Hoffmann-La Roche who studied chlordiazepoxide metabolism also considered diazepam. The results they have obtained [818] originate from experiments with a product ^{3}H-labeled on the phenyl ring. Seven possible metabolites were prepared. A separation procedure was elaborated, using bidimensional thin-layer chromatography on silica gel (primary dimension: heptane : chloroform : ethanol 10 : 10 : 1; secondary dimension: heptane : chloroform : acetic acid : ethanol 5 : 5 : 1 : 0.3). The blood was extracted with ethyl ether at pH 6.7. The urine was hydrolyzed with Glusulase (β-glucuronidase + sulfatase) for 2-4 hours at 37°C and extracted with ethyl acetate at pH 7. The extracts were chromatographed and the nature of the metabolites was established by comparison with authentic samples; their amounts were determined by radioactivity counting. Moreover, reverse isotopic dilution confirmed the presence of one of the metabolites (oxazepam).

Thus, three urinary metabolites were identified in man and dog: 3-hydroxydiazepam, N-desmethyldiazepam, and oxazepam or 3-hydroxy-N-desmethyldiazepam, the most important. In the rat, the biotransformation product was not identified. The metabolites were mainly in the conjugated form and no unchanged drug was detected in urine. In blood, only the N-desmethyl derivative was found.

The choice of the combination of bidimensional thin-layer chromatography (which provides a high resolving power) and radioactivity measurement (which enables a precise evaluation at the micro scale) demonstrated a more elegant approach by this research team with respect to this drug than shown in their work on chlordiazepoxide.

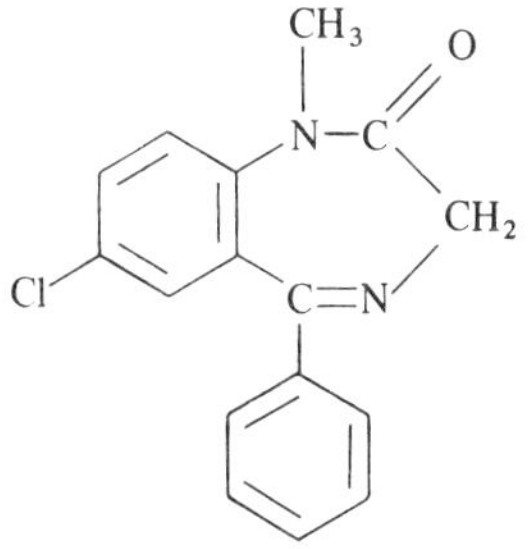

diazepam

The transformation of diazepam into oxazepam was studied by Ruelius et al. [789] in the dog. Urine at pH 7-7.5 was extracted with chloroform to separate the free metabolites and then treated with carbon (Darco G-60). The conjugated metabolites were eluted from the carbon by 80% aqueous acetone and hydrolyzed with β-glucuronidase (2 to 3 days at 37°C, pH 7.5); the liberated metabolites were extracted into chloroform. Thin-layer chromatography on silica gel (chloroform : acetone : ethanol 8 : 1 : 1) revealed the presence of free N-desmethyldiazepam in small amounts. After hydrolysis, oxazepam and 3-hydroxydiazepam were found. Oxazepam was isolated by preparative thin-layer chromatography and identified by IR and NMR spectrometry.

Jommi et al. [467, 468] have published their results from the rabbit. Unfortunately, their investigations were of limited interest. They injected intraperitoneally only very high doses of the drug and applied to urine a very drastic hydrolysis (36% hydrochloric acid, 12 hours at 100°C). Under these conditions diazepam is hydrolyzed into the corresponding benzophenone. Thus, by column and thin-layer chromatography, Jommi et al. found four benzophenones from diazepam and the latter's desmethyl and hydroxyl derivatives in urine. However, it is not possible to ascertain whether these metabolites preexisted as conjugates. The only positive finding of this work was to confirm the demethylation and hydroxylation processes.

Nevertheless, the hydrolysis into benzophenone enabled Silva et al. [845] to develop a gas-liquid chromatographic method for determining diazepam and its N-desmethyl metabolite in blood. Actually, it is possible to chromatograph diazepam directly by electron capture detection, but the sensitivity is increased tenfold by going to the benzophenone. Blood was extracted with ethyl ether and the extracted products transferred to 6N hydrochloric acid. After 1 hour of hydrolysis at 100°C, the benzophenones were extracted in ethyl ether and separated on a column of Gas-Chrom P silanized and coated with 2% of Carbowax 20 M at 190°C. The electron capture detector was able to detect 0.02 to 0.03 μg/ml of blood. The retention times were 5.5 minutes for 2-methylamino-5-chlorobenzophenone and 11.2 minutes for 2-amino-5-chlorobenzophenone. The authors have not considered the hydroxybenzophenones corresponding to hydroxydiazepam, as there were no sizable amounts of hydroxylated metabolites in blood, but have considered only N-desmethyldiazepam [818].

The study of the blood levels of diazepam and its major metabolite by Silva et al. [844] required the development of convenient determination methods. Ultraviolet spectrophotometry makes it possible to simultaneously determine diazepam and its N-desmethyl metabolite after extraction with ethyl ether at pH 7.0. Diazepam is separated from its derivatives by chromatography of this extract on fluorescent silica gel (chloroform : acetone 9 : 1). A more precise identification of the metabolites was achieved by bidimensional thin-layer chromatography (chloroform : heptane : ethanol 10 : 10 : 1, followed by chloroform : heptane : acetic acid : ethanol: 5 : 5 : 1 : 0.3 — chloroform : acetone 9 : 1, followed by chloroform : heptane : ethanol 10 : 10 : 1).

Diazepam and its N-demethylated metabolite can be separated and determined by gas-liquid chromatography after they have been hydrolyzed by 6N hydrochloric acid into the corresponding benzophenones. The column is filled with silanized Gas-Chrom P coated with 2% of Carbowax 20 M-terephtalic acid polyester (CBW-20 M TPA — Wilkens Instruments). It is possible to determine 5 to 30 ng of product with an electron capture detector. Gas-liquid chromatography has been applied to a complete study of the blood levels; the absence of any metabolite other than the N-demethylated derivative has been verified by bidimensional thin-layer chromatography.

The metabolism of oxazepam, which is itself a metabolite of diazepam, has been elucidated by Walkenstein et al. [970]. After administration of the 2-^{14}C-labeled compound, the paper chromatography of urine (butanol : ethanol : water 17 : 3 : 20, or butanol : pyridine : water 6 : 4 : 3) revealed, beside the unchanged drug, only one metabolite: its hydrolysis with β-glucuronidase gave oxazepam.

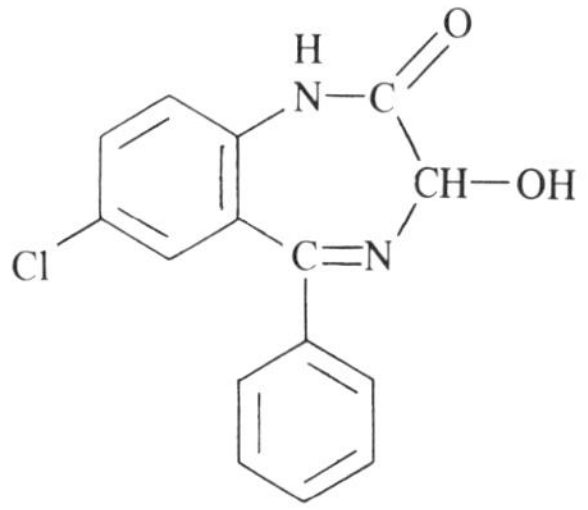

oxazepam

Nitrazepam is an analog of diazepam. Rieder [763] published a series of data concerning its metabolism. In blood and urine, beside the unchanged drug (I), two metabolites were found: the 7-amino (II) and 7-acetamido (III) derivatives of diazepam. They were detected by thin-layer chromatography on silica gel compared to reference compounds (Table 24).

The amino derivative (II) was directly determined after extraction by diazo-coupling with α-naphthylethylenediamine. The drug, after hydrosulfite reduction, and the acetamido derivative (III), after acid hydrolysis, were titrated in the same way.

Pribilla [736] separated nitrazepam from its hydrolysis product, which appeared to be 5-nitro-2-aminobenzophenone on the basis of its elemental analysis and UV and IR spectra. This author applied thin-layer chromatography on silica gel (chloroform : acetone 9 : 1). Four other metabolites were detected in human urine.

TABLE 24

Solvent systems	R_f of the compound:		
	I	II	III
Ethyl acetate : propanol : diethylamine 70 : 30 : 1	0.91	0.76	0.68
Ethylene dichloride : methanol : ammonium hydroxide 90 : 10 : 1	0.66	0.38	0.22
Toluene : acetone : ammonium hydroxide 50 : 50 : 1	0.52	0.30	0.23

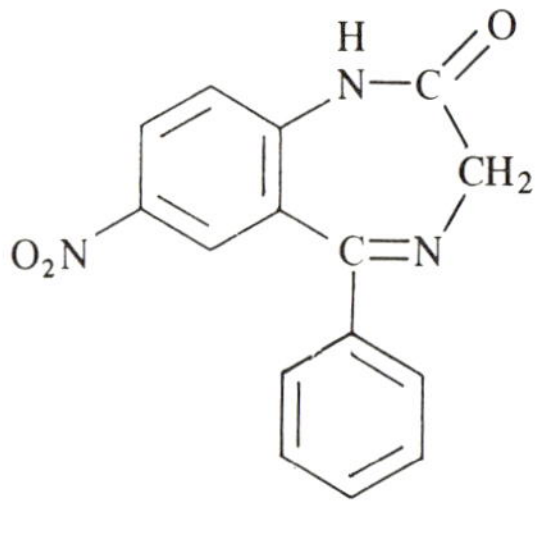

nitrazepam

Chapter VI

CARBAMATES

The discovery of meprobamate, 2-methyl-2-propyl-1,3-propanediol dicarbamate, represented the conclusion of the work that began with the study of mephenesine metabolism.

Mephenesine had a short duration of action because of the rapid oxidation of its primary alcohol group, which could be stabilized by conversion to a carbamate. Extension was made to the propanediol dicarbamates, and the 2-disubstituted derivatives appeared to be the most active.

Similarly, alcohols closely related to mephenesine and their carbamates also appeared interesting and shall be considered in this chapter, although strictly speaking, we may not be rigorous in a chemical classification.

I. MEPHENESINE AND RELATED COMPOUNDS

Wyngaarden et al. [1026] in 1947 had colorimetrically observed that mephenesine was actively metabolized, and a conjugate was assumed. Riley and Berger [768] isolated two urinary metabolites by concentration on Darco G-60 and subsequent acetone elution. These two metabolites were then separated by chromatography (conditions not reported): the most abundant was identified as β-(o-tolyloxy)-lactic acid on the basis of its elemental analysis and UV spectrum as well as by its comparison with the authentic sample.

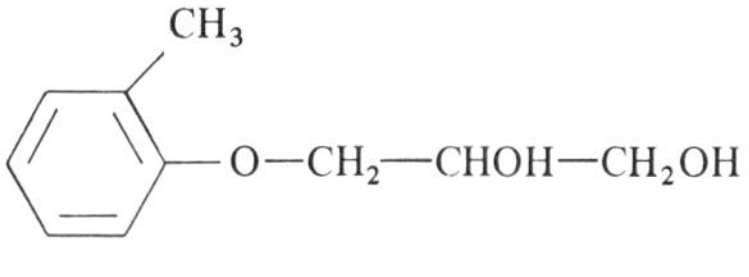

$$\text{Ar-CH}_3\text{-O-CH}_2\text{-CHOH-CH}_2\text{OH}$$

mephenesine

This major metabolite was also isolated at the same time by Graves et al. [382] by the simple extraction from urine at pH 3 into ethyl ether.

Riley [766] then succeeded in elucidating the nature of the second metabolite, β-(2-methyl-4-hydroxyphenoxy)-lactic acid. After elimination of a part of β-(o-tolyloxy)-lactic acid with isopropyl ether at pH 4.5, the residual mixture was extracted with ethyl ether at pH 3. The mixture was fractionated by two successive counter-current distributions (I, saturated phosphate buffer, pH 4.5 : ethyl ether — 21 transfers; II, water : (isobutanol + benzene 35 + 65) — 21 transfers). The second metabolite, purified by recrystallization, was identified by its elemental analysis, UV spectrum, optical rotation, and by comparison with the authentic product.

Riley [767] also applied counter-current distribution to follow the excretion of the major metabolite and was able to account for 33-35% of the administered dose; the secondary metabolite was not very abundant.

Ludwig et al. [566] studied the synthesis of the secondary metabolite in detail and confirmed the identification proposed by Riley. Maas et al. [571] investigated the in vitro metabolism of mephenesine. After incubation with hepatic tissue, one extraction with ethyl ether at pH 6-8 followed by a second at a pH less than 3 yielded β-(o-tolyloxy)-lactic acid, assayable by UV spectrophotometry.

Several authors have examined blood levels of mephenesine. Huff et al. [439] used the colorimetric reaction with diazotized 2,4-dinitroaniline, whereas Titus et al. [908] compared it with the periodic oxidation technique followed by a formol determination with chromotropic acid. This latter method was applied by London and Poet [558] to compare mephenesine and its carbamate, and by Morgan et al. [642], who also introduced methocarbamol into the comparison.

Coppi et al. [220] reported that meprophendiol metabolism was very similar to that of mephenesine. The metabolites were detected by silica gel thin-layer chromatography (butanol : water : ammonium hydroxide 40 : 5 : 1) and compared to the specially synthesized metabolites. They were isolated by radial chromatography under the same conditions and their structures were confirmed by UV and IR spectra. The two were oxidation products of the glyceryl chain: α-hydroxy-β-(o-methoxy-p-propionylphenoxy) propionic acid and o-methoxy-p-propionylphenoxyacetic acid.

The corresponding carbamates were only subject to limited studies. Dresel and Slater [277] reported that mephenesine carbamate, which exhibits a longer duration of action than the diol, was not metabolized into mephenesine. For this purpose they used the analytical procedure of Riley [766] and they determined β-tolyloxylactic acid in urine. Campbell et al. [190] employed ^{14}C-labeled methocarbamol and found six radioactive fractions by paper chromatography (butanol : water 100 : 15). β-Glucuronidase hydrolysis followed by other chromatographies enabled the establishment of the presence of at least three glucuronides, one of which corresponded to the unchanged drug.

II. MEPROBAMATE AND RELATED DRUGS

The metabolism of meprobamate was examined in the course of studies on its pharmacology by Berger [77]. Three differently treated urines were continuously extracted with ethyl ether. The urines were used as is, after acidification and after boiling acid hydrolysis. Ten percent of the unchanged

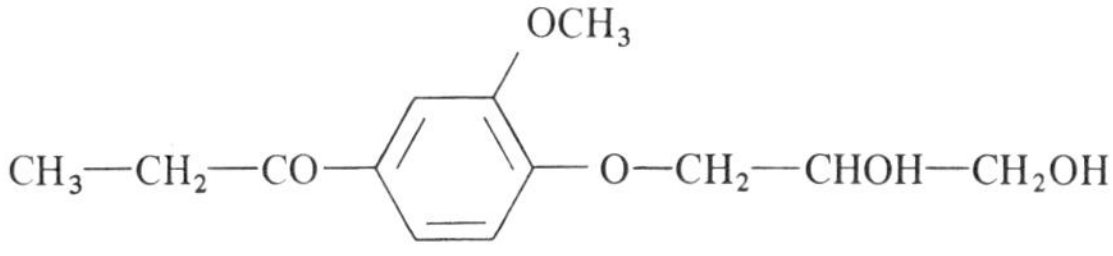

$CH_3-CH_2-CO-\underset{}{\underset{}{\bigcirc}}(OCH_3)-O-CH_2-CHOH-CH_2OH$

meprophendiol

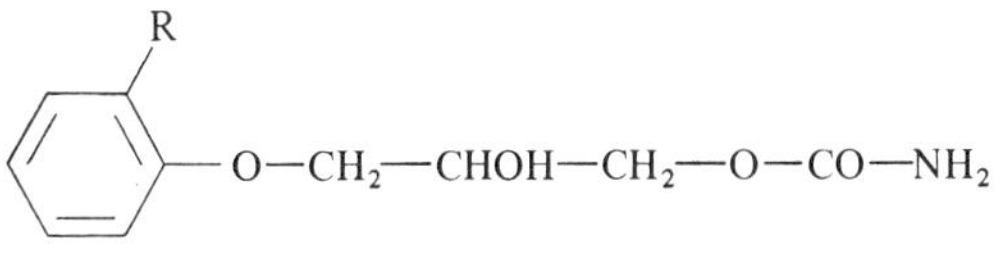

$\underset{R}{\bigcirc}-O-CH_2-CHOH-CH_2-O-CO-NH_2$

R = CH$_3$: mephenesine carbamate
R = OCH$_3$: methocarbamol

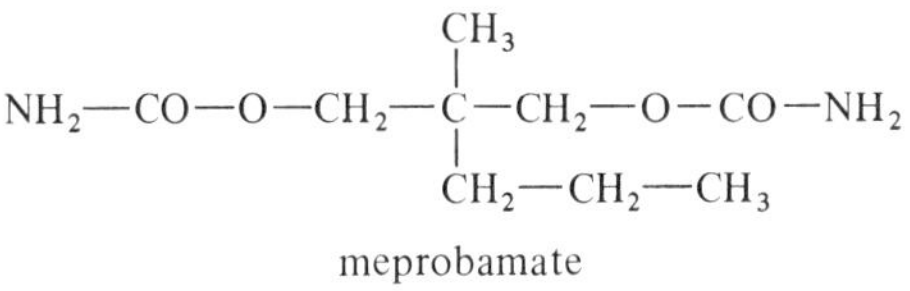

$NH_2-CO-O-CH_2-\underset{CH_2-CH_2-CH_3}{\overset{CH_3}{C}}-CH_2-O-CO-NH_2$

meprobamate

drug was thus detected, plus an equivalent amount of conjugate and one unidentified glucuronide.

These results were confirmed by Agranoff et al. [12]. Meprobamate was extracted into ethyl ether from the alkaline urine and spectrophotometrically determined after heating with sulfuric acid. That the extraction was specific was shown by a solvent partition study. After this extraction, the urine was incubated with β-glucuronidase to liberate a substance similar to meprobamate. Subsequent acid hydrolysis resulted in another similar compound.

The more extensive investigation was due to Walkenstein et al. [967], who utilized the product ^{14}C-labeled on the two carbamate groups. Paper chromatography (butanol:water 100:15) revealed four metabolites beside the unchanged drug, which was identified by isotopic dilution.

The major metabolite, isolated in amounts sufficient for analysis, appeared to be 2-hydroxymethyl-2-propyl-1,3-propanediol dicarbamate. Confirmation was obtained by synthesis of the metabolite, R_f comparison, and reverse isotopic dilution. Two other metabolites were glucuronides and one corresponded to the above-mentioned hydroxylated derivative, as shown by β-glucuronidase hydrolysis followed by paper chromatography. A procedure for meprobamate determination where hydroxymethyl meprobamate does not interfere was described.

Similar results were reported by Emmerson et al. [309], who also worked with the ^{14}C-labeled drug, but they analyzed blood and brain by the Walkenstein paper chromatography method [967].

Wiser and Seifter [1014], using the same technique as the two preceding authors, tried to identify the three yet undefined metabolites. One of them reacted with propionic anhydride, indicative of an hydroxylated compound. The two glucuronides on acid hydrolysis gave meprobamate and hydroxymethylmeprobamate. Electrophoresis showed that the carboxyl groups of these two glucuronides were free.

The chromatographic behavior (butanol:water 100:15) of these metabolites is summarized in Table 25.

Ludwig et al. [565] also synthesized the major metabolite (hydroxymethylmeprobamate). They recorded a constant, but slight, R_f difference (0.75 against 0.78) between the authentic compound and the metabolite isolated from urine, which caused them to doubt the structure proposed by Walkenstein [967].

TABLE 25

	R_f		
	Walkenstein [967]	Emmerson [309]	Wiser [1014]
Meprobamate	0.80	0.77	0.80
Hydroxymethyl meprobamate	0.60	0.60	0.60
Hydroxylated metabolite	0.30	0.43	0.40
Meprobamate glucuronide	0.15	0.09	0.10
Hydroxymethyl meprobamate glucuronide	0.07	0.04	0.05

The unconjugated metabolite was extracted from urine. After meprobamate had been eliminated by ethyl ether from the alkaline medium, the metabolite was separated with ethyl acetate. The solution was evaporated and the metabolite was recrystallized from acetone. Its elemental analysis confirmed its gross formula, but its melting point and IR spectrum showed it was distinctly different from synthetic hydroxymethyl meprobamate.

On hydrolysis, an hydroxylated derivative of meprobamate resulted in a triol but there are three possible triols. The one obtained on hydrolysis of the metabolite was identified by IR spectrophotometry and permitted the conclusion that the product extracted from urine was β-hydroxypropyl meprobamate. This metabolite was prepared by unequivocal synthesis and confirmed the identification.

Ludwig et al. also endeavored to obtain the glucuronide in a pure state, but they only succeeded in isolating its lead salt. Impurities were discarded by lead acetate at pH 2.6, and the glucuronide was precipitated by lead acetate at pH 9.0. The glucuronide was liberated by treatment of the aqueous suspension of its lead salt with hydrogen sulfide. It was then retained on a Dowex-1 (H+) column and eluted by 12N formic acid. The eluate was concentrated under vacuum to discard formic acid and was extracted with ethyl ether and hydrolyzed (4.5N hydrochloric acid, 1 hour under reflux). The ether extract of the hydrolysis solution, after distillation under vacuum and recrystallization was identified as meprobamate by its melting point and IR spectrum.

Douglas et al. [273] determined meprobamate and its hydroxylated metabolite in the urine of various animals by paper chromatography of the crude urine (butanol : acetic acid : water 4 : 1 : 5). The chromatogram was exposed to chlorine gas and sprayed with the starch-potassium iodide reagent and the compounds were determined by visual evaluation.

Paper and thin-layer chromatography of meprobamate and its metabolites were considered for toxicological purposes by Hynie et al. [444]. Alkaline urine was extracted with chloroform or ethyl acetate. Paper chromatography required the following systems:

- Butanol:acetic acid:water, 4:1:5
- Benzene:butanol:acetic acid:water, 3:1:1:5
- Benzene:acetic acid:water, 2:2:1
- Chloroform:acetic acid:water, 100:2:5

Thin-layer chromatography on silica gel dried for 1 hour at 105°C required:

- Chloroform:acetic acid; 20:1
- Ethyl ether
- Benzene:acetone, 2:1
- Chloroform:acetone, 4:1

Detection was achieved by a 0.2% p-dimethylaminobenzaldehyde solution in hydrochloric acid, or by a chlorine solution in carbon tetrachloride followed, after drying, by a benzidine-sodium iodide solution in diluted acetic acid. In the thin-layer plate development, the chlorine solution was replaced by exposure to chlorine gas.

The Japanese authors — Yamamoto, Yoshimura, and Tsukamoto — have published a series of remarkably extensive researches on meprobamate metabolism wherein [1028] the first paper described the isolation and identification of six metabolites in addition to unchanged drug from the rabbit. Urine at pH 2.5 was continuously extracted for 20 hours with ethyl acetate. After evaporation, the products were redissolved in water and the solution was alkalinized by sodium bicarbonate and extracted with ethyl acetate. The substances extracted in this manner were chromatographed on alumina using a series of eluents: benzene:ethyl acetate, ethyl acetate, ethyl acetate: acetone, acetone, methanol.

Meprobamate and two metabolites (I and II) were thus separated.

The alkaline aqueous solution was acidified and extracted again with ethyl acetate. On evaporation and recrystallization, metabolite III was obtained.

The glucuronides were precipitated by lead acetate from the continuously extracted urine. The precipitate was suspended in methanol and the glucuronides were liberated by hydrogen sulfide and, after evaporation, redissolved in water. Metabolite IV was obtained by ethyl acetate extraction. The residual aqueous solution of glucuronides was evaporated and the dry extract was redissolved into methanol. A mixture of metabolites V and VI precipitated on the addition of ethyl ether.

These various metabolites were characterized by paper chromatography and electrophoresis (Table 26).

The metabolites were then identified by their elemental analysis, their various chemical reactions, their IR spectra, and, in the case of the glucuronides, by their acid or enzymic hydrolysis.

(I) Ketomeprobamate
(II) Hydroxypropyl meprobamate
(III) Carboxy meprobamate
(IV) Meprobamate glucuronide
(V) Glucuronide of II
(VI) Unidentified glucuronide

These important investigations were confirmed later [1029] by syntheses that showed that the three metabolites resulted from an oxidation of the propyl chain, supporting Ludwig's [565] contradictions of Walkenstein's conclusions [967].

A third paper [933], one year later, completed elucidation of the meprobamate glucuronide structure. This conjugate, isolated according to the previously described procedure, was purified by chromatography on a silica gel column (ethyl acetate, ethyl acetate : methanol). Its purity was checked by paper chromatography and its identity established by comparison with the synthetic glucuronide, obtained by direct reaction of meprobamate with glucuronic acid. The corresponding glucurone was also prepared. The metabolite and the reference compound had the same chromatographic behavior on paper (same systems as above) (Table 27).

The IR spectra of the methyl and acetyl derivatives of the metabolite and the reference compound were the same.

The glucuronide was also prepared enzymically in small amounts by the procedure of Dutton and Storey [286]. A suspension of ground hepatic cells was mixed with a liver extract obtained at the boil (active factor). This mixture was capable of synthesizing glucuronides (glucuronyl transferase). The R_f of the resulting product was the same as the synthetic compounds.

TABLE 26

Compound	R_f in solvent system: I[a]	II[a]	III[a]	Electrophoretic mobility[b]
Metabolite I	0.72	0.62	0.63	− 0.9
Metabolite II	0.72	0.62	0.63	− 0.9
Metabolite III	0.70	0.60	0.05	+32.0
Metabolite IV	0.57	0.47	0.15	+23.0
Metabolite V	0.31	0.08	0.03	+20.0
Metabolite VI	0.20	0.03	—	—
Meprobamate	0.85	0.81	0.80	− 0.9
Urea	0.40	0.25	0.23	− 0.9

[a] I, butanol:acetic acid:water 4:1:5; II, 3% ammonium hydroxide saturated butanol; III, butanol:water 100:15.
[b] Electrophoresis in a 1% sodium borate solution for 1 hour under 400 V.

TABLE 27

	R_f in solvent system: I	II
Meprobamate-N-mono-β-D-glucopyranosiduronic acid	0.57	0.17
Meprobamate-N-monoglucurone	0.67	0.29
Meprobamate	0.84	0.83
Glucuronic acid	0.12	0.00
Glucuronolactone	0.32	0.00

Further confirmation of the structure of meprobamate glucuronide [934] was obtained by purifying either the natural or synthetic compound by countercurrent distribution (butanol:water — 7 transfers) and by transforming it into the methylated and acetylated derivative with repurification by chromatography on a silica column. On reaction with sodium methoxide, it was converted into the sodium salt and crystallized. The IR spectrum and the melting point were the same for the synthetic compound and the metabolite.

These investigations are extremely detailed in the original papers, particularly in regard to the glucuronide synthesis.

The metabolism of sodium meprobamate N-glucuropyranosiduronate [891] was studied by hydrolyzing the glucuronide present in urine by diluted hydrochloric acid and determining it by gas-liquid chromatography (1.5% of silicone SE-30 on Chromosorb W at 180°C). The glucuronide and free meprobamate were separated by paper chromatography (butanol : acetic acid : water 4 : 1 : 5; 3% ammonium hydroxide saturated butanol). The glucuronide was excreted unchanged in rabbit urine.

Carisoprodol is N-isopropyl meprobamate. The Douglas and Ludwig team also studied its metabolism [271].

Blood metabolites were investigated after administration of the ^{14}C-labeled drug. They were extracted in chloroform : carbon tetrachloride 1 : 1 from the alkaline medium, then in ethyl acetate. Both concentrated extracts were chromatographed on paper (carbon tetrachloride : acetic acid : water 1 : 2 : 1). In the first extract two metabolites were found besides the unchanged drug. One was easily identified as unchanged meprobamate. The other was likely to be a hydroxylated derivative of carisoprodol, but it was not possible to identify it. The second extract revealed a third metabolite, which had the same R_f as hydroxypropyl meprobamate.

Urinary metabolites were extracted in ethyl ether, the solution was dried, and the residue was treated by trichloroethylene, which separated meprobamate from unchanged carisoprodol. The urine previously extracted with ethyl ether was extracted again with ethyl acetate. The compounds thus removed were chromatographed on a cellulose column with butanol, which isolated hydroxypropyl meprobamate according to the already cited procedure [565].

$$NH_2-CO-O-CH_2-\underset{\underset{CH_2-CH_2-CH_3}{|}}{\overset{\overset{CH_3}{|}}{C}}-CH_2-O-CO-NH-CH\underset{CH_3}{\overset{CH_3}{<}}$$

carisoprodol

The structure of the hydroxylated metabolite (hydroxycarisoprodol) was elucidated 2 years later by the same workers [296]. It was isolated from the ethyl ether extract of urine by chromatography on a cellulose column with water saturated butanol, followed by a second chromatography using the mixture carbon tetrachloride : acetic acid : water 1 : 2 : 1. The fraction containing this metabolite together with hydroxypropyl meprobamate was separated by paper chromatography (carbon tetrachloride : acetic acid : water 1 : 2 : 1). The pure metabolite was hydrolyzed and the liberated amine was identified as isopropylamine by gas-liquid chromatography. The residual triol was converted into naphtylurethane: its melting point and IR spectrum affirmed that the metabolite was β-hydroxypropylcarisoprodol which was confirmed by the synthesis of this compound.

Tybamate is the N-butyl derivative of meprobamate and its metabolism was studied by Douglas et al. [272] with the ^{14}C-labeled drug. The paper chromatography of urinary extracts was performed in three solvent systems:

(I) Butanol : acetic acid : water, 4 : 1 : 5

(II) Water saturated butanol

(III) Carbon tetrachloride : acetic acid : water, 1 : 2 : 1

Four radioactive fractions were detected. These fractions were obtained in larger amounts on extraction. Urine treated for 24 hours by carbon tetrachloride gave a substance which, on recrystallization, was identified as unchanged tybamate by its IR spectrum, melting point, and R_f. The residual urine was extracted with ethyl ether and the same assays showed that the crystals obtained on concentration and recrystallization were meprobamate. After these two extractions, the concentrated urine was submitted to chromatography on paper or cellulose powder (system II). The main fraction was purified by counter-current distribution (butanol : water — 50 transfers).

$$NH_2-CO-O-CH_2-\underset{\underset{CH_2-CH_2-CH_3}{|}}{\overset{\overset{CH_3}{|}}{C}}-CH_2-O-CO-NH-CH_2-CH_2-CH_2-CH_3$$

tybamate

The metabolite thus isolated revealed by IR spectrophotometry the probable presence of an hydroxyl group; its alkaline hydrolysis gave butylamine and 2-methyl-2-(β-hydroxypropyl)-propanediol, detectable by gas-liquid chromatography. These data permitted the assignment of the β-hydroxypropyl tybamate formula to the metabolite.

One of the fractions from the cellulose powder chromatography was rechromatographed on paper (system II). It gave a fourth metabolite, the IR spectrum and R_f of which were similar to those of hydroxypropyl meprobamate.

Small amounts of glucuronide were obtained by basic lead acetate precipitation.

The same team from the Wallace laboratories, worked on mebutamate [270] and applied the same techniques. One metabolite was isolated in addition to the unchanged drug. Its elementary analysis and IR spectrum suggested a hydroxylated derivative. The OH position was defined by NMR and by subsequent synthesis of the authentic compound: 2-methyl-2-(β-hydroxyl-α-methylpropyl)-1,3-propanediol dicarbamate.

III. CHLORPHENESINE CARBAMATE

The metabolism of chlorphenesine carbamate, related to mephenesine carbamate, was studied by Buhler [149]. The urine was extracted first as

---

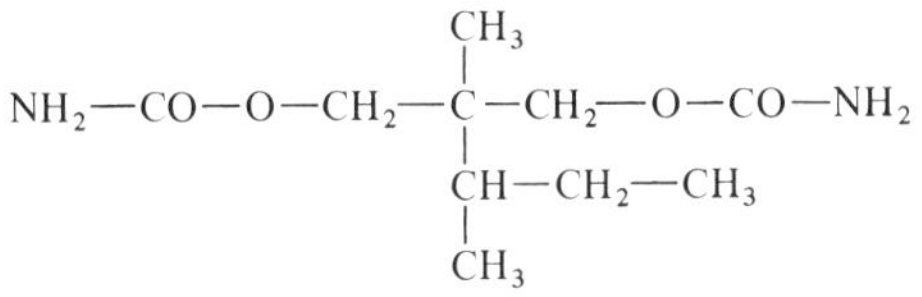

$$NH_2-CO-O-CH_2-\underset{\underset{CH_3}{|}}{\underset{\underset{CH-CH_2-CH_3}{|}}{\overset{\overset{CH_3}{|}}{C}}}-CH_2-O-CO-NH_2$$

mebutamate

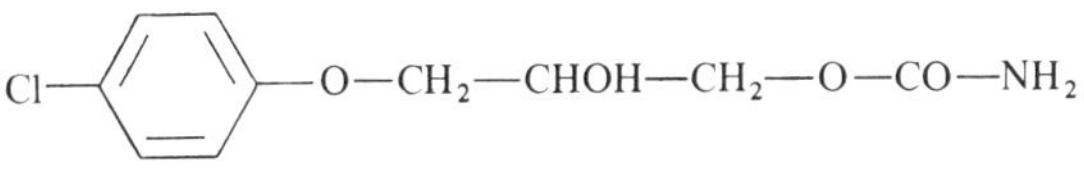

$$Cl-\langle\!\!\!\bigcirc\!\!\!\rangle-O-CH_2-CHOH-CH_2-O-CO-NH_2$$

chlorphenesine carbamate

is at pH 2, then after successive hydrolyses by β-glucuronidase and sulfatase (72 to 120 hours). The extraction was performed with chloroform and the chloroform solution washed with 0.1N sodium hydroxide to separate the neutral and acid fractions.

Neutral metabolites were separated by paper chromatography (benzene : methanol : water 10 : 5 : 5; benzene : Skellysolve B : methanol : water 7 : 3 : 5 : 5; butyl acetate : formamide : water 100 : 5 : 5 on a formamide-impregnated paper). The acid metabolites were chromatographed on paper with other systems (isopropanol : ammonium hydroxide : water 8 : 1 : 1; benzene : propionic acid : water 2 : 2 : 1; chloroform : methanol : formic acid : water 100 : 100 : 4 : 96; benzene : acetic acid : water 1 : 1 : 2).

Thin-layer chromatography on silica gel (benzene : propionic acid : water 2 : 2 : 1; chloroform : methanol : formic acid : water 100 : 100 : 4 : 96) was applied to acid metabolites. Neutral metabolites were also separated on thin layers of silica or alumina (mixtures ethyl acetate : cyclohexane or chloroform : methanol). The metabolites were in some cases analyzed by gas-liquid chromatography (2.6% of Carbowax 20M on Gas Chrom Z at 140°C, flame ionization detector).

Preparative chromatography of the acid metabolites on a silica gel column (stationary phase: 0.05M citrate buffer, pH 3.42; mobile phase: chloroform : ethanol in various proportions) and of the basic metabolites on a Florisil column (mixtures of acetone : Skellysolve B) were performed.

All these experiments were done after administration of ^{3}H-labeled chlorphenesine carbamate. A series of metabolites were detected and it was possible to identify, besides chlorphenesine carbamate, p-chlorophenoxylactic acid, p-chlorophenoxyacetic acid, and p-chlorophenol. The major metabolite was the glucuronide of the drug. The identifications were achieved by comparison with authentic compounds in the various kinds of chromatography, in UV and IR spectrophotometry, and by chemical analysis, elemental and functional.

This investigation is an elegant example of a drug metabolism study: 90% of the excreted radioactivity corresponded to the various identified metabolites!

Buhler [150] completed elucidation of the nature of the glucuronide. It was precipitated from urine by basic lead acetate after the procedure of Kamil et al. [480]. It was methylated by diazomethane and acetylated by acetic anhydride. The esters were purified by chromatography on a silica gel column (mixtures of ethyl ether : benzene and ethyl ether : ethyl acetate of increasing polarity) or on Florisil column (mixtures of acetone : isooctane). The product purified in such a way appeared to contain two components by thin-layer chromatography on alumina gel (0.75% of methanol in chloroform). By recrystallization from methylene chloride, a pure product that was the less polar component was obtained. This metabolite, as studied by NMR, appeared to be methyl-(chlorphenesine carbamate-tri-O-acetyl-β-D-glucoside)-uronate. The UV spectrum confirmed the p-phenoxy structure and the IR spectrum was in agreement. The optical rotation and gross formula were also determined, as well as the molecular weight by mass spectrometry.

It was impossible to purify the second metabolite completely but it was possible to obtain an enriched fraction by recrystallization in the mixture benzene : isooctane. Its study by NMR, IR, and UV spectrophotometry and elemental analysis showed it had a formula similar to that of the first glucuronide. These two metabolites actually were the two diastereoisomers corresponding to the splitting of racemic chlorphenesine carbamate followed by conjugation.

This very remarkable investigation showed that powerful instrumental methods permit the elucidation of structures of complex metabolites without demanding high states of purity.

The study performed on rat and man was extended by Buhler et al. [151] to the dog using ^{3}H-labeled chlorphenesine carbamate prepared by gas exchange, the purity of which had been thoroughly checked. The urine was fractionated according to the procedure shown in Fig. 2 (see page 100).

The paper and thin-layer chromatography systems were the same as given previously [149]. The separated metabolites were identified by the numerous methods already cited.

The results reconfirmed the previous information and made possible a very precise qualitative and quantitative study of chlorphenesine carbamate metabolism. Radioactivity is found in the seven fractions obtained after urine has been treated according to the procedure shown in Fig. 2. The further column chromatographic separations followed by paper chromatography in four different systems permit the separation of four neutral metabolites (one identified as unchanged drug) and 11 acid metabolites (the three main ones identified as p-chlorophenol, p-chlorophenoxyacetic, and

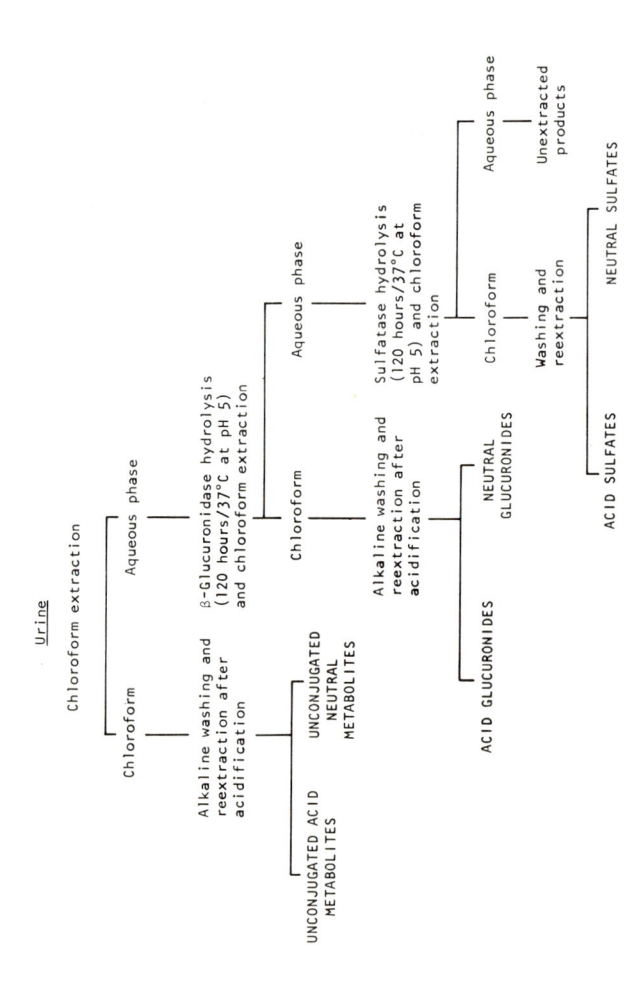

Fig. 2. Fractionation of the urinary metabolites of chlorphenesine carbamate, after [151].

p-chlorophenoxylactic acids). The unidentified metabolites represent individual amounts less than or equal to 4% of the total radioactivity.

The stoichiometric balance, despite the complexity of the metabolic process, is satisfactory. The metabolites identified in the 48 hour urines represent 66% of the administered dose (urines contain 81.7% of the administered radioactivity).

IV. ETHINAMATE

The first studies on ethinamate are somewhat crude and they applied paper chromatography with the following solvents:
(I) Mohrschulz [641] : isopropanol : chloroform : 25% ammonium hydroxide, 45 : 45 : 10
(II) Fischer and Specht [321]: methanol : water, 1 : 2

McMahon [580] prepared ethinamate ^{14}C-labeled on the carbonyl group. He detected two main metabolites by paper chromatography of urine (butanol : 1.5N ammonium hydroxide 1 : 1). A partial extraction of the major metabolite into methylene chloride permitted the observation of ethynyl and carbamoyl groups by IR spectrophotometry. This suggested a likely hydroxyethinamate. A second, more complete procedure involved a methylene chloride extraction of 25 l of human urine at pH 10, followed by a chromatographic purification on a silica gel column (butanol : water), then on an alumina column (benzene : methylene chloride), and recrystallization. The IR spectrum, molecular weight by X-ray diffraction, and chemical assay confirmed the hydroxyethinamate hypothesis. The other main metabolite was hydroxyethinamate glucuronide. The OH position on the ring was not specified, but McMahon [582] demonstrated subsequently that it was 1-ethynyl-4-hydroxycyclohexyl carbamate by converting the metabolite into the corresponding diol by hydrogenation and comparing it with the authentic diols.

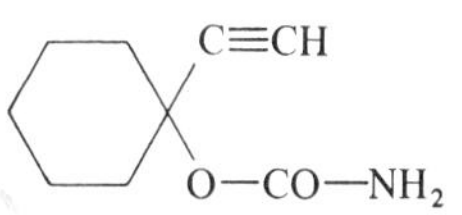

ethinamate

Since 1959, ethinamate metabolism has been the subject of a series of papers by Preuss et al. The first [725] described the extraction of urine at pH 2 by ethyl ether, the treatment of the ether solution by a 5% lead acetate solution, and its purification on a small active carbon column. The metabolites were separated by chromatography on paper impregnated with 1,2-propylene glycol (30% or 45% of dioxane in ligroin). Six metabolites were detected with the silver nitrate reagent in addition to small amounts of the unchanged drug. One of the metabolites was ethynylcyclohexanol.

A second paper [726] attempted to demonstrate that one of the metabolites was McMahon's hydroxyethinamate. Paper chromatography was applied, as well as a series of detection reagents with acid and enzymic hydrolyses.

Paper chromatography (dioxane : ligroin and ethyl ether : pentane) showed that each of the three previously detected metabolites actually consisted of two components [728].

In fact, more extensive studies [733] revealed that most of the so-called detected metabolites appeared to be mixtures up to as many as 15. They were separated into two groups by counter-current distribution [water : (ethyl ether + pentane 3+1) 1 : 1 — 37 transfers]. In the hydrophilic group four metabolites were isolated by preparative paper chromatography [(ethyl ether + pentane 3+1) : water]. They were filtered on a silica gel column and crystallized. Their structure was elucidated by their elemental analysis, IR spectrophotometry, and hydrolysis into the corresponding diol. The same authors had previously performed a metabolic study of 1-ethynyl-1-cyclohexanol [732] which made possible R_f and melting points comparisons. Thus the 2-, 3-, and 4-hydroxy derivatives, the last one in the form of α and β stereoisomers, were found as metabolites. These two isomers were also present as glucuronides, and it was shown by enzymic hydrolysis followed by paper chromatography.

The structure and the conformation of these hydroxylated metabolites have been studied [734]. The behavior on paper chromatography, the NMR spectra of the diols and of their diacetates, and the IR spectra allowed precise conclusions on the structure of the four metabolites:

1a-Ethynyl-4e-hydroxy-1e-cyclohexylcarbamate
1e-Ethynyl-4e-hydroxy-1e-cyclohexylcarbamate
1a-Ethynyl-3e-hydroxy-1e-cyclohexylcarbamate
1a-Ethynyl-2e-hydroxy-1e-cyclohexylcarbamate

Two other metabolites were isolated in 1965 [729]. The same isolation and structure elucidation procedures were applied and the authentic compounds corresponding to these metabolites were synthesized. One of these metabolites was unstable and easily converted into the other: they both gave on saponification 1e-ethynyl-2e-hydroxycyclohexanol. The first one was 1e-ethynyl-2e-hydroxy-1e-cyclohexyl carbamate; the other was 1-ethynyl-1-hydroxy-2-cyclohexyl carbamate. The transition from one to the other was achieved by intramolecular transesterification.

Preuss [730] provided the structure elucidation for the other three metabolites. These could not be separated by chromatography on alumina or silica column because of decomposition. He separated two fractions by preparative paper chromatography (ethyl ether : pentane 1 : 3, water saturated) and studied the decomposition products of these two fractions by paper chromatography to reveal that each one was a mixture of three inseparable components. He did not succeed in elucidating the composition of the first fraction. The second fraction, purified by chromatography on carbon and recrystallization, was examined by IR spectrophotometry. It indicated the presence of an acetoxy group, in agreement with the elemental analysis. The mixture of the hydroxylated metabolites that was obtained from urine by paper chromatography was acetylated and gave a mixture that behaved exactly as this second fraction as analyzed above. Preuss concluded from these data that this second fraction is a mixture of three metabolites:

1a-Ethynyl-4e-acetoxy-1e-cyclohexylcarbamate

1e-Ethynyl-4e-acetoxy-1a-cyclohexylcarbamate

1a-Ethynyl-3e-acetoxy-1e-cyclohexylcarbamate

This is an interesting example of structure elucidation on incompletely separated metabolites. However, this work has been possible only because of the very extensive studies previously performed.

These eight papers did not use labeled derivatives. Sizable amounts of metabolites were isolated from large urine volumes from patients dosed with high amounts of ethinamate (1 g/day and more). Many metabolites remain unidentified and no information is available on the relative amounts.

Preuss studied Dolcental in parallel with the above studies.

Kum-Tatt [530] isolated an ethinamate metabolite from human brain, by treating the organ with alcohol, evaporating, and redissolving the residue in ethyl ether. The recrystallized metabolite was examined by IR spectrophotometry, elemental analysis, and chemical reactions. It contained neither a hydroxyl nor a carbamate group. Exact identification was not possible.

Murata of Japan has been investigating ethinamate metabolism since 1961.

He first discovered three metabolites by paper chromatography [650], one of them coincident with the hydroxyethinamate of McMahon [580] on the basis of its IR spectrum, chemical reactions, hydrogenation, and acetylation.

A glucuronide was isolated from urine by basic lead acetate precipitation at pH 8 after precipitation of the impurities by neutral lead acetate at pH 4. It was regenerated by hydrogen sulfide, reprecipitated by barium acetate, redissolved in water, and submitted to a second cycle: basic lead acetate-hydrogen sulfide. This glucuronide was detected by the paper chromatography of urine, together with the other glucuronides (butanol:acetic acid: water 4:1:5). On β-glucuronidase hydrolysis, hydroxyethinamate was liberated from it [651].

The purified glucuronide was successively methylated by diazomethane and acetylated. This led to a crystallized derivative that was characterized by elemental analysis and IR spectrophotometry [653].

The metabolites extracted in ethylene dichloride were chromatographed on an alumina column; the mixture benzene:dichloromethane eluted the hydroxylated derivative. Subsequently, methanol eluted another metabolite which had an ethynyl group and reacted with periodate. It was a 1,2-diol that was identified as 1-ethynyl-1,2-cyclohexanediol by synthesis of the authentic compound [652].

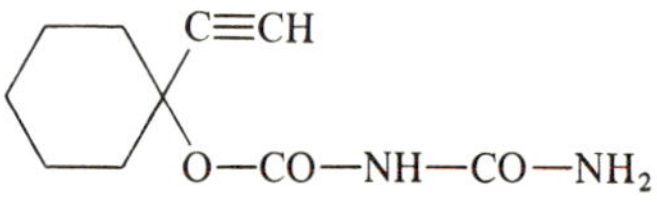

$$\text{cyclohexane} \begin{cases} C\equiv CH \\ O-CO-NH-CO-NH_2 \end{cases}$$

Dolcental

This diol, isolated as above, was purified by vacuum distillation. Hydrogenation into the corresponding ethyl derivative and comparison with authentic products by paper chromatography and electrophoresis identified it with 1-ethynyl-1, 2-transcyclohexanediol [654].

V. OTHER CARBAMATES

The phenprobamate metabolism involved the various steps of side-chain oxidation, and the major metabolite was hippuric acid. There was also a partial parahydroxylation on the ring. The various metabolites were separated by Schatz and Jahn [800] by means of thin-layer chromatography on silica gel (benzene : dioxane : acetic acid 90 : 25 : 4) (Table 28).

TABLE 28

Product	R_f
Benzoic acid	0.80
Phenylacetic acid	0.63
Phenylpropionic acid	0.60
p-Hydroxybenzoic acid	0.54
p-Hydroxycinnamic acid	0.45
p-Hydroxyphenylacetic acid	0.42
Hippuric acid	0.15

They were identified by their R_f's and melting points after extraction. They were determined either by measurement of spot area after chromatography or by various colorimetric methods.

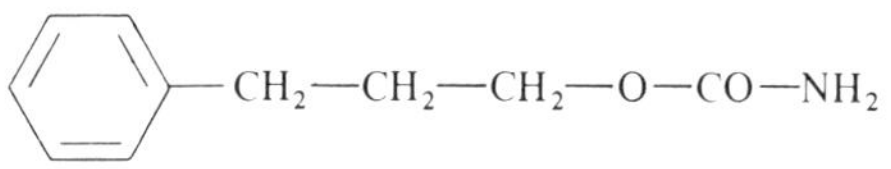

C_6H_5—CH_2—CH_2—CH_2—O—CO—NH_2

phenprobamate

Ethyltrichloramate is a simple carbamate. Its metabolism was studied by Glazko et al. [366]. As chloral hydrate it yielded trichloroethanol which can be determined by steam distillation, conversion into formaldehyde, and colorimetric titration. Most of the drug was excreted unchanged as a glucuronide. The urine was concentrated by partial freezing and fractionated by countercurrent distribution (water : butanol : acetic acid : pyridine 100 : 80 : 8 : 8 — 24 transfers). The determination of ethyl trichloramate and glucuronic acid on every fraction showed the presence of the glucuronide.

The two labeled benzyl-N-benzylcarbethoxyhydroxamates (N- and O-benzyl-α-^{14}C) were prepared by Douglas [269]. The compound labeled on the N-benzyl showed that the carbonyl group was rapidly and completely converted into respiratory CO_2. The O-benzyl derivative led to two radioactive metabolites, which were separated by paper chromatography (propanol : ammonium hydroxide : water 85 : 1 : 14; butanol : acetic acid : water 4 : 1 : 5). These were hippuric and benzoic acids, which were identified by their R_f's and by isotopic dilution.

CCl_3—CHOH—NH—CO—O—C_2H_5

ethyltrichloramate

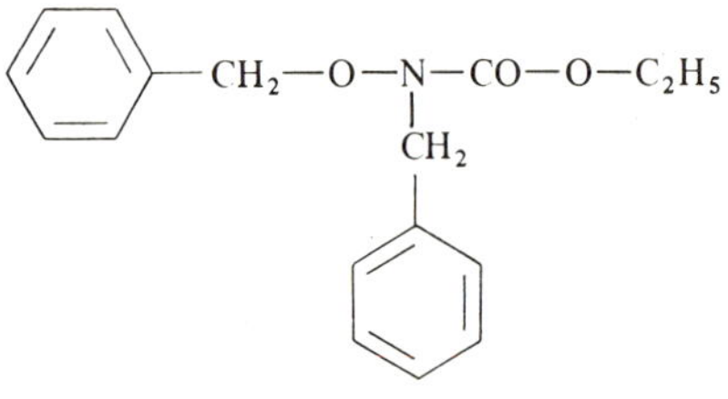

benzyl-N-benzylcarbethoxyhydroxamate

Chapter VII

ANILIDES

One of the most remarkable successes of drug metabolism studies has been in the anilide series. The major metabolite of acetanilide and phenacetin, p-acetamidophenol, has been recognized as the font of the analgesic and antipyretic action of both drugs. It has itself become a significant drug (paracetamol).

I. ACETANILIDE

Acetanilide was introduced into therapeutics in 1887 and it was immediately reported that it was oxidized in vivo and that p-aminophenol could be isolated from urine.

In 1888, Jaffe and Hilbert [454] found 2-benzoxazolone, as well as small amounts of p-aminophenol, in dog urine. It is now evident that 2-benzoxazolone is an artifact, as Bray et al. [112] have demonstrated that it is produced from p-aminophenol during the course of the analysis.

In 1889, Mörner [643] isolated potassium p-acetamidophenyl sulfate as a double salt with potassium ethyl oxalate from human urine. He also isolated a glucuronide which he supposed to be a conjugated p-acetamidophenol. These results were confirmed 59 years later, after a long period of uncertainty as to whether p-aminophenol or p-acetamidophenol was formed.

The first quantitative study was that of Greenberg and Lester [385]. They determined p-aminophenol colorimetrically by reaction with α-naphtol,

---

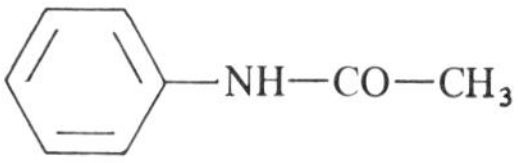

⟨⟩—NH—CO—CH₃

acetanilide

before and after hydrolysis (1N hydrochloric acid — 45 minutes at 100°C). p-Acetamidophenol was separated on the basis of its low water solubility in the presence of dibasic potassium phosphate. After ethylene dichloride extraction it was hydrolyzed and the p-aminophenol colorimetrically determined. Similarly aniline and its conjugates were titrated by diazo coupling with α-naphtol and colorimetry. The presence of 2-benzoxazolone was investigated, after ethylene dichloride extraction of the hydrolyzed urine, by making its benzoate. Azo compounds were treated by zinc powder and the resulting aniline determined by colorimetry.

This study concluded that 70 to 90% of the administered acetanilide was excreted as conjugated p-aminophenol, whereas p-acetamidophenol represented only 4% of the total amount of conjugates, partly as sulfates and partly as glucuronides. Greenberg and Lester did not check the specificity of their analytical procedures, nor did they identify the metabolites.

The same authors extended their investigations to blood metabolites [551] and they developed micromethods based on the preceding principles.

Two fundamental papers on acetanilide metabolism appeared in 1948, the first by Smith and Williams [858] and the second by Brodie and Axelrod [126].

Smith and Williams extracted alkalinized rabbit urine with ethyl ether. On acidification of the concentrated extract, acetanilide crystallized. The alkalinized filtrate was also extracted with ethyl ether and the aniline was isolated as benzanilide.

The urine was also hydrolyzed by 1N hydrochloric acid (20 minutes with boiling; under these conditions only the sulfates were hydrolyzed). The liberated p-aminophenol was extracted into ethyl ether after alkalinization. The amount obtained corresponded closely to the percent of organic sulfate determined separately.

Lastly, the glucuronides were precipitated by basic lead acetate after the impurities had been removed by neutral lead acetate. The glucuronides were regenerated by hydrogen sulfide and the resulting product was hydrolyzed by boiling 5N hydrochloric acid for 1 hour. p-Aminophenol was extracted as described previously and identified by the melting point of its dibenzoyl derivative. The glucuronide was also methylated and acetylated, and then crystallized. Its UV spectrum and its elemental analysis, as well as its comparison with the synthetic derivative (obtained from p-aminophenyl glucuronide prepared according to the Williams procedure [1005]), identified it as p-acetamidophenyl glucuronide.

This glucuronide was precipitated quantitatively after isolation as its lead salt, regenerated, and redissolved in ethanol by addition of benzylamine. The crystallized salt was compared to a synthetic product.

Thus, Smith and Williams demonstrated that acetanilide is metabolized in the rabbit to give 70% of p-acetamidophenyl glucuronide and 12% of p-acetamidophenyl sulfate.

In their parallel paper on phenacetin metabolism [859] Smith and Williams reported the presence of small amounts of a labile p-aminophenol glucuronide in urine (see Section II, Phenacetin).

Brodie and Axelrod applied their usual methodology to this problem. They criticized the lack of specificity of the methods used by Greenberg and Lester and developed new analytical techniques [125], demonstrating their specificity by comparing the partition coefficients of the authentic compound with those of the product extracted from biologic media in various solvent systems.

Aniline was extracted with benzene, transferred into acid solution, and colorimetrically determined by diazo coupling with N-naphtylethylenediamine (sensitivity 0.5 to 2 µg — accuracy 98 ± 5%). Acetanilide was extracted into ethylene dichloride, hydrolyzed, and determined as aniline (sensitivity 4 to 15 µg — accuracy 99 ± 5%). p-Acetamidophenol was extracted by ethyl ether from the sodium chloride saturated aqueous phase. It was then hydrolyzed and determined colorimetrically by diazo coupling with α-naphtol (sensitivity 10 to 50 µg — accuracy 100 ± 4%). Free p-aminophenol was determined without hydrolysis on the ethyl ether extract by reaction with phenol in the presence of sodium hypobromite.

Brodie and Axelrod further applied these techniques to acetanilide in man [126]: 80% of the administered dose was found in urine as the conjugated p-aminophenol; 4% was in the free state, with traces as unchanged drug and aniline. p-Aminophenol, which resulted from the hydrolysis of conjugated products, appeared as a single compound when analyzed by counter-current distribution and it was identical with the reference substance.

The authors reported that p-acetamidophenol was also a metabolite of phenacetin and had an analgesic action equal to that of acetanilide.

Subsequent papers essentially referred to the hepatic metabolism of acetanilide.

Kiese and Renner [506] compared the hydrolytic activity of the microsomes and the soluble fraction: they determined aniline and p-aminophenol by the Brodie and Axelrod procedure [125].

Degkwitz and Staudinger published a series of papers on the role of ascorbic acid on hepatic hydroxylation of acetanilide. The first [250] described the analytical techniques used. The enzymic reaction was stopped by adding water to the incubation medium and heating for 5 minutes at 80-90°C. After centrifugation, it was extracted with ethyl ether and the concentrated extract was analyzed by thin-layer chromatography on silica gel (benzene:amyl alcohol:methanol 85:7:8). This chromatography separated the three o-, m-, and p-hydroxylated isomers but the R_f values were not reported. The separated products were eluted and determined with the Folin and Ciocalteu reagent [331].

Hammar and Prellwitz [395] studied acetanilide metabolism in hepatic or cirrhotic patients. They determined p-acetamidophenol in serum and urine after Brodie and Axelrod [126]. The glucuronide level was evaluated by β-glucuronidase hydrolysis (24 hours at 37°C, pH 6.0) and on the determination of the liberated p-acetamidophenol. The percentage of total conjugated metabolites was determined in the same way after hydrochloric hydrolysis (2 volumes of concentrated hydrochloric acid, 2 hours at 120°C).

II. PHENACETIN

Phenacetin biotransformations were investigated concomitantly with those of acetanilide.

Mörner [644], as early as 1889, isolated from human urine potassium p-acetamidophenyl sulfate and detected the presence of a glucuronide and p-aminophenol, as well as small amounts of phenetidine.

The first significant investigation was reported in 1949 by Smith and Williams [859] who had worked similarly with acetanilide (see Section I)

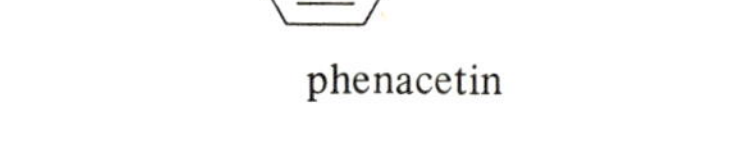

C_2H_5—O—⟨ ⟩—NH—CO—CH_3

phenacetin

and phenacetin in the rabbit. It was published about at the same time as the paper of Brodie and Axelrod [127]. Smith and Williams used the method described previously for acetanilide [858]. Free phenetidine was extracted with acetone from ammonium sulfate saturated urine; p-aminophenol was separated by successive extractions, recrystallized, and identified by elemental analysis and melting point comparison with the authentic compound as the diacetylated derivative. The glucuronide was isolated and identified by the same method as for acetanilide. Moreover a labile p-aminophenol glucuronide was detected in the urine by making a crystalline complex in the presence of p-toluidine and ammonium ions according to the technique described by Smith and Williams [860].

In conclusion, it appeared that 47% of the administered dose was found in rabbit urine as p-acetamidophenyl glucuronide and 7% as the sulfate.

Phenacetin metabolism was studied in man by Brodie and Axelrod [127]. Phenacetin was extracted with benzene, hydrolyzed, and determined by diazocoupling with α-naphtol (sensitivity 1 to 5 μg — accuracy 96 ± 5%). p-Ethoxyaniline was titrated in the same way without hydrolysis. p-Acetamidophenol and p-aminophenol were determined as previously noted. The benzene used for extraction had to be successively washed with 1N sodium hydroxide, 1N hydrochloric acid, and three times with water. A volume of isoamyl alcohol (1.5%) washed in the same manner was frequently added to reduce the adsorption of extracted products on glass surfaces. This was a frequent practice of Brodie and Axelrod.

The major metabolite was conjugated p-acetamidophenol and its analgesic role was confirmed, although its antipyretic action had been known since 1923.

The method of Brodie and Axelrod for determining total p-acetamidophenol was used again by Welch and Conney [994]. They checked that 3-hydroxyphenacetin and phenetidine do not interfere and that 2-hydroxyphenacetin resulted in a slight blank value.

Rüdiger and Büch [788] separated phenacetin and p-acetamidophenol, as well as their metabolites, by thin-layer chromatography and determined them spectrophotometrically.

A secondary metabolite of phenacetin was discovered in 1963 by Jagenburg and Toczko [455]. These authors observed in the urine of phenacetin-dosed patients a compound that reacted with ninhydrin. None of the previously described metabolites had given such a reaction. Six liters of human urine were concentrated by lyophilization and deproteinized; the metabolite was retained on an Amberlite IRC-120 (H$^+$) column. After washing, it was eluted by a 0.05M citrate buffer, pH 4.25. It was further purified by a second chromatography under the same conditions. The chromatographic fractions were examined by paper chromatography (butanol:acetic acid:water 4:1:1 or phenol:water:ammonium hydroxide 800:200:1) and only the purest ones were kept. After ion-exchange desalting, these fractions were concentrated under vacuum, and the residue was extracted with ethanol and then redissolved in water. On concentration of the solution, the pure metabolite crystallized.

It was characterized by its elemental analysis, potentiometric titration, IR and UV spectra, electrophoretic and chromatographic behavior, and chemical properties. It is S-(1-acetamido-4-hydroxyphenyl)-cysteine. The excreted amount represented 2% of the administered phenacetin.

Klutch et al. [510] isolated and identified a supplementary metabolite of phenacetin. The urinary extract, obtained by treatment with chloroform and hydrolyzed with β-glucuronidase, was submitted to paper chromatography. A ring hydroxylated derivative was detected by diazotized sulfanilic acid. It had the same R_f value as 2-hydroxyphenacetin synthetized by an unequivocal procedure. The metabolite was then isolated on a preparative scale by chromatography of the chloroform extract on an alumina column (chloroform). The pure compound obtained was identified on the basis of its UV and IR spectra and its melting point. The excretion product was 2-hydroxyphenacetin glucuronide.

Shahidi et al. [830] briefly reported that they crystallized two compounds from the urine of a patient in whom the therapy resulted in a methemoglobin increase. Elemental analysis and NMR spectrometry demonstrated that the compounds were 2-hydroxyphenacetin and 2-hydroxyphenetidine glucuronides. Synthetized hydroxyphenetidine appeared to be a powerful methemoglobin producer.

Büch et al. [148] described another phenacetin metabolite: p-methoxy-o-sulfonyloxyaniline. They did not report how they identified it. It was determined by colorimetry: acid hydrolysis actually gave p-ethoxy-o-hydroxyaniline, which, in alkaline medium, condensed into 3-amino-7-ethoxy-2-phenoxazone.

These authors also determined p-acetamidophenol, free and conjugated as sulfate and glucuronide, as well as traces of o- and m-hydroxyphenacetin. However, the methods were not described.

Kiese and Renner [506] investigated the hepatic metabolism of phenacetin and its N-methyl derivative. They had to develop methods for identifying and determining N-methylphenetidine. This compound was separated by paper chromatography (96% ethanol containing 1% of petroleum ether) or on silica gel thin layer (methanol). It could be extracted into benzene and separated from N-methylphenacetin by reextraction into 0.1N hydrochloric acid; it was determined by UV spectrophotometry.

Bernhammer and Krisch [78] studied the influence of hepatic esterase on the in vitro metabolism of phenacetin. The produced p-phenetidine was identified by thin-layer chromatography on silica gel (chloroform : toluene : acetic acid 42.5 : 15 : 42.5; benzene : methanol : acetic acid 79 : 14 : 7) and determined by the procedure of Brodie and Axelrod [127].

Welch et al. [995] demonstrated that 3-methylcholanthrene stimulated phenacetin production, dealkylation, and conjugation in the cat. They determined phenacetin, phenetidine, p-acetamidophenol, and p-aminophenol by the procedure of Brodie and Axelrod [125, 127] and aromatic amines by the procedure of Bratton and Marshall [108]. p-Acetamidophenol was identified by thin-layer chromatography on alumina (benzene : toluene : acetic acid : water 2 : 2 : 2 : 1; benzene : ethanol : water : acetic acid 20 : 20 : 20 : 1; chloroform : methanol : water : acetic acid 20 : 10 : 20 : 1).

III. OTHER ANILIDES

Newell et al. [669] investigated the metabolism of the three monochloracetanilides by means of ^{36}Cl-labeled compounds. A series of supposed metabolites was obtained either by synthesis or by purification of commercial products. Rat urine was collected in flasks surrounded by Dry Ice so that the urine could be frozen immediately. The various metabolites were determined by reverse isotopic dilution either directly, or after hydrolysis by either acid (5N sulfuric acid — 1 hour under reflux) or enzymic (β-glucuronidase or sulfatase — 24 hours) methods.

Ethopabate is completely metabolized in the chicken. Buhs et al. [153] showed that it was mostly excreted in urine. They administered the ^{14}C-labeled product (four preparations labeled on carboxyl, ethoxyl, N-acetyl, or the ring).

Urine was chromatographed on Darco G-60 : Celite 1 : 2 and the metabolites eluted by the mixture methanol : 5N ammonium hydroxide 8 : 2. The resulting concentrate was then chromatographed on an Amberlite CG-120 (H$^+$) column. The neutralized effluent was chromatographed on alumina and three fractions were obtained with the mixtures 6N ammonium hydroxide : methanol 1 : 9, 2 : 8, and 5 : 5. The main fraction was chromatographed again under the same conditions, and then transferred to a DEAE-cellulose (acetate) column and eluted by a 0.2M buffer pyridine : acetate. It was again purified on a polyamide column (ethanol containing 0.5% of 6N ammonium hydroxide). The pure product was lyophilized before analysis. Infrared and NMR spectrometry, elemental analysis, and potentiometric titration assigned it the formula of 2-ethoxy-4-amino-5-hydroxybenzoic acid 5-O-sulfate.

Another metabolite was prepared by chromatography of urine on Amberlite CG-120 (ammonium). After elution at pH 3 and 4.5, a fraction was obtained at pH 5.5. It was transferred to a CG-120 (H$^+$) column and eluted at pH 7. The eluate was extracted with chloroform, concentrated, and fractionated on a cellulose powder column impregnated with 0.2M phosphate buffer, pH 7, by means of the mixture butanol : isooctane 7 : 3. The crystallized product was identified by comparing its UV and IR spectra to those of an authentic sample of 2-ethoxy-4-acetamidobenzoic acid.

It was not possible to purify a third fraction, but its chromatographic behavior and its partition coefficients indicated it to be 2-ethoxy-4-aminobenzoic acid.

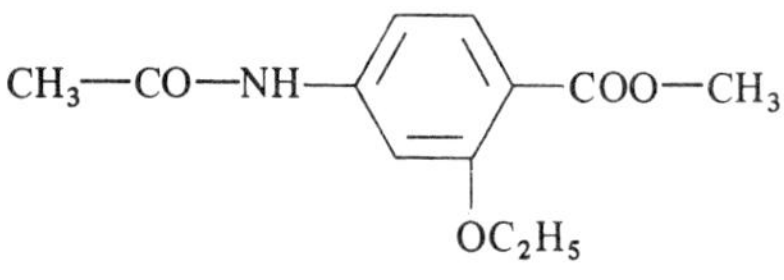

CH$_3$—CO—NH—⟨benzene ring⟩—COO—CH$_3$
                          OC$_2$H$_5$

ethopabate

OTHER ANILIDES

The three metabolites thus detected represented about 60% of the administered radioactivity. Two conjugated derivatives corresponding to the acid metabolites were not identified.

Various benzanilides or phthalanilides have been tried as cytostatic agents. They appeared not to be metabolized. Rogers et al. [781] studied 4',4''-bis-(1,4,5,6-tetrahydro-2-pyrimidinyl)-terephtalanilide (I) and 4-(1,4,5,6-tetrahydro-2-pyrimidinyl)-4'-[p-(1,4,5,6-tetrahydro-2-pyrimidinyl)-phenyl]-carbamoylbenzanilide (II). They applied chromatography on DEAE-cellulose column according to the procedure previously employed by Sivak et al. [848] for 4',4''-bis-(2-imidazolin 2-yl)-terephtalanilide (III).

Kreis et al. [525] studied the 2-chloro homolog of compound III by means of thin-layer chromatography on silica gel (butanol:acetic acid:water 4:1:5).

Lidocaine metabolism has been studied since 1951, but not very extensively. MacMahon and Woods [579] briefly reported an increase of excreted sulfate after lidocaine administration.

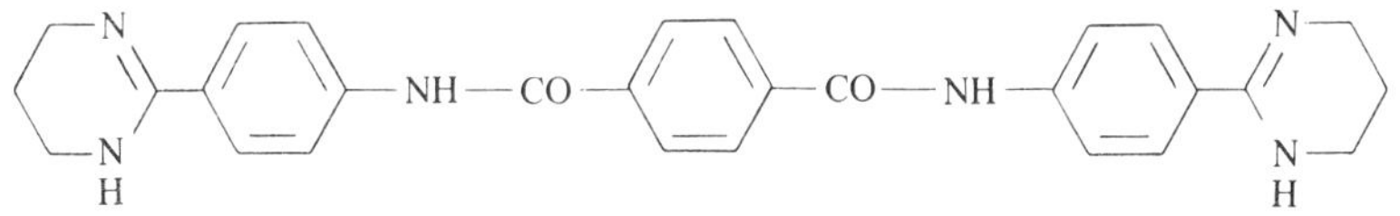

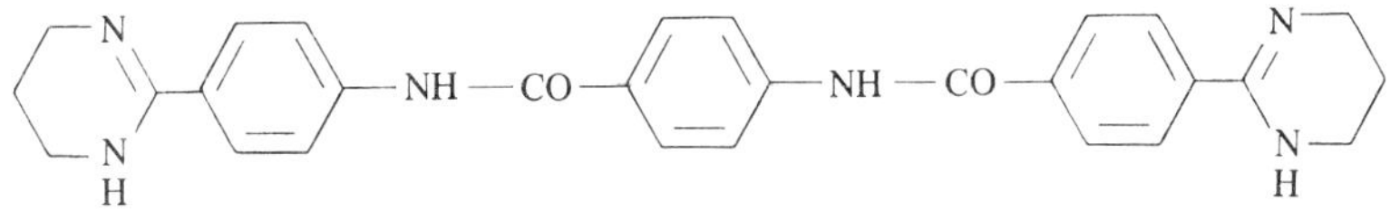

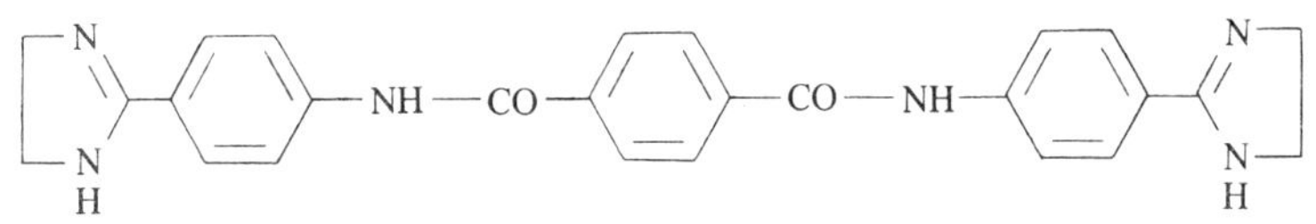

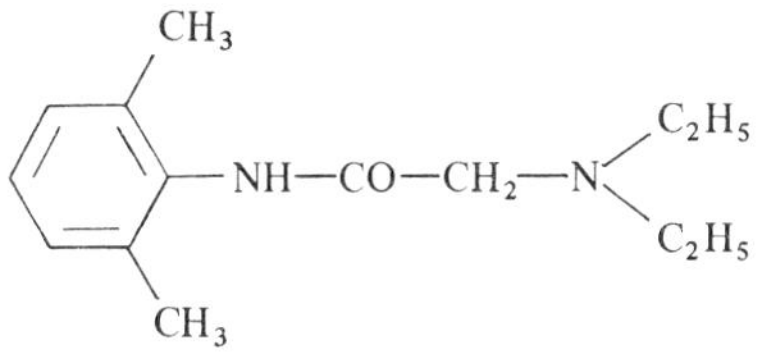

lidocaine

Sung and Truant [882] adapted the methyl orange procedure of Brodie and Udenfriend to lidocaine and checked the specificity of the ethylene dichloride extraction by the partition coefficient method.

Geddes and Douglas [350] studied in vitro metabolism with ^{14}C-labeled lidocaine. They determined the unchanged drug and identified diethylaminoacetic acid by chromatography. Conditions were not defined.

More recently, Hollunger [430, 431] showed that hepatic enzymes transform lidocaine by removing one ethyl group, and then hydrolyzing it to xylidine. The metabolites were analyzed by paper chromatography (methylethylketone : butanol : acetic acid : water 160 : 80 : 2 : 30). This system separates lidocaine and its mono- and didesethyl derivatives well. The monodesethyl derivative was identified as a metabolite by R_f comparison with the authentic compound.

Beckett et al. [57] extracted lidocaine and its metabolite from urine by ethyl ether and from blood by carbon tetrachloride. They analyzed the extract by gas-liquid chromatography according to the technique they had previously published [58]. Akermann et al. [15] investigated, in parallel, lidocaine and prilocaine in vivo and in vitro using the ^{14}C- and ^{3}H-labeled molecules. They used the extraction procedure of Sung and Truant [882], as well as paper chromatography (butanol : acetic acid : water 6 : 1 : 3; 25% acetic acid saturated butanol; methylethylketone : butanol : acetic acid : water 80 : 40 : 1 : 15). Prilocaine gave two metabolites which have same R_f values as o-toluidine and N-propylalanine.

IV. THIOCARBANILIDES AND DERIVATIVES

The metabolism of thiocarbanilide and its derivatives was studied by Smith and Williams in 1961.

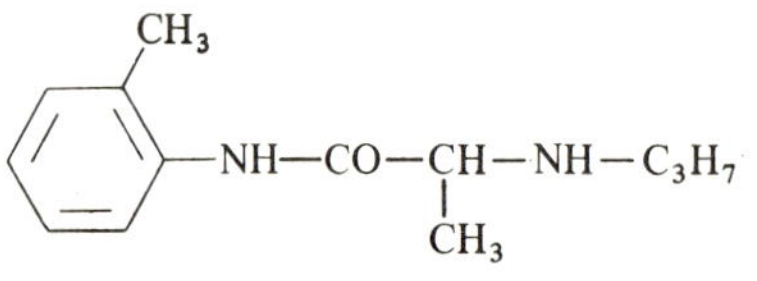

prilocaine

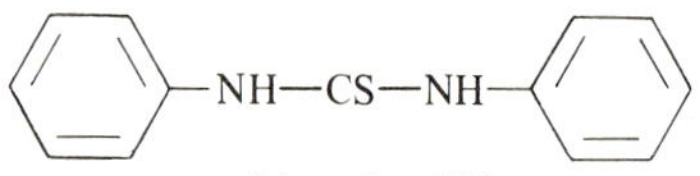

thiocarbanilide

In their first paper [864] the paper chromatography of urine in four solvent systems was described:
(I)   Butanol:acetic acid:water, 4:1:5
(II) 98% formic acid saturated benzene
(III) Benzene:butanol:water 5:1:1
(IV) Benzene:acetic acid:water 1:1:2

Two metabolites were revealed after thiocarbanilide administration, one of which was most likely a hydroxythiocarbanilide glucuronide. It reacted with Grote reagent and naphthoresorcinol. This glucuronide was not precipitated by lead acetate, as it is desulfurized. It was separated by crystallization of the acidified urine at 0°C. After recrystallization, it was identified as p-hydroxythiocarbanilide glucuronide by elemental analysis, spectrophotometry, and comparison with an authentic sample. This sample was prepared from p-aminophenyl glucuronide and phenylisothiocyanate. The same analytical methods were applied to p-hydroxy-, p,p'-dihydroxy- and p,p'-dichlorocarbanilides and thiocarbanilides.

A second paper [803] was devoted to 1-phenyl-2-thiourea biotransformations. The ^{14}C- and ^{35}S-labeled compounds were used. Numerous metabolites were detected by paper chromatography in systems I, II, and IV, to which were added:
(V)   Butanol:pyridine:benzene:water 5:1:3:3
(VI) Propanol:ammonium hydroxide 7:3
(VII) Water saturated isopropyl ether

The different metabolites were characterized and determined by isotopic dilution. Chromatographic assays combined with the use of many detecting reagents, as well as hydrolysis experiments, helped to confirm identifications. The authors thus arrived at a very complete picture of the metabolism.

These investigations were extended to p-chlorophenyl and p-tolylthiourea [865] with related procedures. Smith and Williams examined the fate of

---

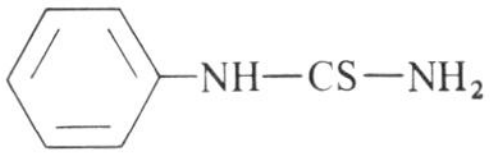

C$_6$H$_5$—NH—CS—NH$_2$

1-phenyl-2-thiourea

thiambutosine which is active as antileprosy agent in the organism [866].
Paper chromatography of urine (water saturated ethyl acetate; ethyl acetate : benzene 1 : 1; ethanol : butanol : 3N ammonium hydroxide in 3N ammonium carbonate 11 : 40 : 19) revealed four metabolites: the main one reacted positively with Grote reagent, naphthoresorcinol, and ferric chloride. It is thus a glucuronide with a thione group and a p-dimethylaminophenylthiourea group. On hydrolysis this metabolite disappeared and it was replaced by a compound of same R_f value as the acid resulting from the oxidation of the terminal CH_3 on the butoxy chain. This acid was isolated from the ammonium sulfate saturated urine at pH 7 by acetone : ether 1 : 1 and was identified by its elemental analysis and the melting points of its derivatives.

These various detailed articles are excellent examples of research based on paper chromatography, with the use of many detection reagents and a complete series of reference compounds. The quantification is achieved by the use of labeled molecules.

In 1965, Shibasaki et al. reported the results of their investigations on 4,4'-diethoxythiocarbanilide [840] and N-methylthiocarbanilide [839]. They fractionated metabolites extracted from urine with ethyl ether by thin-layer chromatography (benzene : ethyl acetate in various proportions), on

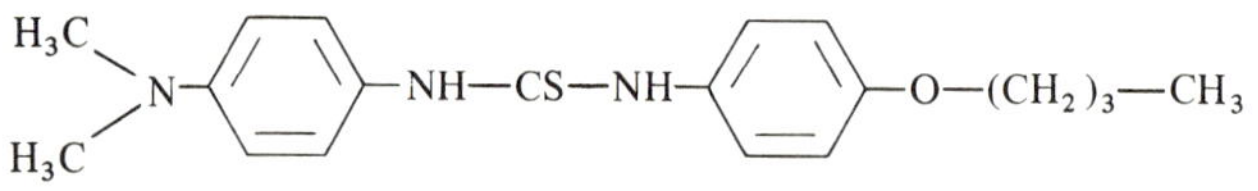

thiambutosine

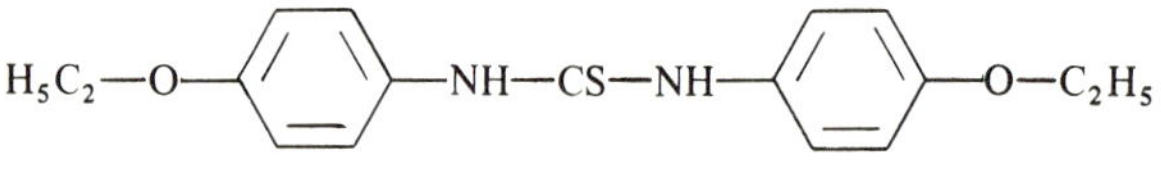

4, 4'-diethoxythiocarbanilide

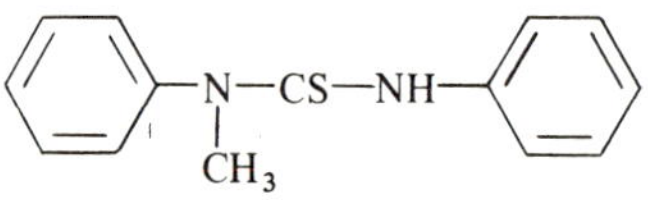

N-methylthiocarbanilide

paper (butanol : acetic acid : water 4 : 1 : 5; ammonium hydroxide saturated butanol), and on alumina column. The same separations were performed after β-glucuronidase hydrolysis. Different desethylation, desulfuration, and conjugation products were thus identified by comparison with synthetic derivatives.

The second paper showed, with similar methods, that N-methylthiocarbanilide is mainly hydroxylated in position 4', then conjugated to glucuronic acid. Nevertheless hydroxylation, N-demethylation, desulfuration, and C—N splitting derivatives were also detected by the thin-layer chromatography of an ethyl ether extract from urine.

NEWER TRACE ELEMENTS IN NUTRITION

ERRATUM

Chapter 12, page 258

line 12 from top of page

9.5 ml should read 0.5 ml

Chapter VIII

BARBITURATES

The barbiturate family is a numerous one but its members are closely related chemically. Structural variations, if we exclude the thiobarbiturates, occur mainly in the 5 position. It is therefore expected, and it has been confirmed, that they show similar metabolism.

Barbiturate metabolism is well reviewed in the literature; but usually the metabolic processes are discussed and the analytical methods that have been employed [568, 895, 896, 622, 747, 759, 338] are not mentioned. Raventos [747] did give methodology some emphasis and Frey [338] limited his discussion to paper chromatography.

The analytical procedures that were developed were applied to series of related compounds. Certain authors, Butler, Bush, Maynert, Van Dyke, Brodie, Richards, and Block, have frequently published studies using similar methods.

The thiobarbiturates differ in that they and their metabolites easily desulfurize on oxidation to give the corresponding barbiturates. This led to many erroneous results when insufficient precautions were practiced during the isolation of metabolites.

I. GENERAL ANALYTICAL METHODS

The study of barbiturate metabolism began when convenient analytical methods were greatly lacking and new general methods had to be devised.

Bush and Densen [166] developed a fractional solvent extraction process to isolate a known product, analyze mixtures, and check purity. The principle of their procedure is related to the counter-current distribution that Craig and his co-workers developed at about the same time. The equipment is not as sophisticated, however, in that the separations were achieved with a series of six to ten separatory funnels.

The partition of one or two substances between two immiscible solvents was mathematically analyzed. The authors developed a series of graphs to define the optimum extraction or separation conditions for a given number of extractions. They also demonstrated that the separation of two compounds A and B with partition coefficients K_A and K_B was optimum for a given number of extractions when the ratio of the volumes, V, of the two solvents X and Y was maintained at:

$$\frac{V_X}{V_Y} = \sqrt{\frac{1}{K_A K_B}}$$

Ten years later, Bush [162] described an apparatus for the extraction of metabolites in a large volume of diluted aqueous solution by partial redistillation of a small portion of the extracting solvent. As an example, 223 mg of barbital dissolved in 5000 ml of water was extracted with 250 ml of ethylene dichloride for 24 hours. The yield of this extraction was equivalent to that of 22 successive extractions with 250 ml of solvent each or of seven extractions with 1000 ml.

Bush [163] in 1961 described in detail the procedure he applied to a series of barbiturates. This procedure involved solvent-selective extraction, concentration in a pH 10 buffer, and determination by a special UV spectrophotometric method that utilized the change in barbiturate absorbance as a function of pH. The absorbance difference at pH 10 and 6.25 was measured and the difference between the two determinations at two wavelengths was calculated. These conditions eliminated a great number of interferences.

Thin-layer chromatography was extensively applied in toxicology for identifying barbiturates, and made possible the detection of a series of metabolites. Eberhardt et al. [291] used silica gel with the system piperidine:petroleum ether 1:5 and separated products were observed by UV light or with a 1% mercurous nitrate solution. They found one to five metabolites for the differing administered barbiturates, but they did not identify them.

Gas-liquid chromatography was also applied to the analysis of barbiturate and metabolite mixtures by Svendsen and Brochmann-Hansen [885]. Acidified urine was extracted with ethyl ether. The ether phase, washed with a

1% sodium bicarbonate solution, was extracted with 0.1N sodium hydroxide, and the barbituric acids were again reextracted into ethyl ether after acidification. The ether was evaporated to dryness and the solutes redissolved in acetone. Chromatography was usually performed on a column of Chromosorb W, acid and alkali washed and coated with 1.7% of neopentylglycol sebacate. p-Hydroxyphenobarbital was not eluted from this column and required Apiezon L. The results obtained confirmed the previous works for five barbiturates. It is to be noted that this method enabled barbital urinary excretion to be followed 6 and 16 days after administration of a single 300 mg dose to man.

Column chromatography on ion-exchange resin (Dowex-1) for identifying and separating urinary metabolites was used by Ledvina and Baladova [542].

II. BARBITURATES WITH ALKYL SUBSTITUENTS

Barbital is the simplest of alkyl-substituted barbiturates. It appears to be different than its analogs in that it undergoes minimum transformation in vivo. Maynert and Van Dyke [626] demonstrated this in 1950 using ^{15}N-labeled barbital. Isotopic dilution showed that the total amount of isotope contained in urine is present as unchanged barbital. Giotti and Maynert [360] confirmed this finding a year later by extracting barbital from urine with ethyl ether and comparing its partition coefficient in ethyl ether : phosphate buffer at four different pH values with that of an authentic sample. They also studied barbital binding to plasma proteins by ultrafiltration.

These experiments were performed with the dog. In the rat, however, Goldschmidt and Wehr [374] used 2-^{14}C-labeled barbital and found three metabolites but the total only represented 5% of the administered drug; the

---

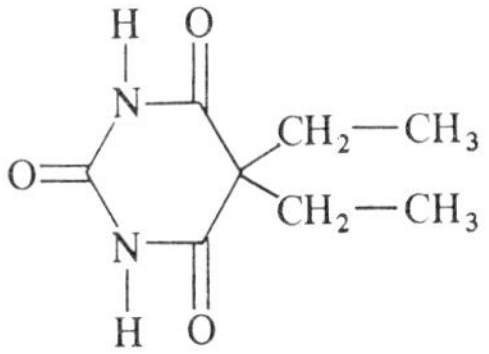

barbital

other 95% was excreted unchanged. Paper chromatography of urine (butanol: ammonium hydroxide:water 340:3:57; butanol:acetic acid:water 4:1:5) was compared to that of the specially synthesized reference compounds. One of these metabolites was identified as 5-ethylbarbituric acid and the other as 5-ethyl-5-β-hydroxyethylbarbituric acid. This identification was confirmed by isotopic dilution. The third metabolite, on hydrochloric acid hydrolysis, gave 5-ethyl-5-β-hydroxyethylbarbituric acid, although glucuronic acid was not detected.

The same authors applied their methods to vinylethylbarbituric acid metabolism. They found four metabolites that represented 8% of the administered dose. One was identified in the manner described above as 5-ethylbarbituric acid.

Seven years later, Ebert et al. [295] compared barbital metabolism in tolerant and nontolerant rats. Three metabolites were detected in urine of animals dosed with 2-^{14}C-labeled barbital by paper chromatography (butanol: water:ammonium hydroxide 340:57:3). They did not identify these metabolites, which were similar to those found by Goldschmidt and Wehr [374].

Butobarbital is excreted into urine as a single metabolite. Maynert [616] extracted this metabolite with ethyl ether, reextracted it into water, and crystallized it after concentration. Elemental analysis, UV and IR spectra, and iodoform reaction characterized it as 5-ethyl-5-(3-hydroxybutyl)-barbituric acid. Confirmation was obtained by synthesis of the metabolite.

When the length of the 5-alkyl chain was increased the complexity of metabolic reactions increased also. In the case of hexetal, Maynert [619] extracted dog urine, concentrated under vacuum and saturated with sodium chloride, with ethyl ether at pH 2. The concentrated ether extract was

---

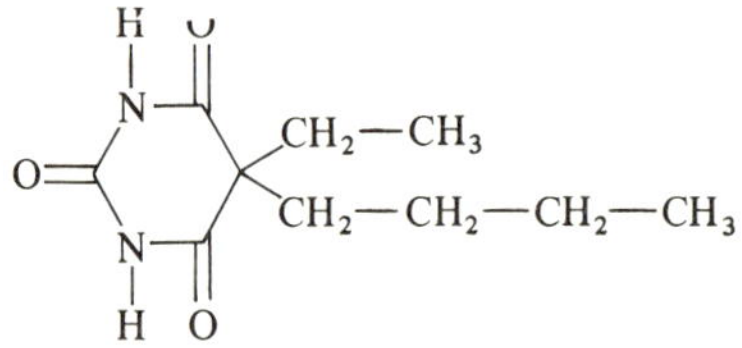

butobarbital

transferred to a column filled with Darco G-60 : Celite 300 : 50. This column was eluted with ethyl ether and the concentrated eluate analyzed by counter-current distribution (0.1M phosphate buffer, pH 7.9 : ethyl ether). Seven components, in addition to the two end fractions, were obtained. The hydrophilic end fraction was adjusted to pH 2, extracted with ethyl ether, evaporated, and submitted to another counter-current distribution (0.01N hydrochloric acid : ethyl ether), which revealed three other metabolites.

The same experiment after β-glucuronidase hydrolysis showed only a few conjugated metabolites. It is to be mentioned that the author treated one half of the sample with the active enzyme and the other half with the denatured enzyme. A gross determination by UV spectrophotometry immediately indicated no significant conjugate.

The ten metabolites thus isolated appeared homogeneous on paper chromatography in four solvent systems:

(I)   5N ammonium hydroxide saturated butanol
(II) 7.4N ammonium hydroxide : isopropanol, 1 : 4
(III) 14.8N ammonium hydroxide : isopropanol, 1 : 8
(IV) Toluene : water : acetic acid : ethylene glycol, 100 : 50 : 40 : 5

The various metabolites were more or less definitely identified by elemental analysis, IR and UV spectrophotometry, and chemical reactions (iodoform, oxime, acetate formation). They are alcohols, ketones, and carboxylic acids resulting from n-hexyl chain oxidation. This 1965 paper of Maynert is much more analytically complete than his 1950-1952 publications on barbital and butobarbital.

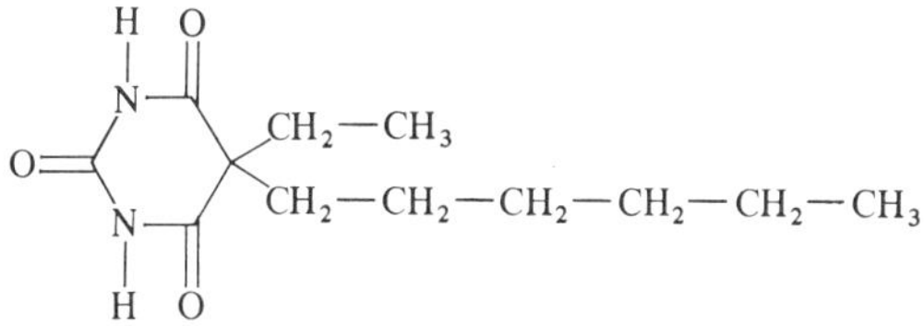

hexetal

Secbutabarbital is a derivative with a branched alkyl chain and was also studied by Maynert [621]. On counter-current distribution (0.5M citrate buffer, pH 2: ethyl ether) one metabolite was isolated; its elemental analysis indicated that it was a carboxylic acid and corresponded to 28-35% of the administered dose with 3-5% of the drug excreted unchanged. The other metabolites were not investigated.

Pentobarbital has been studied in more detail. Van Dyke et al. [954] in 1947 demonstrated by isotopic dilution that the ^{15}N-labeled product was excreted mostly as metabolites. In 1949, Maynert and Van Dyke [624] gave evidence, by isotopic dilution, that ethyl-(1-methylbutyl)-malonuric acid was not a metabolite in itself, but that it coprecipitated with a metabolite. Actually the specific radioactivity of the carrier-metabolite mixture was distinctly reduced by purification. This showed that results obtained from simple isotopic dilution must be accepted cautiously.

The same year, Maynert and Van Dyke [623] announced the isolation of a metabolite corresponding to an hydroxylated derivative on the basis of its elemental analysis, UV spectrum, and acetate formation.

In 1951, this research was completed [620] by identification of the two main metabolites. The ethyl ether extract of urine was reextracted with water as above [619] and a crystalline substance was obtained. On evaporation, the mother liquor gave another product. These two components had the same elemental analysis, but they were different with respect to UV and IR spectra, optical rotation, pK_a, melting point, and solubility. Both gave an acetate on acylation. Reaction with concentrated sulfuric acid showed that the hydroxyl was not on the ethyl group. It was thus concluded that they

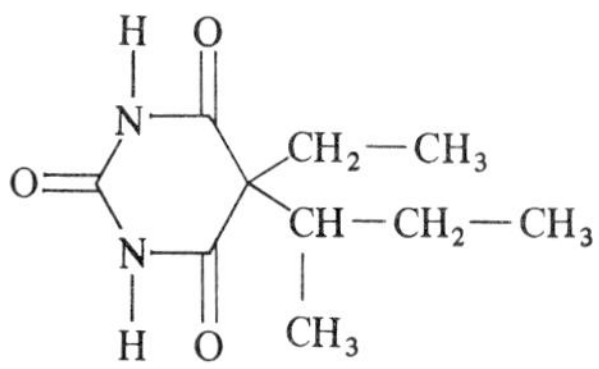

---

secbutabarbital 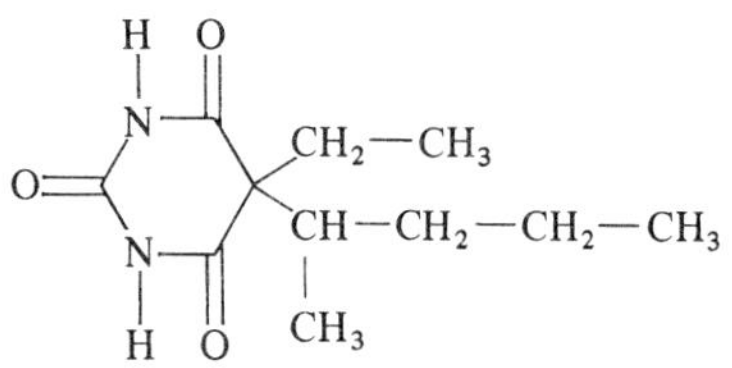   pentobarbital were the two diastereoisomers of 5-ethyl-5-(3-hydroxy-1-methylbutyl)-barbituric acid.

This acid was detected by Algeri and McBay [17] by means of paper chromatography (5N ammonium hydroxide saturated butanol) in addition to another unidentified product. With the same kind of chromatography Roth et al. [786] were able to find five metabolites in the urine of rats dosed with 2-^{14}C-labeled pentobarbital.

In 1953, Brodie and his co-workers also became interested in pentobarbital biotransformations [133]. They confirmed Maynert's results in the dog and extended them to man. Urine was extracted at pH 6 with isoamyl alcohol, and the extracted compounds were transferred into water at alkaline pH after petroleum ether addition. The metabolite was purified by a series of two counter-current distributions (1M citrate buffer, pH 5 : petroleum ether containing 35% of isoamyl alcohol) and by recrystallization. Its physicochemical properties made it possible to identify it as the isomer of 5-methyl-5-(3-hydroxy-1-methylbutyl)-barbituric acid. The authors described a pentobarbital determination procedure in which the metabolite was not extracted.

Titus and Weiss [909], using pentobarbital as an example, demonstrated that labeled molecules are best used after preliminary fractionation of the metabolites. They preferred this to direct isotopic dilution.

The direct paper chromatography of the urine of dogs administered with 2-^{14}C-labeled pentobarbital (1% ammonium hydroxide saturated butanol) revealed nine metabolites in addition to the unchanged drug. A larger volume of urine was extracted with butanol at pH 7 and the concentrated extract fractionated by counter-current distribution (butanol : 0.25M pyrophosphate buffer, pH 9.4 — 10 transfers). The bulk of metabolites was thus isolated from urinary components and it was fractionated again by 124 transfers in the same system. The separation of all metabolites was not achieved but was completed by a preparative paper chromatography.

Some of the components were identified. The R_f and UV spectra of the D and L forms of 5-ethyl-5-(3-hydroxy-1-methylbutyl)-barbituric acid were similar to those given in the literature. Another metabolite was urea.

A third metabolite was isolated from a large volume of urine from a group of dogs which received a total of 50 g of pentobarbital. After addition of the labeled likely metabolite, this urine was extracted with butanol, the butanol solution evaporated, and the residue fractionated by a 12-transfers counter-current distribution followed by a second of 200 transfers (butanol:0.5M borate buffer, pH 8.2). The analysis of this metabolite — UV spectrophotometry, potentiometric titration, pK_a and molecular weight determinations, and elemental analysis — identified it as 5-ethyl-5-(3-carboxy-1-methylpropyl)-barbituric acid. This was confirmed by synthesis of the compound [1018] and by comparison of IR spectra and X-ray diffraction diagrams.

A glucuronide-like metabolite was detected by enzymic hydrolysis. It was not identified because its hydrolysis was very slow.

Maynert [618] studied pentobarbital metabolism quantitatively in man. He administered the ^{15}N-labeled barbiturate and determined the excretion products by isotopic dilution. The alcoholic metabolites were extracted into ethyl ether, recrystallized in water, and transformed into acetates. The two diastereoisomers of 5-ethyl-5-(3-hydroxy-1-methylbutyl)-barbituric acid, respectively, corresponded to 7 and 41%.

Maynert and Van Dyke also investigated <u>amobarbital</u> by the isotopic dilution procedure [625]. They stated that ring opening was not implicated in the metabolism since they did not find acetamide, acetylurea, or malonuric acid in urine. Maynert [615] provided evidence that two thirds of the amobarbital was excreted by the dog as 5-ethyl-5-(3-hydroxyisoamyl)-barbituric acid. The analytical procedure used was the same as given previously [620] for isolation and identification of the metabolite. Its nature was confirmed by synthesis.

Irrgang [446] more recently studied the pharmacological properties of this metabolite as compared to amobarbital. Paper chromatography (chloroform : isopropanol : 25% ammonium hydroxide 45 : 45 : 10) of urine revealed that the amobarbital was rapidly metabolized into the above described metabolite (R_f 0.45-0.51) and two other products, R_f 0.28 and 0.17, which

----

amobarbital

could not be isolated or identified. Irrgang was unable to detect the unchanged drug in most of the cases and he concluded that the drug itself has an immediate, but short, hypnotic effect that is prolonged by the action of the metabolite: hydroxyamobarbital.

This hypothesis was criticized by Frey and Magnussen [341]. They extracted urine at pH 1-3 with ethyl ether and performed paper chromatography (butanol: chloroform: 25% ammonium hydroxide 120:55:25) according to the procedure that Frey et al. [342] had already applied to various barbiturates. The treatment of blood was by a similar procedure already described by Frey et al. [340]. Determinations were performed by the method of Brodie et al. [135] in which the barbiturate was separated from its metabolite by extraction into petroleum ether containing 1.5% of isoamyl alcohol. Both products were determined by spectrophotometry at 255 and 275 mµ as per Frey et al. [339]. Frey and Magnussen showed by these methods that the unchanged amobarbital could be found several days after administration in addition to hydroxyamobarbital and traces of other metabolites.

Maynert [618] studied the quantitative aspect of amobarbital metabolism in man: he used the same isotopic dilution procedure as for pentobarbital. He found 51-73% of the administered dose as 5-ethyl-5-(3-hydroxy-3-methylbutyl)-barbituric acid.

Jackson and Moss [450] detected hydroxyamobarbital in the urine of a suicide. The ethyl ether extract of the acid ammonium sulfate saturated urine was chromatographed on paper (butanol: ammonium hydroxide 3:1). Four spots were observed and one of these (R_f 0.65) was identified as the hydroxylated metabolite by R_f comparison and by its X-ray diffraction diagram.

Kamm and Van Loon [481] developed a procedure for the determination of small amounts of hydroxyamobarbital by gas-liquid chromatography. Urine at pH 6 was extracted with ethyl ether. On treatment with hexamethyldisilazane and chlorotrimethylsilane in the mixture benzene: pyridine, the silyl derivatives were obtained. Chromatography was performed on Diatoport S coated with 3.8% of silicone SE-30 at 160°C. Methyl stearate was used as the internal standard. It was thus possible to determine 5 µg/ml of hydroxyamobarbital.

Secobarbital was the subject of Waddell's investigations from 1962 to 1965 [962]. This author found five metabolites in human urine. Three of these metabolites were first detected by paper chromatography of urinary extracts. The material extracted from the spots was used for spectrophotometric determination of partition coefficients. These served as the basis for the preparative isolation procedure.

The ethyl acetate extract was submitted to a counter-current distribution (0.1M phosphate buffer, pH 6 : ethyl acetate). The two separated metabolic fractions were each subjected to another distribution, then crystallized several times. Their purity was tested by paper chromatography. By elemental analysis, UV and IR spectrophotometry, and various chemical reactions it was possible to identify the two stereoisomeric forms of 5-allyl-5-(3-hydroxy-1-methylbutyl)-barbituric acid (I and II) and 5-(2,3-dihydroxypropyl)-5-(1-methylbutyl)-barbituric acid (III). The identification of this last metabolite was confirmed by synthesis of the authentic compound.

In addition to these three main metabolites, paper chromatography was used to detect and characterize by comparison 5-(1-methylbutyl)-barbituric acid (IV) and 5-hydroxy-5-(1-methylbutyl)-barbituric acid (V). Four other spots were recorded on chromatograms. The UV spectrum after elution suggested 5,5-disubstituted barbituric acid derivatives (Table 29).

Tsukamoto et al. [931] were interested in secobarbital metabolism so that the metabolism of its homolog thiamylal could be better elucidated. The same analytical procedures as in [932] were applied. Two major metabolites were found and compared to the authentic compounds. They were 5-allyl-5-(1-methyl-3-carboxypropyl)- and 5-allyl-5-(1-methyl-3-hydroxybutyl)-barbituric acids.

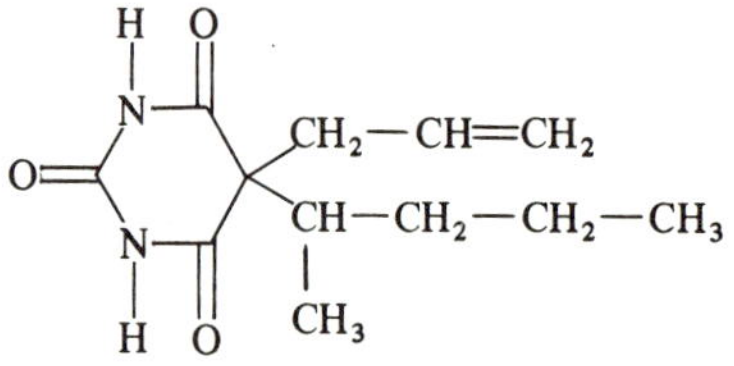

secobarbital

TABLE 29

Product	R_f in:		
	Ethyl acetate(+) [a]	Chloroform	Methylene chloride [b]
Secobarbital	0.95	0.95	0.95
Metabolite I	0.47	0.23	0.36
Metabolite II	0.37	0.18	0.28
Metabolite III	0.2	0	0.03
Metabolite IV	0	0.6	0.5
Metabolite V	0.07	0.04	0.1

[a] Paper moistened with 0.5N sodium hydroxide and dried; stationary phase: water.

[b] Stationary phase: water.

A supplementary metabolite was identified in the rat by Niyogi [673]. It was extracted from acid urine into methylene chloride and, after washing with a pH 7.3 buffer, reextracted into sodium hydroxide and transferred to ethyl ether on acidification. Paper chromatography (butanol : ammonium hydroxide : water, proportions not reported) revealed two spots: one had the same R_f (0.55) as 5-(2,3-dihydroxypropyl)-5-(1-methylbutyl)-barbituric acid; the other was not identified.

Another procedure was used 1 year later by the same author [675]. Preparative thin-layer chromatography on silica gel (chloroform containing 3% of methanol) was applied to rat urine extracts. After elution of the zones, the eluates were injected into a gas chromatograph (5% of DC-200 on Chromosorb at 165°C).

The same author has had an opportunity to analyze the liver of a suicide with secobarbital [674]. The tissue was extracted with hot chloroform in the presence of hydrochloric acid. The washed chloroform phase was extracted with sodium hydroxide and this solution in its turn was acidified and chloroform extracted. The chloroform solution was injected into a gas chromatograph under the above-stated conditions. In addition to the two peaks corresponding to one known metabolite [5-(1-methylbutyl)-barbituric acid], a new compound appeared to be revealed by a third peak. It must be mentioned that the chromatographic separations were poor.

Cochin and Daly [210], in their general study on the thin-layer chromatographic identification of barbiturates in biological media, were able to confirm the presence of 5-allyl-5-(1-hydroxy-1-methylbutyl)-barbituric acid in human urine. The extraction with methylene chloride was performed at pH 5. The chromatography on silica gel dried for 30 minutes at 80°C used the mixture chloroform:acetone 3:1. Moreover, three other metabolites were detected, one of which was identified as 5-(1-methylbutyl)-barbituric acid.

Kalypnon was investigated by Ledvina and Baladova [543]. The urine extract was chromatographed on Dowex-1 (OH⁻) according to the previously described procedure [542].

The separated metabolite was purified by counter-current distribution and recrystallization, and analyzed by paper chromatography in two solvent systems (butanol:5N ammonium hydroxide 1:1; propanol:5N ammonium hydroxide 7:3). The theoretical R_f of the supposed metabolites were calculated by means of the R_M function by comparison with a series of compounds. The identifications were confirmed by IR spectrophotometry and chemical reactions: the metabolite was 5-ethyl-5-(3-carboxyallyl)-barbituric acid. Another metabolite with an alcohol function was also detected.

The N-methylated alkyl-substituted barbiturates will now be considered. Metharbital was studied as early as 1939 by Butler and Bush [182, 164] and confirmed in 1953 [174]. The only metabolite present in large amounts is the N-demethylation product (barbital). It was extracted at pH 6 with ethyl

---

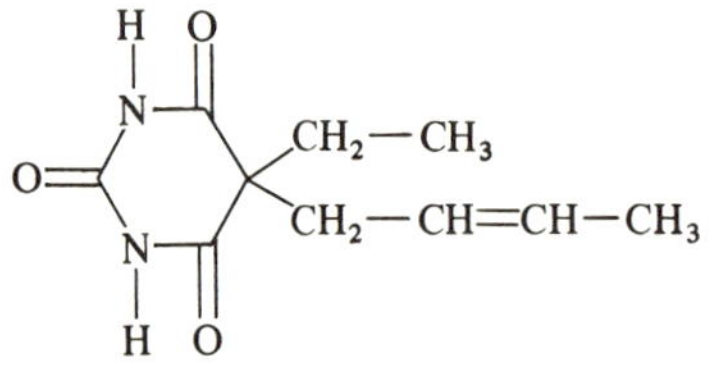

Kalypnon

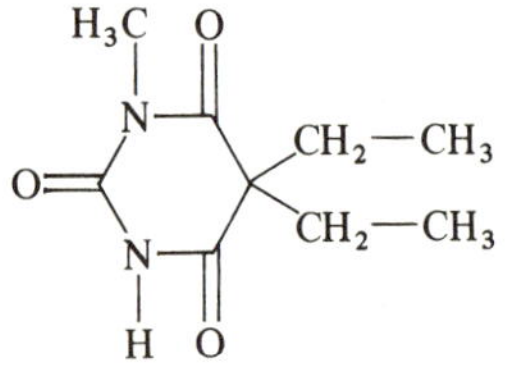

metharbital

ether together with metharbital, and its partition coefficients in ethyl ether: 0.05M borate buffer at pH 8.0, 8.6, and 9.2 were consistent with that of barbital. Barbital was determined in the presence of metharbital by UV spectrophotometry at two wavelengths (241 and 250 mμ).

^{14}C-labeled methohexital was used by Welles et al. [997] on the rat and dog. Urine was treated with ethyl ether at pH 4 and the extract chromatographed on paper. Four metabolites were observed with permanganate or mercurous nitrate in the absence of unchanged drug. The major metabolite (R_f 0.72 in the system butanol:5N ammonium hydroxide 1:1) was isolated by chromatography of the ethyl ether extract on a silica gel column, and methanol elution gave a crystallized product. Its UV spectrum in alkaline medium showed the 246 mμ peak with the 220 mμ shoulder characteristic of N-methyl barbiturates. The IR spectrum presented a strong band at 2.8μ to indicate the presence of an hydroxyl. Catalytic hydrogenation demonstrated that the ethynyl and allyl bonds were intact. Treatment with sodium hypoiodite yielded iodoform to show that the OH was on the penultimate carbon of the chain. The metabolite was 5-(1-methylallyl)-5-(1-methyl-4-hydroxy-2-pentynyl)-barbituric acid and its identification was confirmed by microanalysis.

The urinary metabolites of Pronarcon, according to Frey [338] are very numerous. Paper chromatography (butanol:chloroform:25% ammonium hydroxide 120:55:25) showed at least six, of which one was the demethylated derivative. When the latter was administered, four metabolites were observed in the urine.

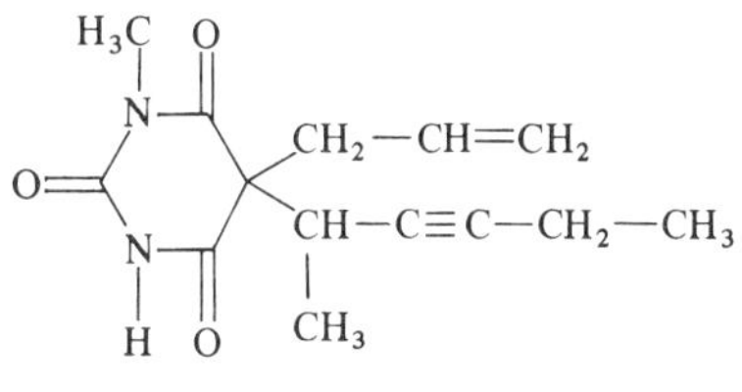

methohexital

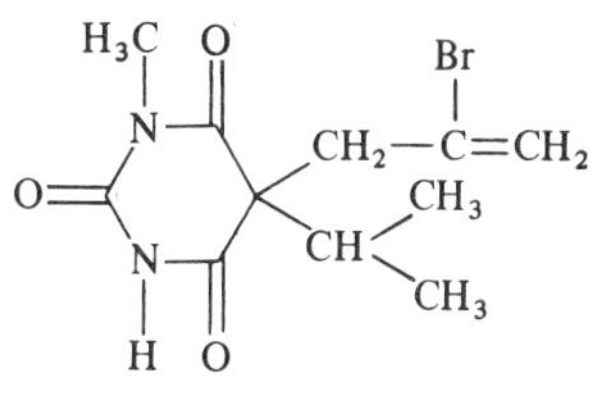

Pronarcon

III. BARBITURATES WITH CYCLIC SUBSTITUENTS

The conversion of phenobarbital into its p-hydroxylated metabolite was discovered by Butler et al. [184]. The analytical procedure was the same as for metharbital [174]. The identification was assured by synthesis of the metabolite [175], and Butler demonstrated by acid hydrolysis (addition of an equal volume of concentrated hydrochloric acid to urine and heating for 3 hours under reflux) that it was mostly conjugated.

Curry [236] detected this metabolite by paper chromatography (5N ammonium hydroxide saturated butanol) and isolated it [237] by chloroform extraction of acid human urine. Its elemental analysis and its UV spectra at pH 2, 10, and 13 were consistent. Acid hydrolysis (conditions not defined) did not increase its yield. Algeri and McBay [18] determined this metabolite in urine after paper chromatographic separation by elution and UV spectrophotometry. Hydrolysis under the conditions described by Butler [175] indicated an important percentage as conjugate.

Butler [177] prepared the m- and p-hydroxyphenobarbitals. The metabolite extracted from dog urine appeared similar to the p-hydroxylated derivative by solubility analysis in water. It was then isolated from urine by counter-current distribution (1M phosphate buffer, pH 8.2:ethyl ether), which made possible its precise determination. The urine was hydrolyzed by an equal amount of hydrochloric acid after a few minutes of boiling and by treatment with β-glucuronidase. The urine was treated by counter-current distribution after both kinds of hydrolysis. It was thus demonstrated that in the dog, p-hydroxyphenobarbital is entirely conjugated, mostly as the glucuronide, whereas in man it is only partly conjugated, probably as sulfate. The fact that β-glucuronidase was responsible for the liberation of the

---

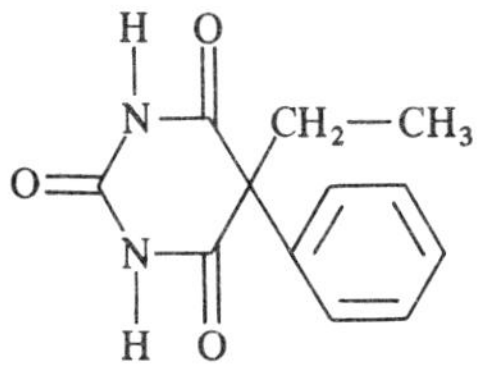

phenobarbital

hydroxylated product was confirmed by the inhibitory effect of potassium saccharate on the enzyme.

Even in a simple problem relative to a single phenobarbital metabolite in large amounts, crude analytical procedures may lead to conflicting results.

Glasson and Benakis [361] investigated radioactive metabolites of ^{14}C-labeled phenobarbital by paper chromatography. The system butanol: methanol:water 45:5:50 revealed five components. One of these, eluted and chromatographed again in the system isoamyl alcohol:ammonium hydroxide:ethylene glycol 85:15:5, was separated into three components. The identities of several of these metabolites were established by their R_f values.

Tompsett [918] used bidimensional paper chromatography (I, isopropanol:88% ammonium hydroxide:water 8:1:1; II, benzene:propionic acid: water 2:1:1) to detect p-hydroxyphenobarbital in urine (R_f 0.60 in I and 0.00 in II).

Kutt et al. [533] applied the methods of Butler [177] and Armstrong et al. [29] to study phenobarbital metabolism in patients suffering from hepatic diseases. After paper chromatography the metabolites were eluted and determined by diazotization.

Methylphenobarbital was examined by Butler [172] with the same techniques as used previously [174, 184] and shown to be simply demethylated.

Hexobarbital was the subject of the investigations of Bush et al. [165], who isolated small amounts of four metabolites, the solubilities and partition coefficients of which were measured to serve as a basis for their final fractionation process. After urine extraction at pH 6 by ethyl acetate, a series of partitions in ethyl acetate:water, and lastly benzene:water, gave the pure metabolites. The knowledge of the partition coefficients permitted the

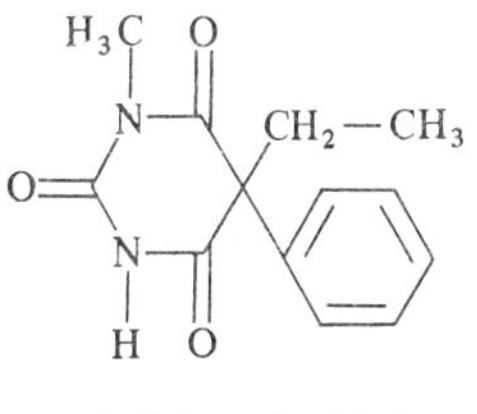

methylphenobarbital

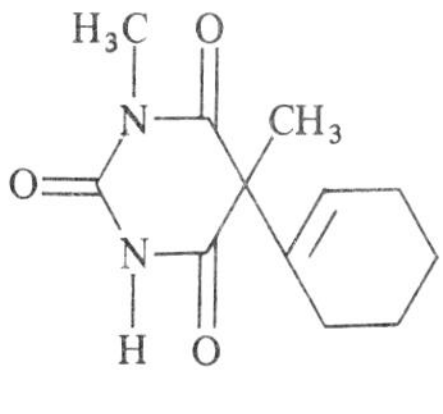

hexobarbital

calculation of the extraction yields. Elemental analysis, molar weight determination by cryoscopy in camphor and ebulliometry in acetone, and chemical reactions were performed on the metabolites. One was identified as a demethylation product; the three others had a ketone function. Unlike other barbiturates, hexobarbital is extensively metabolized. Oxidation appears to precede demethylation, as indicated from a comparative study of desmethylhexobarbital metabolism.

Cooper and Brodie [219] pursued the study of hexobarbital and demonstrated that liver microsomes oxidize the cyclohexenyl chain to yield one of Bush's three ketone metabolites. After incubation with the enzyme preparation, the acidified medium was extracted with chloroform and the extract chromatographed on paper (1% ammonium hydroxide saturated butanol). The four Bush metabolites were found in addition to 25% unchanged hexobarbital. The major metabolite was isolated from the chloroform extract by two countercurrent distributions (0.2M phosphate buffer, pH 7: chloroform containing 30% of heptane). It was a ketonic derivative of undefined structure and its formation required the presence of reduced TPN.

Frey [338] also examined hexobarbital urinary metabolites by paper chromatography (butanol: chloroform: 25% ammonium hydroxide 120:55:25) and confirmed the previous results. Desmethylhexobarbital administration resulted in four metabolites of low R_f values under the same conditions.

Tsukamoto and Yoshimura of Kyushu University have worked on hexobarbital metabolism [930]. They extracted rabbit urine with ethyl acetate and the resulting solution was chromatographed on an alumina column (no experimental details). Three metabolites were obtained, one of which was identical to the product that had resulted from hexobarbital chromic acid oxidation [probably 5-(3-oxo-1-cyclohexen-1-yl)-3,5-dimethylbarbituric acid]. Another product resulted from the opening of barbituric ring.

Tsukamoto et al. [936] prepared chromic acid oxidation products of hexobarbital and its demethylated derivative in an attempt to identify the urinary metabolites.

These metabolites [937] were removed from urine at pH 4-5 by ethyl acetate and chromatographed on an alumina column (acetone). The metabolites thus purified were dissolved in water and extracted into benzene. The benzene fraction was chromatographed on alumina, then recrystallized from methanol. The residual aqueous solution was evaporated, the residue dissolved in acetone, and the solution also chromatographed on alumina. On crystallization a pure metabolite was obtained in the eluate. The first metabolite did not exhibit the UV spectrum of barbiturates and it was identified by melting point comparison with cyclohexenylmethyl-N-methylacetylurea obtained by synthesis. The second one appeared on the basis of IR spectra to be 5-(3-oxo-1-cyclohexenyl)-3,5-dimethylbarbituric acid which was prepared by chromic acid oxidation. The third, according to its UV and IR spectra, was likely to have the formula of 5-hydroxycyclohexenyl-3,5-dimethylbarbituric acid. These three metabolites corresponded respectively to 0.5-1 and 5% of the administered dose.

The UV and IR spectra of these metabolites were studied [938] and the collected data confirmed the proposed identifications.

Yoshimura [1031] identified two other metabolites. The separation procedure was the same as above. By a series of seven chromatographies and crystallizations, several pure fractions were obtained. The new metabolites were characterized by their elemental analyses, their UV and IR spectra, and the formation of derivatives. These were 5-(3-oxo-1-cyclohexen-1-yl)-5-methylbarbituric acid and 5-(3-β-hydroxy-1-cyclohexen-1-yl)-3,5-dimethylbarbituric acid. These metabolites only represented 10% of the administered dose. Yoshimura also reported that the 3-keto and 3-hydroxy metabolites have the same R_f values on paper chromatography in the system used by Cooper and Brodie [219].

Tochino [913] isolated a metabolite from rabbit urine and assigned it the structure of an hydroxyhexobarbital by elemental analysis and IR spectrophotometry. By paper chromatography (5N ammonium hydroxide saturated butanol) he detected the presence of several biotransformation products [914].

Yoshimura in 1958 published [1032] a confirmation for the structure of the hydroxylated metabolite. He prepared it by reducing the 3-oxo metabolite and separating the two diastereoisomers. Tsukamoto, Yoshimura, and Toki [939] extracted acidified rabbit urine with ethyl acetate as previously, chromatographed it on alumina (acetone), and redissolved the eluate into methanol. It was then chromatographed on paper and compared to reference compounds (paper buffered with borate : sodium hydroxide buffer, pH 11, butanol saturated with the same buffer, ascending). The components were identified by their R_f and their UV spectra after elution. Five metabolites were thus detected. They were determined [940] by UV spectrophotometry after elution. Their total represented 25.7% of the administered dose. The major metabolite was the 3-α-hydroxy derivative.

Tsukamoto et al. [927] applied the same extraction and paper chromatography techniques, but with a pH 10 buffer to demonstrate that the 3-hydroxy and 3-keto groups on the cyclohexene ring are interconvertible and that the equilibrium favors the 3-hydroxy compounds.

Tsukamoto and Kuroiwa [923] used for the hydroxylated metabolite of hexobarbital the same methods as for cyclobarbital. They prepared an authentic sample and compared it with the urinary metabolite by paper chromatography.

Toki and Tsukamoto [916, 917] investigated the mechanism of metabolism by in vitro methods. Hexobarbital was incubated with microsomes and soluble enzymes preparations from rabbit liver. Paper chromatography, according to the already described procedure, revealed that microsomes first metabolized the drug into 3-hydroxyhexobarbital, and the soluble fraction yielded 3-ketohexobarbital.

Toki and Takenouchi [915] recently discovered a 3-α-hydroxyhexobarbital glucuronide. It was detected by paper chromatography (butanol:acetic acid:water 4:1:5, ascending) and isolated according to Kamil et al. [480]. The glucuronide thus purified was dissolved in water saturated ethyl acetate and chromatographed on silica gel column (ethyl acetate; ethyl acetate: methanol 1:20, 1:10, and 1:5). Elemental analysis, UV spectrum, and hydrolysis assays established its structure.

Yoshimura and Tsukamoto [1033] studied the metabolism of 5-ethyl-5-(1-cyclohexenyl)-4,6-dioxohexahydropyrimidine, an anticonvulsant agent resulting from the hydrogenation of the 2-thiobarbituric acid corresponding to cyclobarbital. They applied techniques similar to those previously described.

The same group of Japanese researchers investigated the metabolism of cyclobarbital, an hexobarbital isomer. Tsukamoto et al. [926] extracted a cyclobarbital metabolite from rabbit urine adjusted at pH 5 by treating it for 20 hours with ethyl ether. This derivative was identical with the cyclobarbital chromic acid oxidation product on the basis of its melting point and UV spectrum. A certain number of chemical reactions permitted the assignment of the formula of 5-cyclohexenonyl-5-ethylbarbituric acid.

Tsukamoto et al. [935] considered this metabolite, which may show a carbonyl grouping in either the 3 or 6 position of the cyclohexenyl ring. When heated at 180°C in the presence of a palladium-on-carbon catalyst, this metabolite gave, besides other compounds, 5(m-hydroxyphenyl)-5-ethylbarbituric acid, which was identified by comparing its melting point with that of the authentic sample. The carbonyl group was thus assignable to the 3 position.

Tsukamoto and his team [925] recovered about 44% of the administered dose of cyclobarbital as the metabolite on paper chromatography.

Cyclobarbital [889] was incubated with rat liver slices in the Krebs-Ringer medium containing 0.2% of glucose. The known metabolite was detected by paper chromatography. The metabolite was also isolated [890] after acidification of the incubation medium and extraction for 15 hours with ethyl ether. Counter-current distribution (butanol : borate buffer, pH 9 1:1 — 23 transfers) was applied to the extract. The metabolite that was separated from the unchanged drug was identical with the in vivo metabolite on melting point comparison.

Tsukamoto and Kuroiwa [923] detected in rabbit urine the 3-hydroxylated cyclobarbital derivative. This metabolite was obtained from the urine acidified at pH 2-3 and extracted with ethyl acetate. The washed extract was recrystallized in methanol and the 3-oxo metabolite precipitated. The residual solution was chromatographed on alumina column. The metabolite thus purified was treated with Girard's P reagent, reextracted with ethyl ether, and recrystallized in ethyl acetate. After another chromatography and another crystallization it was finally converted into the p-nitrobenzoate. Its identification was obtained by comparison with the synthetic product.

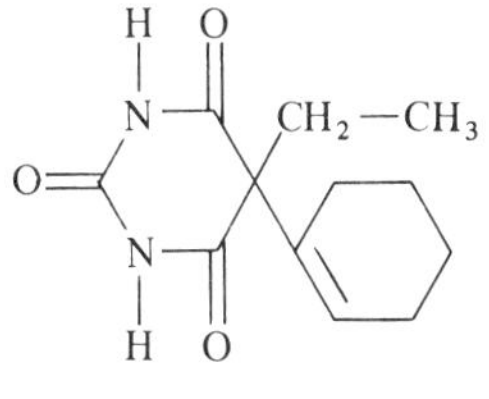

cyclobarbital

IV. THIOBARBITURATES

As early as 1942, it had been reported that the thiobarbiturates appeared to lose their sulfur in vivo and to revert back to the corresponding barbiturate. However, in 1950, Brodie et al. demonstrated a side-chain oxidation without sulfur loss and several authors confirmed and extended their observations. Block and Ebigt showed in 1960 that thiobarbiturates easily lose their sulfur in vitro during the course of the analytical procedures and that metabolic desulfurization appears to be very limited, or even nonexistent.

Thus, the numerous old papers will not be examined in detail.

Williams and Shideman [1009] injected 2-^{14}C-labeled thiobarbital to rats and extracted their urines at pH 2 by ethyl ether for 48 hours. The extract was analyzed by counter-current distribution (phosphate buffer, pH 7.5 : chloroform). Two radioactive fractions were detected, one corresponding to thiobarbital, the other to barbital.

Raventos [747], in fractionating urinary extracts by alumina column chromatography, separated and determined thiobarbital (20%) and barbital (10%).

Thiopental has been most extensively studied. Brodie et al. [135] reported it as completely metabolized and that large amounts of a metabolite can be extracted from urine by very polar solvents. Its UV spectrum was almost similar to that of thiopental in acid or alkaline medium. Its molecular weight was estimated by comparing its absorbance to that of the drug at the wavelength of maximum absorption (305 mμ). Potentiometric titration indicated two acid groups (pK$_a$ 5.2 and 8.2) and confirmed its molecular weight. Its elemental analysis completed the picture and the structure of a carboxylic acid was assigned that corresponded to the oxidation of one of the

---

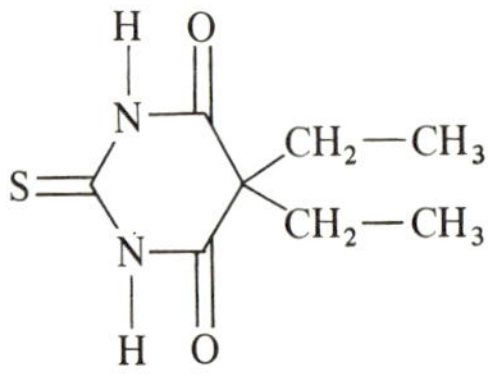

thiobarbital

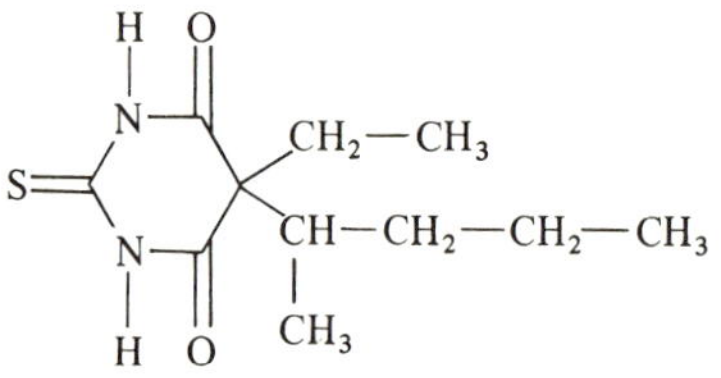

thiopental

CH$_3$ groups on the side chains. The identification was confirmed by synthesis of 5-ethyl-5-(1-methyl-3-carboxypropyl)-thiobarbituric acid [1018] and comparison of UV and IR spectra, partition between solvents, and behavior on paper chromatography.

Schmidt and Arnold [806] found this acid in human urine by paper chromatography (1% ammonium hydroxide saturated butanol).

This metabolite represented 10 to 25% of the administered dose in man. The fate of the remaining part was studied in the rat and monkey with ^{35}S-labeled thiopental by Tailor et al. [897]. They first noticed that an important part of ^{35}S may be precipitated from acid urine by barium chloride. They showed that Ba^{35}SO$_4$ actually precipitated and that the ^{35}S was not trapped by the BaSO$_4$ precipitate by very thoroughly washing the precipitate, redissolving it in an EDTA solution, and precipitating it again. In addition to this inorganic sulfate, they found by acid hydrolysis an hydrolyzable sulfate and by chloroform extraction a small amount of ^{35}S. The paper chromatography of urine was performed with three solvent systems:

- Phenol : water : ammonium hydroxide, 70 : 30 : 4
- Ethanol : water, 77 : 23
- Lutidine : collidine, 1 : 1 (water saturated, 1% of diethylamine added)

Migrations in one and two dimensions revealed the presence of at least 12 unidentified metabolites. Very small amounts of thiourea were determined by isotopic dilution. About 90% of the administered dose was recovered but data on the nature of the intermediary metabolites were not fully provided.

Thiopental metabolism investigations were pursued by Winters et al. [1011]. They used both 2-^{35}S- and 2-^{14}C-labeled thiopental. Experiments were performed in vitro on rat liver slices. The acidified incubation medium was extracted with petroleum ether and ethylene dichloride, both containing 1.5% of isoamyl alcohol. The concentrated aqueous phase was chromatographed on a Dowex-1 (formate) column: the eluent was an aqueous ammonium formate solution of increasing concentration. Several radioactive fractions were thus detected. The main one disappeared on barium chloride precipitation. One of these fractions was identified as thiourea by paper chromatography. The petroleum ether extract — which represented about 50% of the whole radioactivity — was analyzed by counter-current distribution. Several radioactive fractions were observed, one of which corresponded to thiopental. The major metabolite isolated by counter-current distribution was identified as pentobarbital by UV and IR spectrophotometry, elemental analysis, and partition coefficient determination.

Dietz and Soehring [258] studied thiopental as well as various other thiobarbiturates. Their extraction procedure was also applied by Frey [338] and will be described below. Paper chromatography (butanol:amyl alcohol 1:1 saturated with 25% ammonium hydroxide) revealed, besides the unchanged drug, six metabolites.

Spector and Shideman, who had participated in the preceding investigations [1011], restudied them some years later [870]. After incubation with rat liver tissue, pentobarbital was isolated and identified by the same analytical procedure.

Frey [338] was also interested in the metabolism of several thiobarbiturates: thiopental, ethylbutylthiobarbital, methitural, and buthalital. The ethyl ether extract of urine was shaken with a 5% lead acetate aqueous solution and then evaporated. The dry residue was redissolved in ethyl ether, treated with carbon black, evaporated again, and redissolved in ethyl ether for paper chromatography (butanol:amyl alcohol:25% ammonium hydroxide 2:2:1). This extraction procedure was severely criticized by Block and Ebigt [98] as it induced an important sulfur loss.

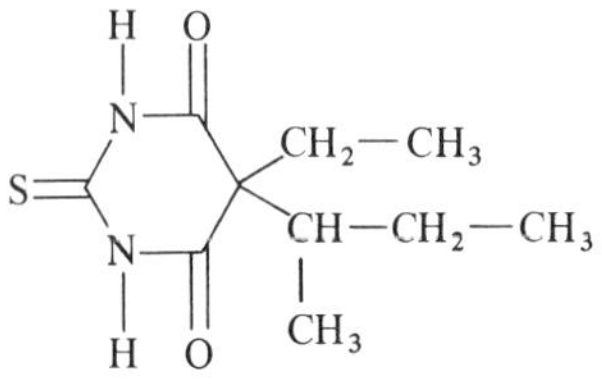

---

ethylbutylthiobarbital

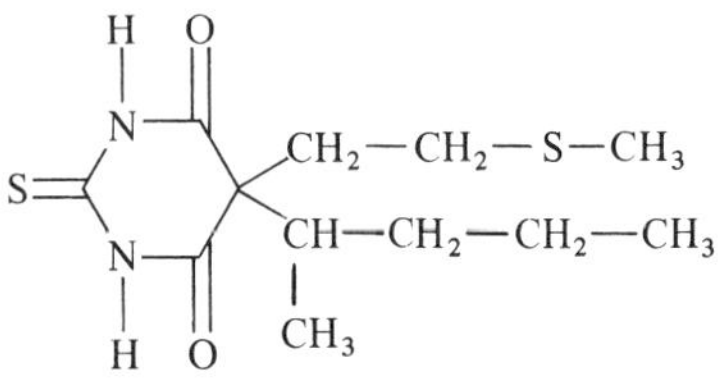

methitural

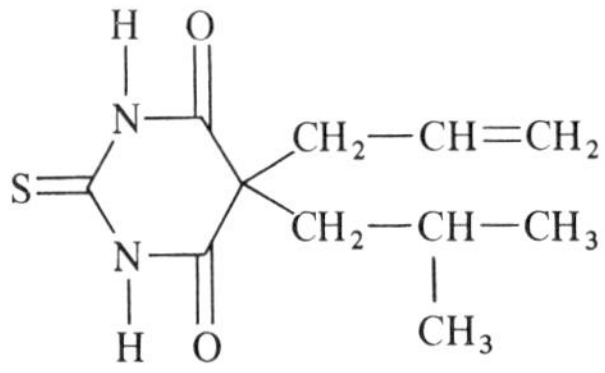

buthalital

In fact, in 1960, Block and Ebigt [98] demonstrated that when a thiobarbiturate, such as ethylbutylthiobarbital or thiopental, was extracted from its acidified aqueous solution with ethyl ether and the extract dried on a water bath at 50°C, paper chromatography revealed a quantitative conversion into the corresponding desulfurized product. Various ethyl ether grades were assayed and only the use of fresh anesthetic ether avoided this degradation. These assays were confirmed by comparing the ethyl ether extractions of thiobarbiturates labeled with ^{14}C or ^{35}S.

These authors consequently studied the metabolism of various thiobarbiturates by direct chromatography of urine without any extraction. Labeled products were synthesized with high specific activity. Metabolite identifications were made possible by preparing a series of side-chain oxidation products (alcohols and carboxylic acids). Two solvent systems were used:

- Butanol : isoamyl alcohol, 1 : 1 saturated with 25% ammonium hydroxide
- Butanol : acetic acid : water, 4 : 1 : 5

Even in these two systems some metabolites still exhibited a tendency to lose sulfur.

However, Block and Ebigt detected and determined ten ethylbutylthiobarbital metabolites and followed their variation as a function of time. They identified most of them: the major metabolite was the secondary alcohol resulting from 1-methylpropyl chain oxidation. Study of the glucuronides was not possible since β-glucuronidase completely desulfurized the metabolites.

Similarly, thiopental showed nine metabolites, but the authors still insisted on the possibility of artifacts due to the instability of the metabolites. In all cases, they were able to definitely contradict the quasiquantitative transformation of the thiobarbiturates into the corresponding desulfurized derivatives in vivo.

One year later, Bush et al. [167] tackled the same thiobarbital problem. The determinations were performed by UV spectrophotometry according to a procedure previously published by Bush [161]. Since all boric acid : potassium hydroxide buffers of various pH actually have the same absorbance, it is possible to use the final alkaline solution resulting from urine or tissues extraction as a reference solution. For this purpose a part of the solution was adjusted to pH 10 and another to pH 6.25 and the differential spectrum was recorded. The determinations were performed at a wavelength of maximum difference. It is thus possible to determine barbital in the presence of thiobarbital.

This spectrophotometric procedure made possible the study of thiobarbital stability in alkaline aqueous solution and during the course of its extraction. Thiobarbital solution in ethyl ether, without peroxides, stored at 21°C showed a slight desulfurization, but ether extraction by a pH 10 solution resulted in higher instability. Butyl chloride, in which thiobarbital is stable, was substituted for ethyl ether. A series of fractional extractions according to the Bush procedure [166, 162, 163] was performed in which thiobarbital was first extracted into butyl chloride, then barbital into ether. The spectrophotometric determination showed that thiobarbital was excreted mostly unchanged, with some barbital.

Spector and Shideman studied the in vitro metabolism of thiamylal [869, 870]. The acidified incubation medium was extracted with ethyl ether in a continuous extraction apparatus for 24 to 48 hours. After evaporation the residue was redissolved in chloroform and, on partition with a 0.2M borate buffer pH 8.8, thiamylal was separated from its supposed metabolite: secobarbital. The chloroform residue was then analyzed by counter-current distribution (chloroform : borate buffer, pH 10.5 — 14 transfers). Two components were thus detected: one corresponded to thiamylal. The other was identified as secobarbital by UV and IR spectrophotometry and elemental analysis.

Tsukamoto et al. [920] detected eight metabolites and the unchanged drug in rabbit urine. After extraction with ethyl acetate, the metabolites were chromatographed on paper. Subsequently [932], the ethyl acetate extract

---

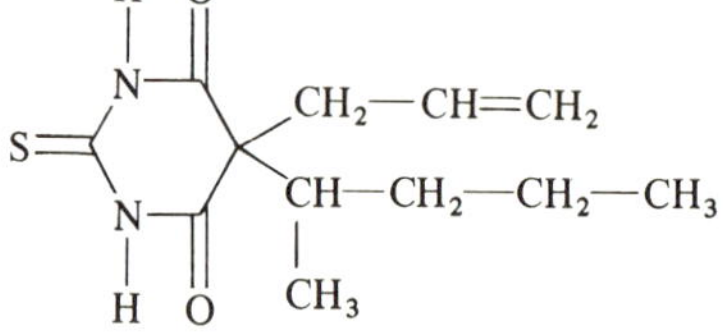

thiamylal

was fractionated into two parts, one insoluble and the other soluble in a sodium bicarbonate saturated solution.

The soluble fraction, in benzene solution, was chromatographed on an alumina column (benzene; 0.5% and 1% of acetone in benzene). Two metabolites that appeared pure on paper chromatography (butanol : ethanol : ammonium hydroxide 4 : 2 : 1.2) were eluted first. The next fraction was chromatographed again on a silica gel column (chloroform; chloroform : methanol 49 : 1; chloroform : methanol 19 : 1). Three other metabolites were thus separated. The insoluble fraction was extracted with ethyl acetate and chromatographed on a silica gel column (chloroform : methanol 19 : 1, then 4 : 1, then 1: 1): two other metabolites were obtained.

These eight metabolites were identified by their IR spectra, R_f on paper (Table 30), chemical reactions and, for the most abundant, by elemental analysis.

A rough study of methitural metabolism was performed by Blake and Perlman [95]. The unchanged drug was determined by UV spectrophotometry. Counter-current distribution (buffer, pH 8 : heptane or buffer, pH 10.7 : chloroform — 8 transfers) indicated that methitural was stored unchanged in the lipid depots. On the other hand, the urine contained an unidentified metabolite.

Preuss and Kopsch [727] extracted a large volume of urine from dogs dosed with methitural by chloroform and they separated six metabolites by paper chromatography according to the procedure of Hübner and Pfeil [433]

TABLE 30

Product	R_f
1. Acetylurea derivative (?)	0.84
2. Thiamylal	0.80
3. Secobarbital	0.72
4. 5-Allyl-5-(1-methyl-3-hydroxybutyl)-2-thiobarbituric acid	0.69
5. 5-Allyl-5-(1-methyl-3-hydroxybutyl)-barbituric acid	0.62
6. 3-Carboxypropylallylacetylthioureide	0.28
7. 5-Allyl-5-(1-methyl-3-carboxypropyl)-2-thiobarbituric acid	0.22
8. 5-Allyl-5-(1-methyl-3-carboxypropyl)-barbituric acid	0.14

(amyl alcohol : butanol 1 : 1 saturated with ammonium hydroxide). One of these showed two components by thin-layer chromatography on silica gel (methylene chloride : acetone 1 : 1). It was impossible to fractionate these two metabolites by chromatography on silica gel column and the authors applied preparative paper chromatography. The isolated metabolites were purified by recrystallization and identified by comparison of their IR spectra, melting point, and chromatographic behavior on paper and thin layer with those of synthetically prepared reference compounds.

Chapter IX

DERIVATIVES OF UREA, GUANIDINE, AND OTHERS

I. OPEN CHAIN UREIDES

Hydroxycarbamide, an anticancer agent, is extensively metabolized in man. Adamson et al. [5] administered ^{14}C-labeled hydroxycarbamide and separated it from its metabolite, urea, by high voltage paper electrophoresis (0.05M borate buffer, pH 12 — 50 V/cm).

Carbromal is metabolically dehalogenated. 2-Ethylbutyrylurea was isolated from the liver of a suicide, crystallized, and identified by its melting point and IR spectrum (Tamminen and Alha [892]).

Curry [238] also isolated this metabolite from liver, purified it by sublimation, and identified it by its IR spectrum.

Butler [181] prepared ^{14}C-labeled carbromal in addition to 2-ethylbutyrylurea and 2-bromo-2-ethyl-3-hydroxybutyrylurea. This last compound was isolated from urine by extraction with chloroform and then with ethyl ether, purified by counter-current distribution (chloroform : water — 30 transfers), and crystallized. It was identified by elemental analysis and IR spectrophotometry. The two metabolites were also separated by paper chromatography (stationary phase: water) (Table 31).

The metabolism of various antidiabetic sulfamylureas was also studied.

$NH_2-CO-NHOH$
hydroxycarbamide

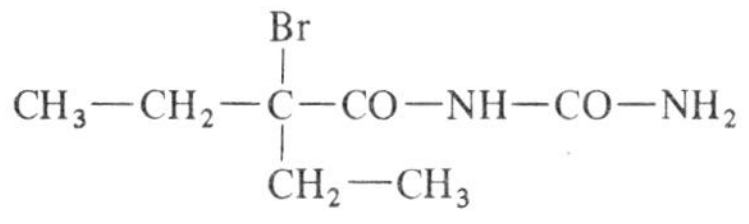

$$CH_3-CH_2-\underset{\underset{CH_2-CH_3}{|}}{\overset{\overset{Br}{|}}{C}}-CO-NH-CO-NH_2$$
carbromal

TABLE 31

Product	R_f in:		
	Benzene	Ethylene dichloride	Chloroform
Carbromal	0.81	—	—
2-Ethylbutyrylurea	0.58	0.80	0.90
2-Bromo-2-ethyl-3-hydroxybutyrylurea	0.24	0.60	0.72

As early as 1952, Wittenhagen and Mohnike [1015] obtained a crystalline metabolite after the acidification of the urine of patients dosed with tolbutamide and recognized the corresponding 4-carboxybenzene sulfonylurea. This identification was confirmed by elemental analysis and hydrolysis [640].

Wittenhagen et al. [1016] separated tolbutamide from its metabolite on paper impregnated with a M/15 phosphate buffer, pH 7 (butanol:ethanol: phosphate buffer 40:11:19). Louis et al. [563] also isolated this metabolite by acidification and analyzed it thoroughly: elemental analysis, potentiometric titration, UV and IR spectrophotometry, and hydrolysis. They also concluded that it was the carboxylic acid.

Dorfmüller [266] also isolated this metabolite from urine and identified it by its IR spectrum compared to the synthesized compound. He published several procedures to determine the drug and its metabolite in biological media [267, 268].

Miller et al. [633] administered rats with ^{35}S-labeled tolbutamide and extracted the urine by the mixture methanol:ethylene dichloride 1:1. The unchanged drug and its metabolite, 1-butyl-3-p-carboxyphenylsulfonylurea, were separated by paper chromatography (butanol:water:piperidine 81:19: 2 by weight).

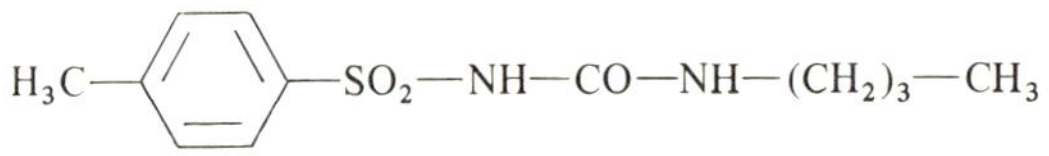

$H_3C-\langle\!\!\!\bigcirc\!\!\!\rangle-SO_2-NH-CO-NH-(CH_2)_3-CH_3$

tolbutamide

A complete study of tolbutamide metabolism in the rat and man was published by Thomas and Ikeda [900] in 1966. They demonstrated the presence of 1-butyl-3-p-hydroxymethylphenyl sulfonylurea in addition to the already mentioned carboxylic metabolite. They used tritiated tolbutamide. Urine was chromatographed on paper (butanol : piperidine : water 81 : 2 : 17) and on silica gel thin layer (chloroform : formic acid 92 : 8; chloroform : methanol : formic acid 95 : 4 : 1). The hydroxylated metabolite was extracted from 1 l of rat urine saturated with ammonium sulfate by the mixture ethyl ether : ethanol 6 : 1. After evaporation, the dry residue was redissolved in water and extracted with ethyl ether. The ether fraction was analyzed by counter-current distribution (chloroform : 0.05M phosphate buffer, pH 5.1 — 200 transfers). The hydroxylated metabolite was characterized by its UV, IR, and NMR spectra, its R_f on paper, and its melting point. In man the hydroxylated metabolite represents 33% of the tolbutamide excretion product. The only other metabolite present is the carboxylated derivative (52%).

The procedure of Nelson et al. [667] permits the determination of the carboxylated metabolite after extraction of the unchanged tolbutamide from the urine at pH 5.5. This procedure does not extract more than 20% of the hydroxylated metabolite, which is lumped together with the carboxytolbutamide.

Chlorpropamide has a different metabolism. Welles et al. [998] extracted urine at pH 2 by methylene chloride and chromatographed the extract on paper (5N ammonium hydroxide saturated butanol). Three components were thus detected and separated; one corresponded to the unchanged drug and another to p-chlorobenzene sulfonamide. The identification was confirmed by IR spectrophotometry and X-ray diffraction on the crystals obtained after chromatography. The third excretion product was identified as p-chlorobenzene sulfonylurea.

$$Cl-\langle\rangle-SO_2-NH-CO-NH-(CH_2)_2-CH_3$$

chlorpropamide

The same authors investigated acetohexamide metabolism [999] by the same methods. They isolated and identified 1-[p-(1-hydroxyethyl)phenylsulfonyl]-3-cyclohexylurea. McMahon et al. [586] elucidated acetohexamide metabolism in the rat and man. After administration of the drug ^{14}C-labeled on the urea carbonyl to the rat, urine was extracted at pH 4 by 2-butanone. Thin-layer chromatography on silica gel (benzene:acetic acid 9:1) revealed three radioactive fractions in this extract. The first corresponded to the unchanged drug, and the second to the hydroxylated metabolite of Welles [999]. The third fraction in its turn was separated into two components on thin layer (chloroform:acetic acid 4:1). These were eluted with 2-butanone. After recording of their UV spectra, they were degraded by heating for 1 hour at 130°C in a 1% fluorodinitrobenzene solution in amyl acetate. The reaction products were the sulfonamide and the dinitrophenyl derivative of the corresponding cyclohexylamine. The dinitrophenyl derivatives obtained were compared by thin-layer chromatographies with the six isomers of synthetically prepared N-hydroxycyclohexyl-2,4-dinitroaniline. McMahon et al. thus demonstrated that one of the two components was a mixture of four isomers of the cyclohexyl ring hydroxylation product of acetohexamide: they were the 3'-trans, 3'-cis, 4'-trans, and 4'-cis derivatives. The second component was also a mixture of the same four isomers of Welles' hydroxylated metabolite.

In man, urinary acetohexamide was isolated and purified by a series of extractions and by preparative thin-layer chromatography. This metabolite exhibited the L(-) configuration. Its hypoglycemic activity was 2.4 times that of acetohexamide, whereas the racemic mixture had equivalent activity.

An antidiabetic sulfonylurea related to acetohexamide is 1-(1-II-hexahydro-1-azepinyl)-3-(p-acetylphenylsulfonyl)-urea. The chromatography of chloroform extracts from urine or plasma (Smith et al. [856]) showed that

---

CH_3-CO-⟨phenyl⟩$-SO_2-NH-CO-NH-$⟨cyclohexyl⟩

acetohexamide

its metabolism was similar to that of acetohexamide and yielded a p-α-hydroxyethyl derivative. The drug and its metabolite were separated on silica gel thin layer (chloroform : formic acid 92 : 8; water saturated ethyl acetate : formic acid 97 : 3). The unchanged drug and its metabolite may be determined in the presence of each other by spectrophotometry at 228 and 247 mμ.

A general analytical procedure for studying sulfamylurea and sulfonylurea metabolism has been proposed by Wiseman et al. [1012]. The unchanged drugs were degraded to the corresponding primary amines and analyzed colorimetrically after reaction with fluorodinitrobenzene. The sulfamide metabolites were hydrolyzed to their corresponding secondary amines and determined by the methyl orange procedure. The other metabolites were directly titrated with methyl orange.

Preuss, in the course of his long series of investigations on ethinamate (see page 102) considered Dolcental metabolism. He detected [725] four excretion products by paper and thin-layer chromatography and reported [722] that these four products were in themselves mixtures. He has succeeded in identifying five of the seven excretion products by comparison with ethinamate metabolites. Two of these were obtained crystallized after chromatography on a column of silica gel (ethyl ether : pentane 3 : 1) or alumina (methylene chloride : benzene 1 : 1) or by paper chromatography (ethyl ether : pentane 3 : 1 water saturated). They were studied by elemental analysis and IR spectrophotometry.

II. HYDANTOINS

These products are related both to barbiturates and phthalimide derivatives. Butler [171] published in 1951 a study on the demethylation of

---

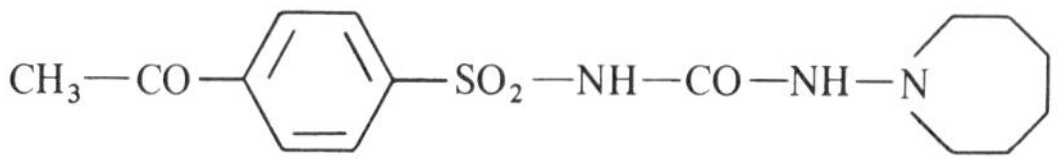

$$CH_3-CO-\langle\!\!\bigcirc\!\!\rangle-SO_2-NH-CO-NH-N\bigcirc$$

1-(1H-hexahydro-1-azepinyl)-3-(p-acetylphenylsulfonyl)-urea

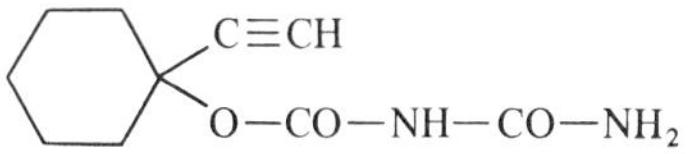

$$\bigcirc\!\!<^{C\equiv CH}_{O-CO-NH-CO-NH_2}$$

Dolcental

IX. DERIVATIVES OF UREA, GUANIDINE, AND OTHERS

mephenytoin into Nirvanol by analogy with that of methylphenobarbital. After determination of the partition coefficients of both products, Nirvanol was extracted from urine at pH 6 by ethyl acetate with the application of fractional extraction procedures. It was separated from the residual mephenytoin by partition between benzene and water, then between water and ethyl ether. Nirvanol thus extracted and crystallized was identified by its melting point and its UV spectrum.

Five years later [178], Butler demonstrated that mephenytoin and Nirvanol were metabolized into 5-ethyl-5-p-hydroxyphenylhydantoin. This derivative was synthesized. It was isolated and determined in urine by counter-current distribution (buffer, pH 9.5 : butanol — 30 transfers).

Butler also demonstrated that 1-methyl-Nirvanol and 3-ethyl-Nirvanol are dealkylated into Nirvanol [185]. The same analytical methods as above were applied. Similarly 1,3-dimethyl-Nirvanol was demethylated into Nirvanol [963]. In addition to the preceding methods, Butler applied paper chromatography (2,2,4-trimethylpentane : butanol : ethanol : 0.1N hydrochloric acid 50 : 1 : 1 : 1; 2,2,4-trimethylpentane saturated with ethanol, ascending).

Phenytoin is also transformed in man into 5-phenyl-5-p-hydroxyphenylhydantoin [179] and Butler used the same methods as in the two preceding papers. He was unable to detect the glucuronide, although the metabolite was conjugated.

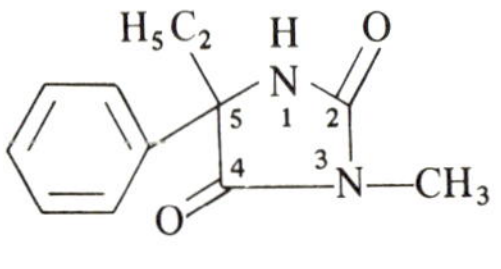

---

mephenytoin

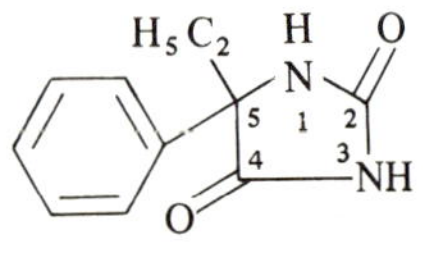

Nirvanol

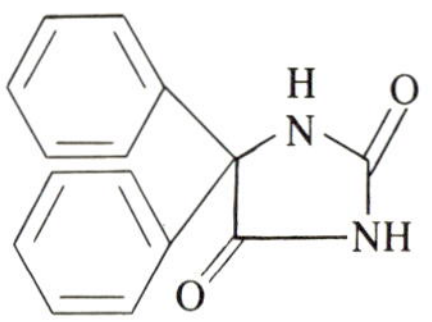

phenytoin

Gorvin and Brownlee [380] obtained the same metabolite and crystallized it after they had extracted it from rabbit urine at pH 4 with ethyl ether.

Kozelka and Hine [521] had isolated α-amino diphenylacetic acid from urine in amounts that represented 10 to 27% of the administered phenytoin. This metabolite was extracted from the concentrated urine with butanol under reduced pressure, recrystallized, and purified by sublimation. It was identified by its melting point. The assigned structure, however, appears to be doubtful on the basis of all other investigations.

Noach et al. [676] prepared 4-^{14}C-labeled phenytoin. They fractionated urinary and biliary metabolites by the procedure of Armstrong et al. [29] for separating urine phenolic acids. The three fractions obtained (acid and neutral compounds and hydantoins) were analyzed by bidimensional paper chromatography [(I) isopropanol:ammonium hydroxide:water 8:1:1; (II) benzene:propionic acid:water 2:2:1]. Only 5-phenyl-5-p-hydroxyphenyl-hydantoin was identified.

Maynert employed ^{15}N-labeled phenytoin [617]. He showed by isotopic dilution that the product was very slightly degraded by hydrolysis and that the major metabolite was a glucuronide of 5-phenyl-5-p-hydroxyphenylhydantoin. This hydroxylated derivative, free or freed by acid or enzymic hydrolysis, was separated by paper chromatography (isopropanol:15% ammonium hydroxide 4:1) and determined colorimetrically. This investigation demonstrated the difficulties to be encountered in the study of the metabolism of compounds not readily water soluble. The small amounts of labeled phenytoin excreted after intraperitoneal injection were due to crystallization within the peritoneal cavity. On oral administration the labeled drug was less readily absorbed than a commercial preparation, probably because the crystals used were larger.

Nakamura et al. [659] administered phenytoin or 3-ethoxycarbonyl-phenytoin to dogs for 1 year and investigated metabolites in the various tissues. These were extracted with ethyl acetate. 5-Phenyl-5-p-hydroxyphenylhydantoin was separated by counter-current distribution (butanol:carbonate buffer, pH 9.5 — 50 transfers). The other possible metabolites were examined by chromatography on paper [(I) benzene:hexane 1:1] and on a thin layer of cellulose Hyflo-super Cel [(II) benzene:acetone 3:1; (III) benzene:acetone 5:1]. Only the above-mentioned two products were found (Table 32).

TABLE 32

Product	R_f in system:		
	I	II	III
Phenytoin	0.05	0.50	—
5-Phenyl-5-p-hydroxyphenylhydantoin	—	0.05	0.10
Diphenylhydantoic acid	—	0.22	—
3-Ethoxycarbonylphenytoin	0.90	0.78	0.69

The p-hydroxylated metabolite was identified by its IR spectrum and its melting point as compared with those of an authentic product. Its glucuronide was hydrolyzed by hydrochloric acid for 1 hour at 120°C.

Yoshimura and Tsukamoto [1034], in an extension of their researches on hexobarbital, studied the metabolism of the 5-methyl- and 5-ethyl-5-(1-cyclohexenyl)-hydantoins. After hydrochloric acid hydrolysis of urine (5 hours at 100°C) the metabolites were extracted into ethyl ether, chromatographed on alumina (chloroform : methanol 9 : 1), and crystallized by the addition of hydrochloric acid. They were identified by their melting point and those of their dinitrophenylhydrazones, and by UV and IR spectrophotometry compared to the synthetic products. It was thus demonstrated that the major metabolic pathway is by the 3-oxidation of the cyclohexene ring.

III. GUANIDINE DERIVATIVES

N,N-Dimethylbiguanide is a hypoglycemic agent whose fate has been studied in mice by Cohen and Costerousse [212] using the ^{14}C-labeled drug. Paper chromatography (butanol : water : formic acid 75 : 15 : 10; pyridine : amyl alcohol : water 84 : 42 : 74) only revealed the unchanged drug.

$R = CH_3, C_2H_5$

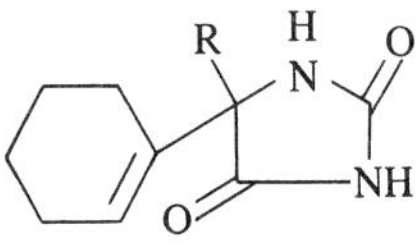

5-methyl- and 5-ethyl-5-(1-cyclohexenyl)-hydantoins

N-butylbiguanide metabolism was studied by Beckmann, who reported [68] the presence, after administration of the drug ^{14}C-labeled on the two biguanidine carbons, of a urinary metabolite. It appeared on paper chromatography (pyridine:amyl alcohol:water 7:7:6). He isolated a metabolite [69] in the urine of various animals by chromatography on a Dowex 50-X8 (H$^+$) column. After elution with water, 1N hydrochloric acid, and 4N hydrochloric acid, a radioactive compound was obtained. It was identified, by its elemental analysis and by comparison of its R_f on paper and UV spectra, as 3-hydroxybutylbiguanide.

N-benzyl-N',N''-dimethylguanidine was investigated by Boura et al. [104], who had the ^{14}C-labeled drug. Paper chromatography of urine in various systems only revealed the unchanged compound. The excretion was very slow and extended beyond one week.

Moroxydine biotransformations were analyzed by Melander et al. [629]. The greatest portion of the drug was recovered unchanged by paper high voltage electrophoresis assays (60 V/cm at pH 4.5).

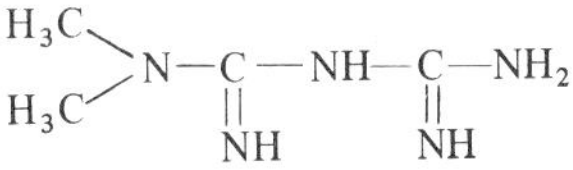

$$\begin{array}{c} H_3C \\ \end{array}\!\!\!\!\!\!N-\underset{\underset{NH}{\|}}{C}-NH-\underset{\underset{NH}{\|}}{C}-NH_2$$

N,N-dimethylbiguanide

$$CH_3-CH_2-CH_2-CH_2-NH-\underset{\underset{NH}{\|}}{C}-NH-\underset{\underset{NH}{\|}}{C}-NH_2$$

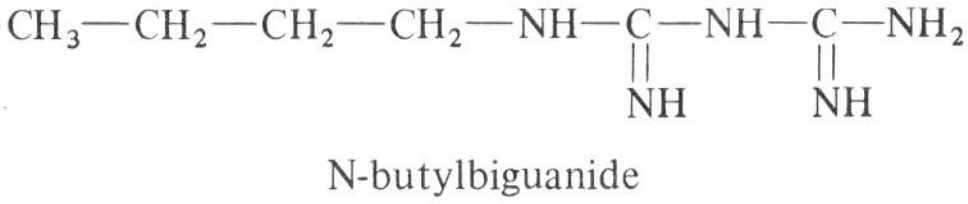

N-butylbiguanide

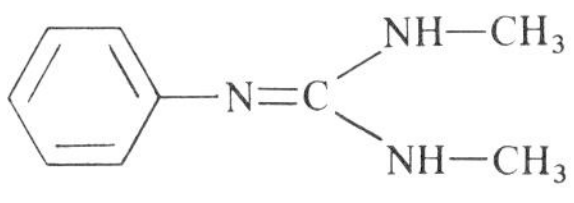

N-benzyl N',N''-dimethylguanidine

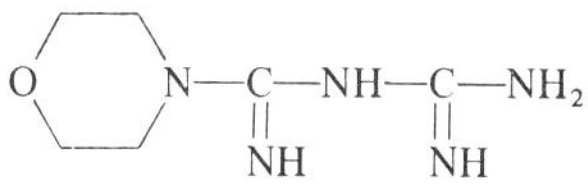

moroxydine

IV. OTHERS

Scott et al. [820] extracted two metabolites of <u>neostigmine</u> with absolute ethanol from dried human urine. These metabolites were purified by chromatography on a column of Amberlite CG-50 buffered at pH 6.86. After methanol and water washing, they were eluted by 0.2N hydrochloric acid. They were then separated by paper chromatography (butanol : ethanol : water : acetic acid 8 : 2 : 3 : 0.25). One of these was identified by its R_f value as m-hydroxyphenyltrimethylammonium bromide. Confirmation was obtained by paper electrophoresis.

The same authors studied the hydrolysis of neostigmine by plasma in vitro [679]. The metabolites and the unchanged drug were separated from plasma proteins on Sephadex G-25 and chromatographed as above. Cholinesterase was inhibited by Dyflos (diisopropyl fluorophosphate) in the blank assay.

Roberts et al. [773] used ^{14}C-labeled neostigmine [772] and separated the unchanged drug from its metabolite by paper electrophoresis (0.1M borate buffer, pH 9.2). A small amount of another metabolite of very low mobility was found by paper electrophoresis [774].

These investigations were extended to ^{14}C-labeled <u>pyridostigmine</u> by Birtley et al. [94]. They mainly used paper electrophoresis, as well as paper chromatography (methylethylketone : acetone : water : ammonium hydroxide 5 : 4 : 1 : 0.01) to separate the drug from its metabolite, 3-hydroxy-N-methylpyridinium.

<u>Benzomethamine</u> metabolism was studied by Levine and Clark [552] who determined the unchanged compound by bromophenol blue colorimetry and its tertiary homolog by methyl orange.

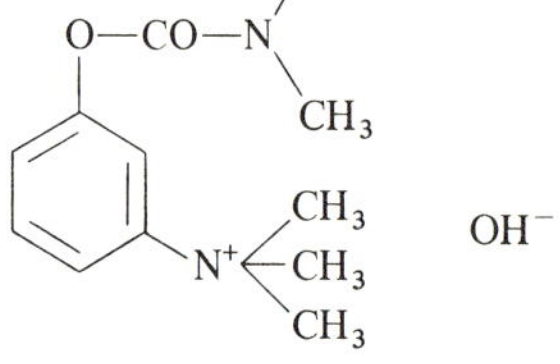

neostigmine

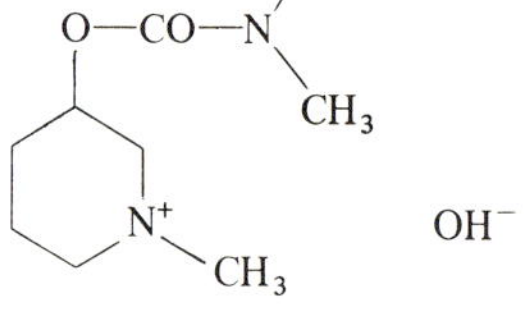

pyridostigmine

OTHERS

Braun et al. [109] investigated the excretion of ^{14}C-labeled <u>etymide</u> and determined primary or secondary amine metabolites by making their dinitrobenzyl derivatives.

Kraml et al. [523] briefly reported they had noticed a metabolic hydroxylation in position 10 of the anticonvulsivant <u>AY-8682</u> and developed a colorimetric determination of its metabolites.

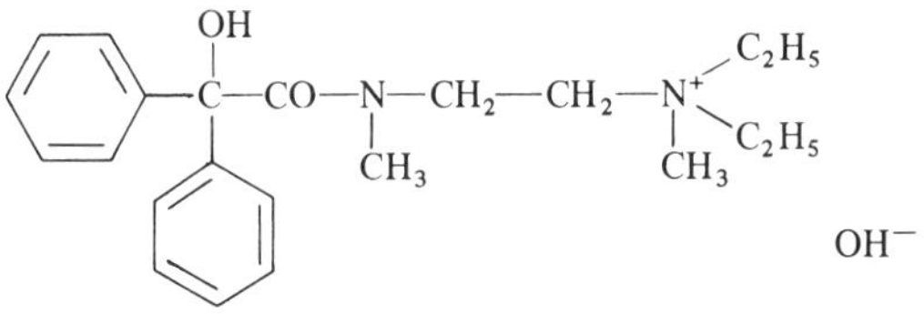

benzomethamine

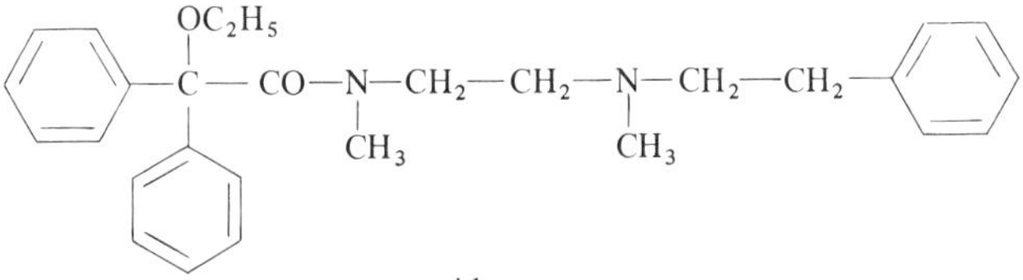

etymide

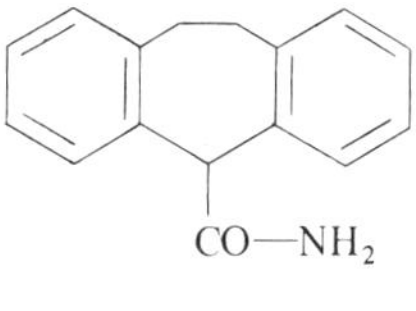

AY - 8682

Chapter X

SULFONAMIDES

Although the discovery of sulfonamides may be considered as a consequence of the studies on the mechanism of action of Prontosil, it cannot be said truthfully that this is a case where a new drug has originated from a metabolism study. Actually Fourneau and Trefouel proceeded along classical microbiological pathways, testing large series of derivatives until the bactericidal action was localized in the sulfanilamide structure. It was only later that Fuller [349] isolated the crystalline molecule from the urine of patients.

The metabolism of the simple sulfonamides was the subject of a long series of papers by Williams and his co-workers from 1940 to 1950. The analytical methodology was relatively unsophisticated. The metabolites were extracted from large volumes of urine and identified by the classical methods of organic analysis, by comparison with specially synthesized authentic samples. It is interesting to compare this long series of investigations to recent research by the same team [119] which applied modern analytical methods in the solution of problems of metabolite identification.

The Bratton and Marshall [108] procedure for sulfonamide determination by diazo coupling of the primary amine function has been used extensively.

Some papers [857, 359] refer to the metabolism of many sulfonamides so that the analytical methods need not be reviewed repetitively for the several compounds.

I. PARAAMINOBENZENESULFONAMIDES

In 1937, Marshall et al. [610] reported that part of the administered sulfanilamide was excreted in a conjugated form that required a hydrolysis to make the amine function titrable by diazo coupling. They later isolated [609] this metabolite by spontaneously crystallizing it from urine. After recrystallization, it was identified as N-acetylsulfanilamide by its elemental analysis, melting point, and acetic acid characterization. Colorimetric determinations showed [607] that the total of sulfanilamide and its acetyl derivative represented a maximum of 90% of the administered dose.

James [456] observed in 1940 that sulfanilamide treatment increased organic sulfate and glucuronic acid excretion. The patients' urine was evaporated, and the dry residue was redissolved in hydrochloric acid and hydrolyzed for 2 hours at 120°C. After ethyl ether extraction at pH values of 5, 7, and 9, p-hydroxylaminobenzenesulfonic acid and N-acetylsulfanilamide were isolated and characterized by elemental analysis. p-Aminophenol was also found in unhydrolyzed urine.

Shelswell and Williams [832] detected the presence of a phenolic metabolite — as an ethereal sulfate — in showing that an increase of hydrolyzable sulfate level followed sulfanilamide administration. They were nevertheless unable to find p-aminophenol or N-hydroxysulfanilamide.

Sammons et al. [793] especially studied p-hydroxybenzenesulfonamide conjugation. This compound was eliminated both as the sulfate and the glucuronide with sulfate detected as previously [832]. The glucuronide was precipitated as the lead salt, regenerated by hydrogen sulfide, and reprecipitated as the barium salt. This salt was purified by repeated precipitations from its aqueous solution on addition of alcohol. Hydrochloric acid liberated p-hydroxybenzenesulfonamide.

Purification of the glucuronide was attempted by methylation with methyl sulfate in the presence of sodium hydroxide, followed by a methylation with

---

H_2N—⟨⟩—SO_2—NH_2

sulfanilamide

methyl iodide in the presence of silver oxide. Thus, the sulfonamide group as well as the glucuronic moiety were methylated with a resultant p-hydroxybenzenedimethylsulfonamide trimethylglucuronic acid methyl ester. When this ester was treated with methanolic ammonium hydroxide, the result was the corresponding glucuronamide, which was crystallized.

The authors noted the presence in small amounts of a catecholic metabolite as revealed by its reactions with ferric chloride and ammoniated silver nitrate. This catecholic metabolite was extracted with ethyl ether from hydrolyzed urine [1004] and converted into the methylated derivative. Comparison of its melting point with that of the synthesized methylated derivative of the authentic product permitted assignment of the formula of 1,2-dihydroxy-4-sulfonamidobenzene.

Four years later Williams determined the major oxidation metabolite of sulfanilamide [1006]. Ten liters of rabbit urine were first concentrated under vacuum, hydrolyzed, and extracted by ethyl ether for 40 hours. The evaporated extract was redissolved in water and the sulfanilamide crystallized at 0°C. The residual solution was lyophilized. The residue was chromatographed on an alumina column, first washed with ethanol until complete sulfanilamide elimination. The column was then sectioned and the metabolite extracted into water. After recrystallization it was identified as 3-hydroxysulfanilamide by its elemental analysis, melting point, and with chemical reactions in comparison with an authentic compound.

This metabolite was further detected in human urine by a specific color reaction [1007]. The urine fractionation process was similar to that discussed above and the purified metabolite was redissolved in water. In the presence of ferric chloride, ethyl ether extracted a colored complex which has a characteristic absorption spectrum.

In 1955, Boyer et al. [105] paper chromatographed the urines of various species (butanol:ammonium hydroxide:water 12:3:15; butanol:acetic acid:water 12:1:12; butanol:ethanol:water 4:1:1). By comparison with synthetic products these authors detected the N^1- and N^4-acetyl and the N^1- and N^4-diacetyl derivatives.* The N^1-acetyl metabolite was separated by adsorption on carbon and acetone elution. It was purified by chromatography

---

*The nitrogen of the sulfonamide function is N^1 and that of the amine function N^4.

on an alumina column (water saturated butanol), eluted with water, and crystallized. Its formula was confirmed by elemental analysis and acid hydrolysis [632].

Thauer et al. [898] investigated sulfanilamide transformations in suspensions of liver microsomes from man and the rat. The metabolites were extracted by ethyl acetate after the proteins were precipitated first by heating at 90°C and then by acetone. These extracts were chromatographed on silica gel thin layer (ethyl ether). Only N-hydroxysulfanilamide was found in addition to the unchanged drug. This metabolite was characterized by comparison with the authentic compound and by oxidation into the corresponding nitroso derivative by means of ferric chloride. It was able to oxidize hemoglobin into methemoglobin.

Robinson and Crossley [778] studied the metabolism of N^1-acyl derivatives of sulfanilamide. Large urine volumes were continuously extracted for several dozen hours with ethyl ether. The pH values for selective extraction of the N^1-acyl derivative and the respective sulfanilamide were determined by preliminary experiments. The separated products were crystallized and characterized by their melting points.

Although metanilamide is therapeutically inactive, it was studied as an extension of sulfanilamide.

Dobson and Williams [263] used the previously described methods to determine the free and total sulfonamide and the ethereal sulfate. The first metabolite, acetylmetanilamide, was extracted from urine by ethyl ether and recrystallized. The second one, 4-hydroxymetanilamide, was liberated by acid hydrolysis and extracted into ethyl ether on prolonged percolation (50 hours). It was purified by chromatography on alumina column: after metanilamide elution by the mixture benzene : ethanol 1 : 1, followed by ethanol, 4-hydroxymetanilamide was extracted from the convenient column section by

---

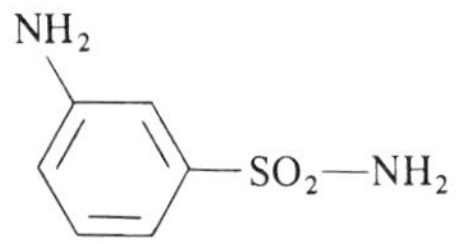

metanilamide

boiling water. It was identified by its elemental analysis, melting point, and color reactions. 4-Hydroxymetanilamide did not give the specific color reaction of 3-hydroxysulfanilamide [1007].

Stubbs and Williams [881] isolated two aminohydroxybenzenesulfonamides similarly from the urine of animals dosed with orthanilamide. They prepared the corresponding products by synthesis [700].

Mafenide is a sulfanilamide homolog, but its biological properties differ. Hartles and Williams [406] showed that it is rapidly converted into inactive metabolites. In fact, on acidification of the urine, a product identical to p-carboxybenzenesulfonamide precipitated. The search for conjugated metabolites, such as hippurate, sulfate, and glucuronide, by means of the usual color reactions gave negative results.

The same authors studied the metabolism of a series of N^4-acyl derivatives of mafenide [407].

McChesney [573] studied the excretion of 3', 5'-dibromosulfanilanilide by analyzing for free and conjugated (after hydrolysis) compound by the Bratton and Marshall procedure. The conjugated metabolite was extracted with ethyl ether from urine at pH 2. After washing of the ether solution, the metabolite was precipitated by benzylamine, redissolved into methanol, and reprecipitated by ether. Elemental analysis and hydrolysis indicated a hydrated salt

---

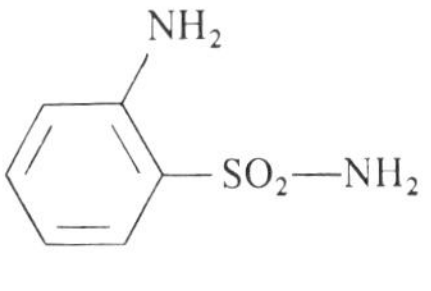

orthanilamide

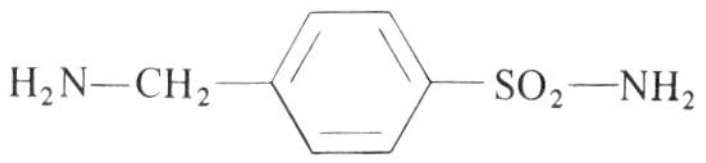

mafenide

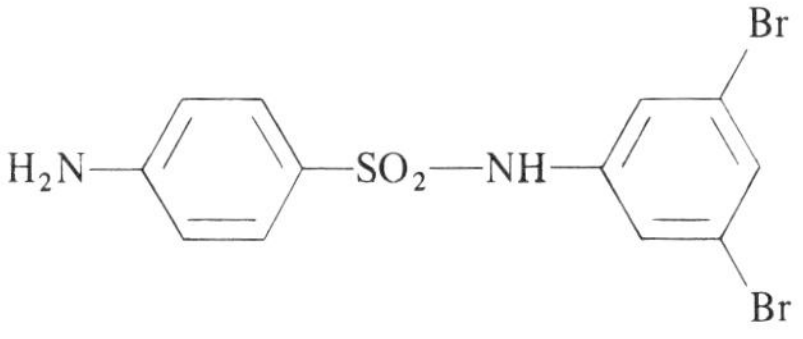

3', 5'-dibromosulfanilanilide

of 3',5'-dibromosulfanilanilide glucuronide. This metabolite contained four hydroxyl groups in addition to the carboxyl. It appeared to be a N-glucuronide.

In 1940, Scudi [821] briefly reported the isolation from dog urine of a monohydroxylated derivative of <u>sulfapyridine,</u> identified by its elemental analysis and color reactions. The glucuronide of this derivative was precipitated as the silver salt.

Weber et al. [988] investigated this hydroxylated metabolite further. Dog urine was extracted for nine hours with ethyl ether to remove the unchanged drug. The aqueous phase was then precipitated by lead acetate in the presence of acetic acid, and the precipitate was discarded. The solution was further alkalinized and a second precipitate was collected, which was treated with hydrogen sulfide. Silver nitrate was added to the aqueous solution of the regenerated glucuronide and the silver salt of the glucuronide precipitated. It was then hydrolyzed and extracted by ethyl ether to yield a crystalline derivative. The presence of an hydroxyl group was revealed by reaction with ferric chloride. Acid hydrolysis permitted the isolation of sulfanilic acid and a pyridine residue which gave a positive reaction with ferric chloride. It must be concluded that the hydroxyl was thus fixed on the pyridine ring.

Bray et al. [114] obtained the same hydroxylated metabolite in the rabbit by applying Weber's method [988]. The lead salt of the metabolite was simply treated with hydrogen sulfide and the aqueous solution concentrated under vacuum. After alkaline and subsequent acid hydrolysis, a product was extracted with ethyl ether which was recrystallized as the picrate. Elemental analysis confirmed that it was an hydroxyaminopyridine derivative.

Four years later the same authors [110] prepared a ^{35}S-labeled sulfapyridine for the purpose of studying its transformation into a derivative which cannot be analyzed by diazo coupling. Urine was fractionated by

---

H_2N—⟨ ⟩—SO_2—NH—⟨N⟩

sulfapyridine

extractions and hydrolyses, and the radioactivity, as well as the diazotizable products was determined in every fraction. The isolated metabolites revealed nothing but the unchanged drug and an hydroxylated derivative.

In 1950 these authors [115] succeeded in identifying the hydroxylated metabolite. They first synthesized two of the four possible aminohydroxypyridines and demonstrated that hydroxysulfapyridine did not give any of these two derivatives on hydrolysis. On the basis of the negative test with dichloroquinonechloroimide, they concluded that the metabolite was 2-(p-aminobenzenesulfonamido)-5-hydroxypyridine (5'-hydroxysulfapyridine).

In 1956, Scudi and Childress [822] succeeded in synthesizing 2-amino-5-hydroxypyridine, and from this compound 5'-hydroxysulfapyridine. The metabolite extracted from urine by lead salt precipitation and hydrolyzed by sodium hydroxide led to an hydroxysulfapyridine. It was recrystallized and identified with the synthetic product by its elemental analysis, UV spectrum, and color reactions.

Uno et al. undertook an extensive study of sulfadiazine metabolism and described the ascending paper chromatography [951] of concentrated human urine with the following solvent systems:

(I)   Water saturated butanol
(II) Butanol : acetic acid : water, 5 : 1 : 4
(III) 3% Ammonium hydroxide saturated butanol
(IV) Butanol : propanol : water, 2 : 1 : 1
(V)   Butanol : propanol : 0.1 N ammonium hydroxide, 2 : 1 : 1
(VI) 1 N hydrochloric acid saturated butanol

The five urinary excretion products were identified by R_f comparison with authentic compounds (Table 33).

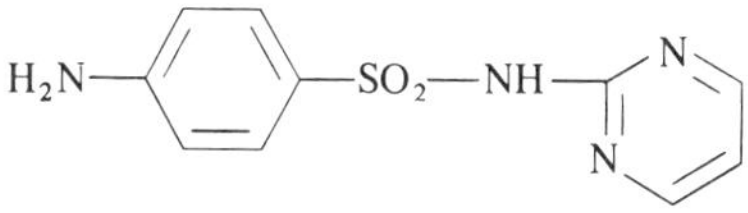

H_2N—⟨ ⟩—SO_2—NH—⟨N⟩

sulfadiazine

TABLE 33

Product	R_f in system:					
	I	II	III	IV	V	VI
Sulfadiazine	0.68	0.76	0.10	0.74	0.11	0.72
N^4-Acetylsulfadiazine	0.73	0.84	0.18	0.82	0.18	0.78
Sulfadiazine-N^4-glucuronide	0.005	dec.	0.00	0.01	0.00	dec.
Sulfadiazine-N^4-sulfonate	0.045	0.21	0.00	0.09	0.01	dec.
Sulfanilamide	0.50	0.60	0.49	0.53	0.55	0.54

The chromatographic identifications were confirmed by hydrolysis assays of the compounds and by paper electrophoresis.

A chromatographic separation that enabled the determination of the metabolites was also developed [952]. Urine was applied as a streak on two paper sheets. One was developed with system II, the other one with system III. After cutting the sheets, the fractions were eluted by 0.25N ammonium hydroxide, hydrolyzed by hydrochloric acid, and determined by Ehrlich reagent.

Bray et al. [113] previously reported in their paper on sulfadimerazine that sulfadiazine was studied by similar methods and its glucuronide was detected. Similarly Smith and Williams [857] found its N-acetyl derivative.

The preliminary studies of Gilligan [359] demonstrated that the administration of sulfadimerazine was followed by an increase in the amount of excreted glucuronic acid. This was determined after hydrolysis by the naphtoresorcinol method [614]. Smith and Williams [857] also determined glucuronic acid by a variant of the naphtoresorcinol method [400]. The ethereal sulfate groups and the free and conjugated diazotizable functions were

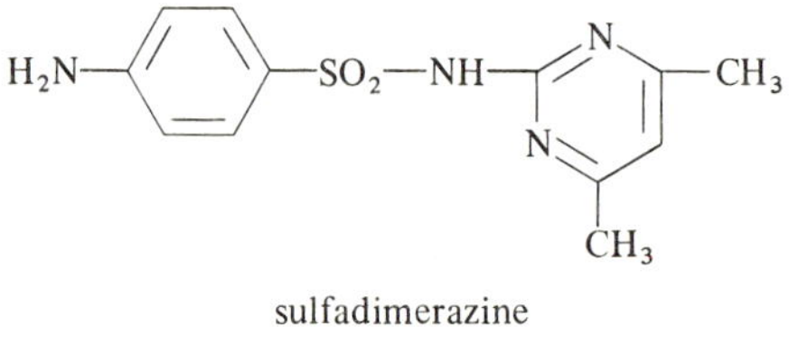

sulfadimerazine

determined similarly to give the proportion of N^4-acetyl metabolites by difference. However, they were not able to isolate the metabolites.

Bray et al. [113] studied this problem further with paper chromatography. The glucuronide was precipitated as the lead salt according to the usual procedure, but it was not possible to crystallize it. On alkaline hydrolysis it gave an hydroxysulfadimerazine, which in its turn was hydrolyzed in acid medium. Paper chromatography identified the following compounds in the reaction medium (nine solvent systems, see Table 34): sulfanilic acid, 3-hydroxysulfanilic acid, 2-hydroxy-4,6-dimethylpyrimidine, and 2,5-dihydroxy-4,6-dimethylpyrimidine. When hydrolysis was performed in 2N rather than 5N hydrochloric acid, 2-amino-5-hydroxy-4,6-dimethylpyrimidine was also detected.

Krebs et al. [524] compared in vitro and in vivo acetylation by means of the Bratton and Marshall method. They observed that N-acetylsulfonamides slowly hydrolyze at room temperature in the presence of the trichloroacetic acid used for the deproteinization. In this particular case, conjugation by acetylation in vitro was limited by the action of an enzyme which is almost inactive in vivo.

Sulfadimethoxine is a more recently discovered sulfonamide and exhibits a long duration of action and a low toxicity. It is characterized metabolically by the fact that its major excretion product is a soluble glucuronide, whereas many other sulfonamides give relatively insoluble acetylated derivatives which can induce renal lithiasis.

This glucuronide was studied in 1965 by Bridges et al. [119]. Human urine was concentrated under vacuum (40 times) and the inorganic salts precipitated by alcohol addition. The solution was then passed through a strongly basic anion-exchange resin (Deacidite FF-BDH). The column was

----

H_2N—⟨benzene⟩—SO_2—NH—⟨pyrimidine⟩—OCH_3
                                    |
                                  OCH_3

sulfadimethoxine

TABLE 34

Product	R_f in system:[a]								
	I	II	III	IV	V	VI	VII	VIII	IX
Sulfanilic acid	0.63	0.51	0.10	0.44	0.50	0.21			
Sulfanilamide	0.96	0.95	0.77	0.88	0.90	0.71			
2-Hydroxysulfanilic acid	0.73	0.74	0.08	0.18	–	0.27			
2-Hydroxysulfanilamide	0.81	0.93	0.06	0.28	0.21	0.61			
3-Hydroxysulfanilic acid	0.62	0.57	0.04	0.13	0.31	0.26			
3-Hydroxysulfanilamide	0.86	1.00	0.22	0.41	0.40	0.63			
Sulfadimerazine	0.80	0.95	0.55	0.60	0.69	0.90			
2-Hydroxy-4,6-dimethylpyrimidine	0.69		0.45				0.60	0.72	1.00
2,5-Dihydroxy-4,6-Dimethyl-pyrimidine (hydrochloride)	0.36		0.10				0.10	0.23	0.15
2-amino-5-hydroxy-4,6-dimethyl-pyrimidine (hydrochloride)	0.74		0.35				0.91	0.83	0.90

[a] Solvent systems: I, butanol:pyridine:sodium chloride saturated solution:ammonium hydroxide 4:8:5:3; II, water saturated collidine; III, butanol:ammonium hydroxide 20:3; IV, isopropanol:ammonium hydroxide: sodium chloride saturated solution 4:2:3; V, 96% ethanol:ammonium hydroxide 20:1; VI, butanol:acetic acid: water 5:1:4; VII, ethyl acetate:acetic acid:water 10:3:7; VIII, 96% ethanol:acetone 1:3; IX, butanol:pyridine:sodium chloride saturated solution 1:1:2.

washed with water, and the glucuronide was eluted by a sulfur dioxide solution. After degassing under vacuum, the solution was transferred to an Amberlite IR-120 (H⁺) column, which was eluted by 2N ammonium hydroxide. A glucuronide mixed with amino acids was thus obtained. It was chromatographed on cellulose MN thin layer (butanol:acetic acid:water 4:1:2). After recrystallization from methanol a 99% pure glucuronide was obtained. It was also purified by crystallization of its S-benzylthiouronium salt. When treated by β-glucuronidase (pH 4.2 – 24 hours at 38°C) the glucuronide gave glucuronic acid and sulfadimethoxine, which were characterized by paper chromatography, as well as its acid and alkaline hydrolysis products (Table 35).

The urinary glucuronide was identified with the synthetically prepared N^1-glucuronide. The N^4-glucuronide was also prepared but in an impure state. The synthetic method was derived from Okumura [690]. The Tsukamoto et al. procedure [934], as it has been applied to meprobamate, gave poor results when used in this case. The structure of the synthetic products was elucidated by UV and IR spectrophotometry and the same methods confirmed the natural glucuronide formula. This very complete investigation was described in detail.

TABLE 35

Product	R_f in system:[a]		
	I	II	III
Sulfadimethoxine	0.40	0.95	0.95
N^4-Acetylsulfadimethoxine	0.65	0.95	0.95
Sulfadimethoxine N^4-glucuronide (Na)	0.00	0.25	—
Sulfadimethoxine N^1-glucuronide (NH₄)	0.15	0.55	0.80
Glucuronid acid	0.00	0.37	0.23
Glucuronolactone	0.03	0.65	0.39
Sulfanilic acid	0.11	0.57	0.32

[a]Solvent systems: I, butanol:ammonium hydroxide:water 10:1:1; II, 80% aqueous ethanol; III, butanol:acetic acid:water 4:1:2.

Similar research on the same compounds was published by Uno et al. [947] with the same results.

After administration of <u>sulfisomidine</u> to man, DiCarlo et al. [257] paper chromatographed the urine (butanol:acetic acid:water 5:1:4, upper phase to which 5% of butanol is added; butanol:ammonium hydroxide:water 4:1:3, upper phase to which 5% of butanol is added, ascending). The fluorescent fractions were eluted from 4-12 sheets and the free and total sulfonamides and glucuronic acid were determined. Six metabolites were thus found, all of which contained glucuronic acid. They were very labile glucuronides that easily liberated the aglycone. Two of these were identified by chromatography and UV spectrophotometry as sulfisomidine and its N^4-acetyl derivative.

DiCarlo et al. [256] also studied <u>sulfametomidine</u>, which is closely related to sulfisomidine. The same procedure was applied to paper chromatograph the urine that had been concentrated by dialysis. Two metabolites were detected in addition to the unchanged drug. One was identified by comparison with the N^4-acetyl derivative and the other was a glucuronide which behaved differently from the synthetic sulfametomidine N^4-glucuronide on paper electrophoresis and chromatography. It was partially purified, after Bratton and Marshall reaction, by chromatography on Florisil column, and then submitted to an enzymic hydrolysis. It was not possible to identify it.

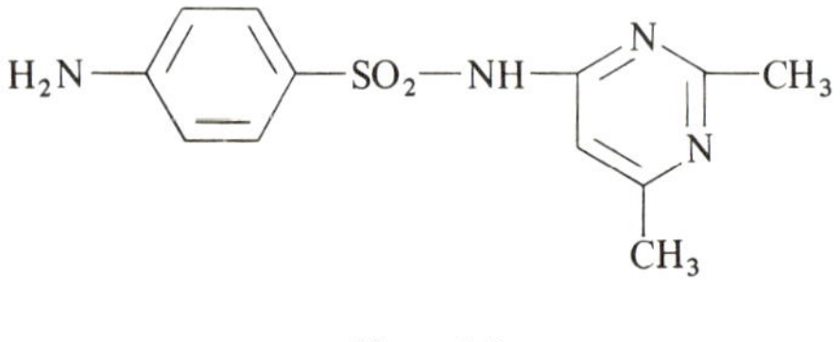

sulfisomidine

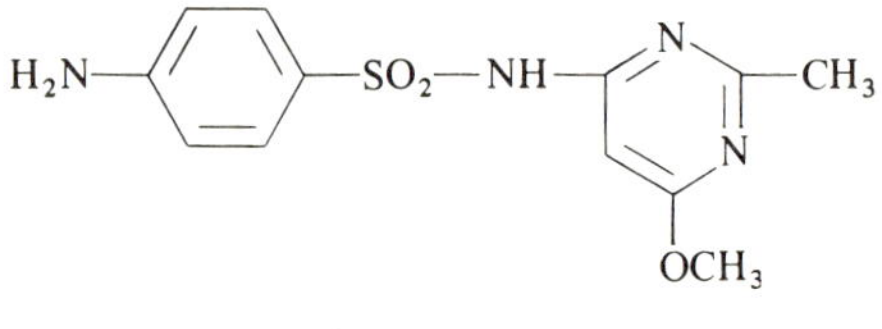

sulfametomidine

Linkenheimer et al. [556] investigated sulfaethoxypyridazine metabolism in the heifer. They separated the sulfonamide from its N^4-acetyl derivative by distribution between ethyl acetate and water. Paper chromatography of the metabolites (butanol : ammonium hydroxide : water 4 : 1 : 5) was performed after acetone precipitation of urine or plasma proteins, eventually followed by a column chromatography purification. Five fractions were detected: one corresponded to the unchanged drug, another to the N^4-acetyl derivative; a third contained a glucuronide as was shown by spraying the chromatogram with naphtoresorcinol. A cellulose column chromatograph was used with the same solvent as was used with paper chromatography.

Riess et al. [765] worked with sulfaphenazol ^{14}C-labeled on the phenyl. The lyophilized plasma was extracted with methanol and the extract separated by paper electrophoresis (carbonate buffer pH 10.4 — 2,000 V — 105 minutes). None of the sulfaphenazol hydrolysis products, located by means of reference compounds, was detected. Urine, similarly extracted, was analyzed by paper chromatography (butanol : acetic acid : water 750 : 75 : 210), by electrophoresis as above, and by isotopic dilution. It contained only sulfaphenazol glucuronide, which was isolated by precipitation of its lead salt with hydrogen sulfide regeneration. It was methylated and acetylated, then chromatographed on an alumina column. A radioactive fraction was eluted by a 1% methanol solution in chloroform to give a crystallized derivative

----

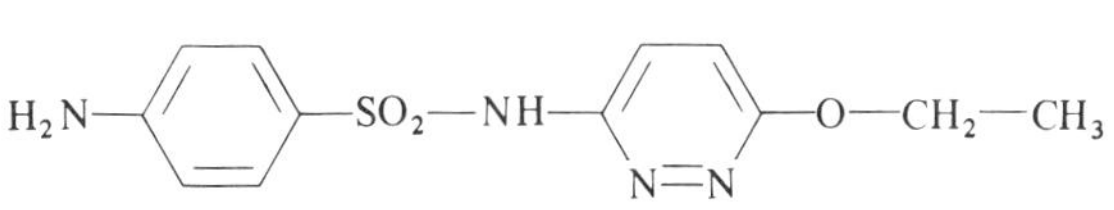

sulfaethoxypyridazine

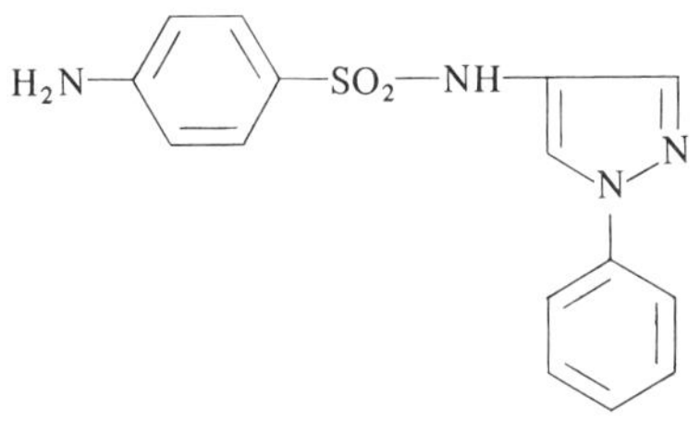

sulfaphenazol

whose elemental composition and UV and IR spectra were determined. This derivative was hydrolyzed and the hydrolysis products analyzed by thin-layer chromatography to reveal sulfaphenazol. The entire analytical data led to the conclusion that the single urinary metabolite was the glucuronide resulting from glucuronic acid attachment on the 2-nitrogen of the pyrazole ring.

Bruck et al. [147] followed free and conjugated sulfamoxol excretion by colorimetry after diazo coupling with and without previous hydrolysis. They also investigated the urinary metabolites. The ethyl acetate extract was fractionated by paper chromatography (methanol : benzene : amyl alcohol : water 32:45:15:8) and the resulting fractions were examined by paper electrophoresis (borate buffer, pH 8.6, 5 hours, 145 to 310 V). The residual extracted urine was treated by lead acetate to detect glucuronides. In addition to the unchanged drug the authors found its acetyl derivative and sulfanilamide, as well as various other metabolites that they were unable to identify definitely. They emphasized that sulfamoxol was unstable in air.

Sulfamoxol also deteriorates in aqueous solution. Seydel et al. [829, 186] separated several degradation products by silica gel thin-layer chromatography (chloroform : methanol 80:15 and 85:15; methylethylketone : acetic acid 75:15; chloroform : methanol : acetic acid 80:15:5; propanol : 25% ammonium hydroxide 70:30). Two of the products, sulfanilcarbamide and sulfanilamide, were characterized by chromatography and by comparing their IR spectra with known compounds. The same procedure applied to blood and urine revealed as many as eight metabolites, of which two were the degradation products already identified. All of the analytical operations, including chromatography, were performed under nitrogen and are described with more detail in [186].

Nevertheless Serick and Loop [828] stated they were unable to positively detect substances other than unchanged drug that reacted with the Bratton

---

sulfamoxol

and Marshall reagent. They applied the paper chromatographic procedure of Bruck et al. [147] and worked in the absence of light.

Uno and Kono published a series of papers on sulfafurazol metabolites. In the first [944] they reported the isolation of several urinary metabolites by paper electrophoresis and chromatography (Table 36).

The metabolites were identified by their R_f values. The acetylated derivative was isolated.

The same authors [945] developed a procedure to determine unchanged sulfafurazol by isolation on a cellulose column (butanol:propanol:water 2:1:1). This method was further applied to the determination in blood [946].

TABLE 36

Product	R_f in system:[a]				
	I	II	III	IV	V
Sulfafurazol	0.88	0.86	0.40	0.86	0.51
Acetylsulfafurazol	0.91	0.90	0.44	0.92	—
Sulfanilamide	0.56	0.58	0.55	0.54	0.56
Sulfafurazol-N-sulfonate	0.30	0.39	0.00	0.23	0.20
Sulfafurazol-N-glucuronide	0.005	dec.	0.002	0.05	0.05

[a]Solvent systems: I, water saturated butanol; II, butanol:acetic acid:water 5:1:4; III, 3% ammonium hydroxide saturated butanol; IV, butanol:propanol:water 2:1:1; V, butanol:propanol:0.1N ammonium hydroxide 2:1:1.

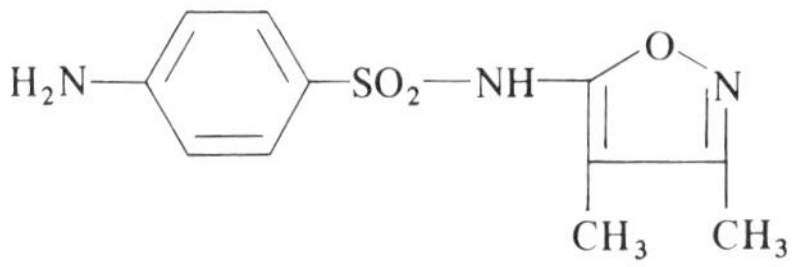

sulfafurazol

After <u>sulfamethoxazol</u> administration to man, Ueda and Kuribayashi [943] were able to isolate the N^1-glucuronide. Their procedure involved the adsorption of the urinary glucuronide on activated carbon at pH 4 and elution by the mixture butanol:methanol:ammonium hydroxide:water 1:1:0.4:8 at 40°C. The product in aqueous solution was then transferred to a Dowex 50W-X8 (H⁺) column and eluted by water. It was purified by paper chromatography (butanol:acetic acid:water 5:1:4), extracted by water, and treated by neutral lead acetate at pH 4 and then by basic lead acetate at pH 8. After lead removal by hydrogen sulfide, the glucuronide was again chromatographed on Dowex 50W-X8, eluted into 1N ammonium hydroxide, and crystallized. It was identified by comparing its melting point and UV and IR spectra with the authentic compound, the preparation of which was described.

After administration of <u>sulfathiazol</u>, Thorpe and Williams [904] collected acetylsulfathiazol crystals in urine and they detected the presence of a glucuronide, the barium salt of which precipitated. Uno and Ueda [949] separated and characterized by their R_f values, compared to authentic samples, three sulfathiazol metabolites (Table 37).

A quantitative balance of sulfathiazol metabolism was established by the same authors [950]. Urine was directly chromatographed on paper (butanol:methanol:0.1N ammonium hydroxide 3:1:1) and the separated products were determined after elution by the Bratton and Marshall procedure.

The main excretion product of <u>sulfasomizol</u> is its N^4-acetylated derivative. Bridges and Williams [120] by paper chromatography (propanol:ammonium

----

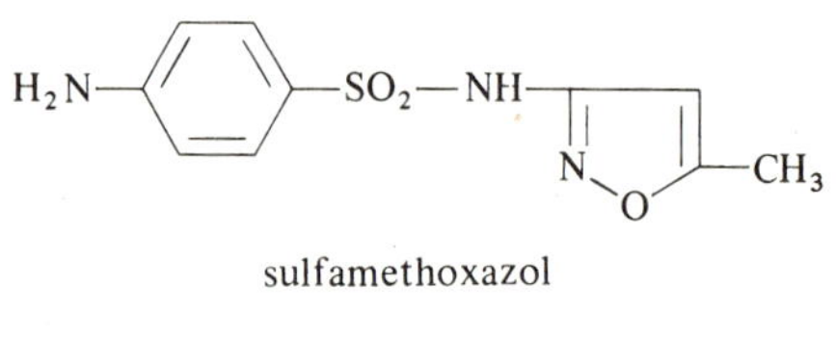

sulfamethoxazol

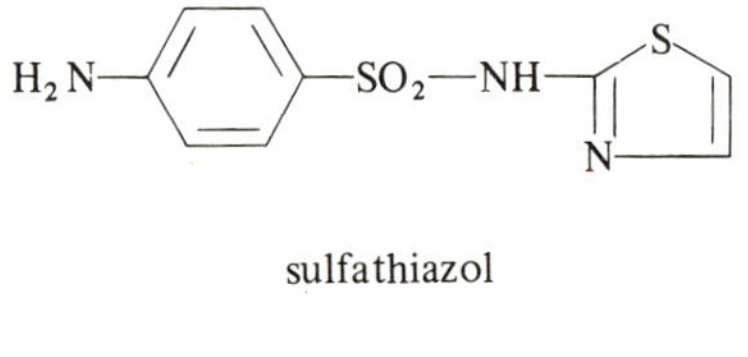

sulfathiazol

TABLE 37

Product	R_f in system:[a]				
	I	II	III	IV	V
Sulfathiazol	0.66	0.68	0.50	0.47	0.35
Acetylsulfathiazol	0.78	0.78	0.57	0.52	0.44
Sulfathiazol N^4-sulfonate	0.13	0.28	0.26	0.21	0.05
Sulfathiazol N^4-glucuronide	0.02	0.10	0.08	0.09	0.00

[a] Solvent systems: I, water saturated butanol; II, butanol : propanol : water 2 : 1 : 1; III, butanol : propanol : 0.1N ammonium hydroxide 2 : 1 : 1; IV, butanol : isopropanol : 0.1N ammonium hydroxide 2 : 1 : 1; V, 2N ammonium hydroxide saturated butanol.

hydroxide 7 : 3) of urine also found three minor metabolites: the N^4-glucuronide, the N^4-sulfate, and an oxidation derivative (Table 38). The separated products were eluted and studied by UV and IR spectrophotometry and by various chemical reactions.

Uno and Okasaki [948] analyzed the excretion products of sulfamethizol by paper electrophoresis, paper chromatography, and a combination of both techniques (Table 39).

These products were also separated by counter-current distribution (butanol : water — 50 transfers).

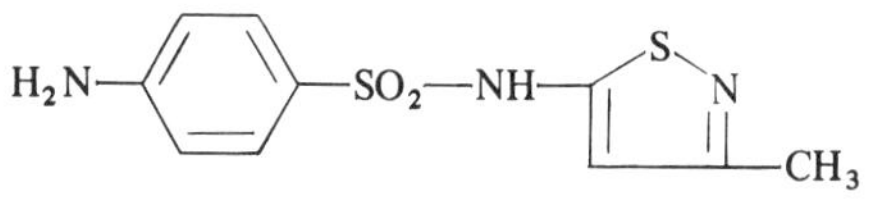

sulfasomizol

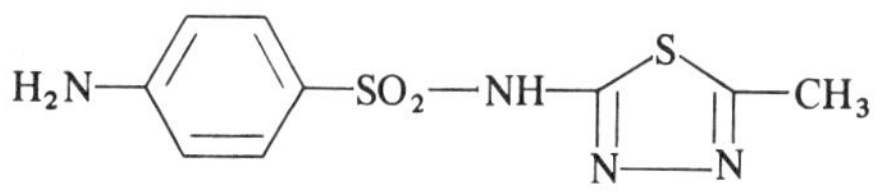

sulfamethizol

TABLE 38

Product	R_f in system:[a]					
	I	II	III	IV	V	VI
Sulfasomizol	0.69	0.72	0.84	0.84	0.94	0.89
N^4-Acetylsulfasomizol	0.79	0.83	0.82	0.86	0.93	0.90
N^1-Acetylsulfasomizol	0.86	0.74	0.80	—	0.91	0.90
Sulfasomizol N^4-glucuronide	0.08	0.10	—	0.08	—	0.09
Sulfasomizol N^4-sulfate	0.36	0.32	0.22	—	0.43	0.01
Sulfanilic acid	0.39	0.52	0.09	0.87	0.89	0.83

[a] Solvent systems: I, propanol:ammonium hydroxide 7:3; II, ethanol: butanol:ammonium hydroxide:water 12:4:1:1; III, butanol:acetic acid: water 4:1:5; IV, butanol:water 1:1; V, methylethylketone:water:acetic acid 200:100:1; VI, methylethylketone:ammonium hydroxide 200:1.

TABLE 39

Product	R_f in system:[a]			Electrophoretic migration (cm) in:[a]	
	I	II	III	IV	V
Sulfamethizol	0.80	0.78	0.71	+4.8	-1.6
N^4-Acetylsulfamethizol	0.85	0.83	0.77	+4.8	-0.5
Sulfamethizol N^4-sulfonate	0.33	0.29	0.44	+6.7	+6.3
Sulfamethizol N^4-glucuronide	0.05	dec.	0.09	+5.2	dec.

[a] Solvent systems: I, water saturated butanol; II, butanol:acetic acid: water 4:1:1; III, butanol:ethanol:water 4:2:1; IV, McIlvaine buffer, pH 7.0 — 200 V for 6 hours; V, 0.2M acetic acid, pH 3 — 900 V for 3 hours.

Marshall et al. [608] showed that sulfaguanidine and its acetylated derivative easily crystallized from urine on cooling. They were identified by their melting points.

II. OTHER SULFONAMIDES

Clapp [206] administered 2-sulfonamidobenzothiazole to a dog and retained its metabolite on an activated carbon (Darco G-60) column and then eluted it by methanol. The crude product was redissolved in water and precipitated with acetone. Elemental analysis and acid hydrolysis, followed by UV and IR spectrophoremetry, permitted the identification of this metabolite as the 2-mercaptobenzothiazole glucuronide. Confirmation was obtained by the positive reaction with naphtoresorcinol. An electrometric titration revealed an acid group of pK_a close to 3.3, which showed that the glucuronic acid carboxyl was not esterified and was likely to be the sodium salt. The binding of 2-mercaptobenzothiazole with glucuronic acid was of the thioether type, as was shown by UV spectrophotometry at various pH values.

This metabolic process, which involves a reduction of the sulfonamide group into sulfhydryl, was confirmed by Colucci and Buyske [216]. Their assays were performed with ^{35}S-labeled products. Urine was extracted with chloroform and the extract analyzed by UV spectrophotometry and paper chromatography, and then compared to synthetic compounds (Table 40).

TABLE 40

Product	R_f in system:[a]		
	I	II	III
2-Sulfonamidobenzothiazole	0.59	0.93	0.40
2-Mercaptobenzothiazole	0.66	0.93	0.75
2-Benzothiazolemercapturic acid	0.54	0.88	0.10
2-Mercaptobenzothiazole glucuronide	0.20	0.65	0.00

[a]Solvent systems: I, butanol:ammonium hydroxide:water 4:1:5; II, butanol:acetic acid:water 4:1:5; III, benzene:petroleum ether:methanol:water 12:8:13:7.

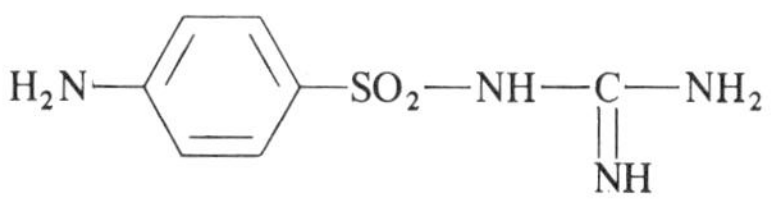

sulfaguanidine

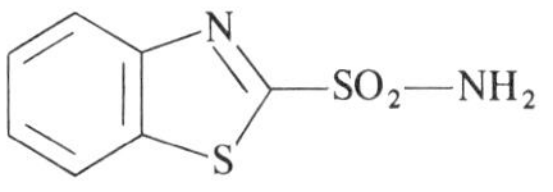

2-sulfonamidobenzothiazole

Caronamide was determined by Earle and Brodie [287] in biological media after specific extraction and nonspecific precipitation. The difference between the results of the two methods revealed the presence of at least one metabolite. The binding of Caronamide to plasma proteins was studied by dialysis.

Probenecid was only excreted in very small amounts, even after large doses had been administered. Tillson et al. [905] administered it to dogs for several weeks until a reducing nonfermentable substance appeared in their urines. The urines were then collected and chromatographed on Celite: Superfiltrol 2:1 column. Water eluted two fractions. One fraction exhibited reducing power and gave a positive test for glucuronic acid. The glucuronide thus detected was very unstable and spontaneously hydrolyzed.

Furosemide is a diuretic sulfonamide. Haüssler and Wicha [410] chromatographed on paper an ethyl acetate extract from urine (butyl acetate: isopropanol: water: ammonium hydroxide 30:50:15:5). They thus detected, besides the intact drug, 4-chloro-5-sulfamoylanthranilic acid, the identity of which was confirmed by UV spectrophotometry.

Corbett et al. [221] administered clorexolone, in some cases ^{14}C-labeled, and fractionated its urinary metabolites on silica gel thin layer (acetone).

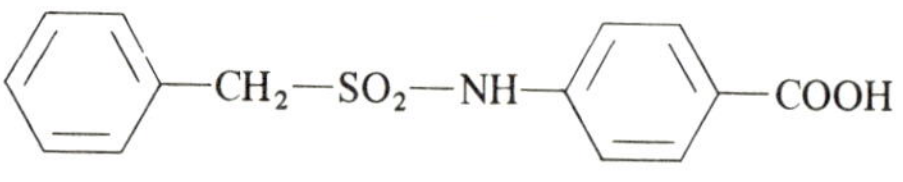

Caronamide

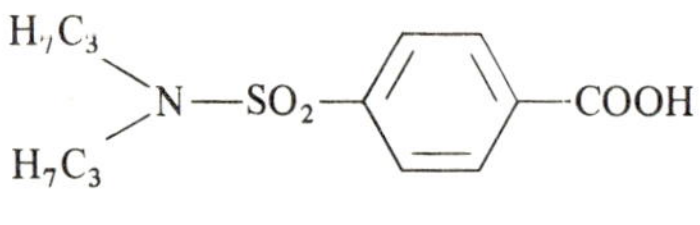

probenecid

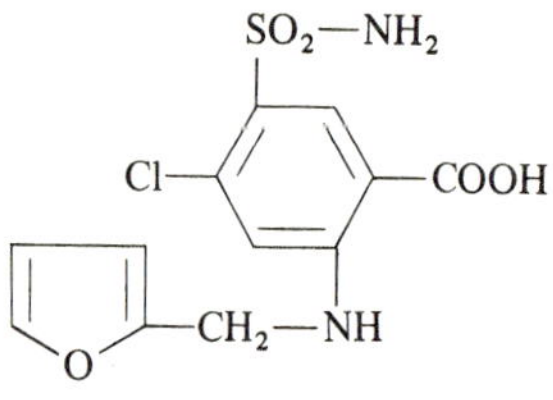

furosemide

One of these was identified as 5-chloro-2-(3'-hydroxycyclohexyl)-1-oxo-6-sulfamoylisoindoline by chromatography and by IR and mass spectrometry.

The metabolism of the antidiabetic sulfonamide glymidine was studied in man with the ^{35}S-labeled compound. Gerhards et al. [351] detected three radioactive compounds in urine by chromatographing the ethyl acetate extract on silica gel thin layer (toluene : ethyl acetate : acetic acid 48 : 50 : 2). After elution and paper chromatography (isopropanol : 1.5N ammonium carbonate solution in 3N ammonium hydroxide 3:1), these compounds were characterized by their melting points, UV and IR spectra, and electrophoretic mobilities, as compared with reference compounds. One compound was unchanged glymidine. The two others were the alcohol that resulted from its demethylation and the carboxylic acid obtainable from the oxidation of this alcohol.

The same authors applied the same procedure [354] to follow the transformations of the titriated alcoholic metabolite in man. They further [352] studied the relationship between the plasma levels of the drug and its two metabolites and the blood glucose content. They [355] undertook in vitro and in vivo investigations in the rat to elucidate phenobarbital influence on glymidine metabolism. Paper chromatography, as above, was used.

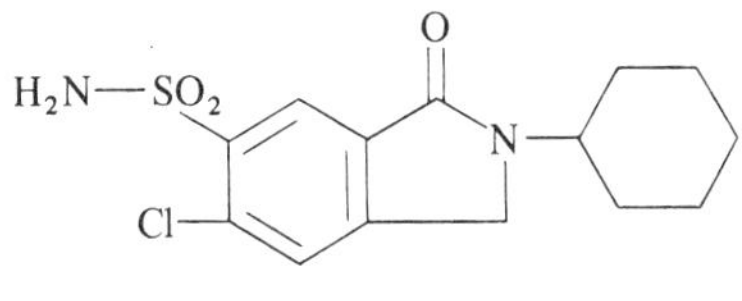

clorexolone

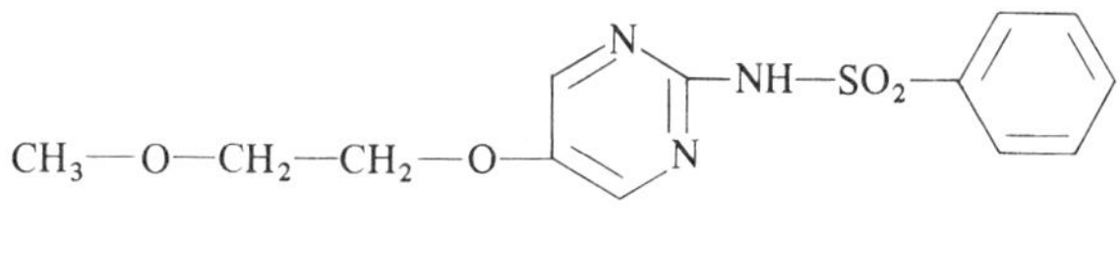

glymidine

Smith et al. [855] examined the metabolism of a series of N-alkyl-4-bromobenzene sulfonamides in connection with their anticonvulsant properties. They separated the drug from its dealkylated derivative by silica gel thin-layer chromatography (chloroform : formic acid 19 : 1) and they determined both of them separately by spectrophotometry after ethanol elution.

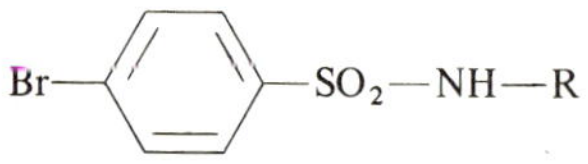

Br—⟨⟩—SO$_2$—NH—R

R = H, CH$_3$, C$_2$H$_5$, CH$_2$—CH=CH$_2$, CH(CH$_3$)$_2$, (CH$_2$)$_2$CH$_3$, (CH$_2$)$_3$CH$_3$

N-alkyl-4-bromobenzenesulfonamides

Chapter XI

IMIDES

The two main representatives of this class of drugs are both glutarimide derivatives. Although they are structurally and pharmacologically analogous, they behave very differently in vivo. One is considerably teratogenic whereas the other has practically no such secondary effects. The grievous accidents due to thalidomide promoted extensive investigations to elucidate the mechanism of its modifications in the organism and to study the teratogenic activity of its metabolites. The other imide is glutethimide, which has two optical isomers that do not undergo the same biotransformations.

I. THALIDOMIDE

The first investigations on the metabolism of this sedative originated in the laboratory of its development (Grünenthal).

In 1962 Beckmann [66] described the synthesis of a labeled thalidomide from 7-^{14}C-labeled phthalic anhydride. After administration of this compound, paper chromatography (system IV*) revealed only traces of the unchanged drug and β-glucuronidase hydrolysis gave no evidence of conjugated derivatives.

---

*The data pertinent to the paper chromatography of thalidomide metabolites are collected in Table 41, page 184.

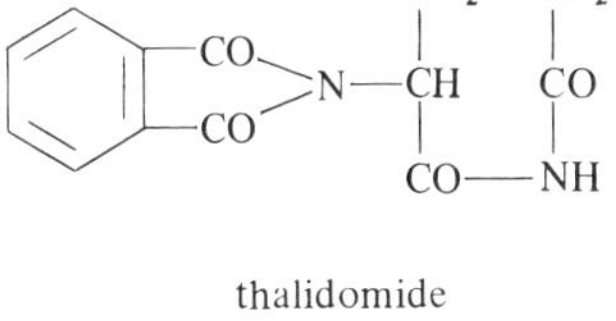

thalidomide

Beckmann synthesized [67] a series of derivatives to allow paper chromatographic identification (systems IV and V). β-Glucuronidase hydrolysis did not modify the chromatograms. Hydrolysis in situ with sodium rhodizonate detection for sulfate conjugation was negative. The amounts of each metabolite were determined by reverse isotopic dilution. The reference compound was dissolved in the alkalinized urine, precipitated with the corresponding metabolite by acidification, and crystallized at 4°C. The cycle was repeated until constant activity was achieved.

Larger amounts of some metabolites were obtained by column chromatography on Amberlite IR-4B (formate) and elution by formic acid solutions of increasing concentrations (from 0.001N to 4N). The metabolites thus separated were identified by their retention volume, their R_f on paper, their melting points, and their IR spectra.

At the same time, Mackensie and McGrath [578] also administered labeled thalidomide to rats and examined urine and organs extracts by paper chromatography (systems I and II). The identity of the metabolites was confirmed by reverse isotopic dilution.

Faigle et al. of CIBA [317] prepared a thalidomide ^{14}C-labeled on the two phtalimide carbonyls in 1962. They demonstrated that 70% of the drug was excreted unchanged in the feces in the dog, as shown by isotopic dilution. The residual 30% were metabolites in the urine. These were identified, after synthesis of the authentic compounds, by paper and thin-layer chromatography (conditions not reported) and by counter-current distribution (0.01N hydrochloric acid:butanol — 100 transfers). Their concentration was determined by reverse isotopic dilution.

Smith et al. [863] characterized one of the metabolites of thalidomide as phthaloylisoglutamine. It was extracted from urine by acetone and crystallized on concentration. It was identified by comparison with the authentic compound and it was possible to separate it from phthaloylglutamine by paper electrophoresis (0.2M citrate buffer, pH 4).

In 1964—1965, the Schuhmacher, Smith, and Williams team published a series of papers [808, 809] on the hydrolysis of thalidomide at various pH values. They elaborated a paper chromatographic method derived from Beckmann (solvents IV and V, mono- and bidimensional chromatography). They subsequently [810] fractionated the urinary metabolites by administration of unlabeled drug. The metabolites were isolated by two procedures.

The first was by column chromatography on alumina of the strongly concentrated urine. The eluting solvent was the system benzene : chloroform 3 : 1, to which was added increasing amounts of a mixture of dioxane, methylethylketone, acetone, methanol, and chloroform in equal parts. The various fractions were examined by paper chromatography. Some were analyzed by alumina thin-layer chromatography (solvent IV) for preparative purposes. The second fractionation procedure involved a series of extractions and crystallizations and is summarized in Fig. III, page 185. The qualitative identifications were completed by color reactions, IR spectra, melting points, and elemental analysis.

Fabro et al. [316] used a ^{14}C-labeled thalidomide and were able to quantitatively interpret the metabolism.

Paper chromatography enabled Nicholls [670] to study the transfer of labeled thalidomide and its metabolites from the mother to the foetus. This author used the mixture isopropanol : water 4 : 1 but did not report his R_f values.

II. GLUTETHIMIDE AND DERIVATIVES

Glutethimide metabolism was studied at CIBA, which developed this hypnotic agent.

Keberle* et al. [495, 496] described the isolation of the major metabolite in dog urine. This was hydrolyzed for 6 hours under reflux at pH 1 and

---

*H. Keberle published until 1963 under the name of J. Kebrle (see [497]). We have adopted the second spelling.

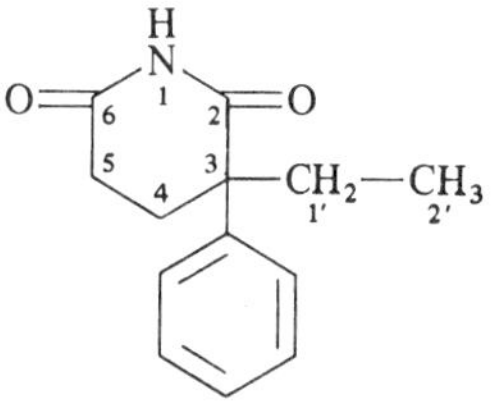

---

glutethimide

TABLE 41

Paper Chromatography of Thalidomide Metabolites

Component	R_f in system:[a]					
	I	II	III	IVa	IVb	V
Thalidomide	0.80	0.80	0.77	0.90	0.89	tailing
N-Phthalylglutamine			0.72	0.41	0.37	0.59
N-Phthalylglutamic acid			0.81	0.22	0.26	0.83
N-Phthaloylglutamine	0.80	1.00	0.57	0.16	0.17	0.50
N-Phthaloylglutamic acid	0.80	0.04	0.64	0.10	0.18	0.46
N-Phthalylisoglutamine			0.75	0.45	0.47	0.63
N-Phthaloylglutarimide			0.60	0.40	0.38	0.63
N-Phthaloylisoglutamine			0.58	0.17	0.16	0.47
Phthalic acid	0.86	0.20	0.71	0.52	0.57	0.74
Isoglutamine					0.13	0.13
Glutamine					0.10	0.05
Glutamic acid					0.06	0.09
Phthalimide	0.91	0.90				
α-Aminoglutarimide					0.41	0.12

[a] Solvent systems: I, dimethylformamide : methanol : water 25 : 70 : 5 [578]; II, 2% ammonium hydroxide saturated butanol [578]; III, butanol : acetic acid : water 4 : 1 : 1 [67]; IV, pyridine : amyl alcohol : water 7 : 7 : 6 IVa [67], IVb [809]; V, butanol : acetic acid 10 : 1, water saturated [809]. Bidimensional chromatography: solvent IV, then solvent V [809, 810].

extracted for 70 hours with ethyl ether. The extract was evaporated and redissolved in benzene, and chromatographed on an alumina column (activity I). It was developed with benzene and then with the mixtures benzene : chloroform and chloroform : methanol. A crystallized product was obtained in the last fraction. Its melting point, elemental analysis, and UV and IR spectra corresponded to the desethyl derivative, α-phenylglutarimide, in an amount

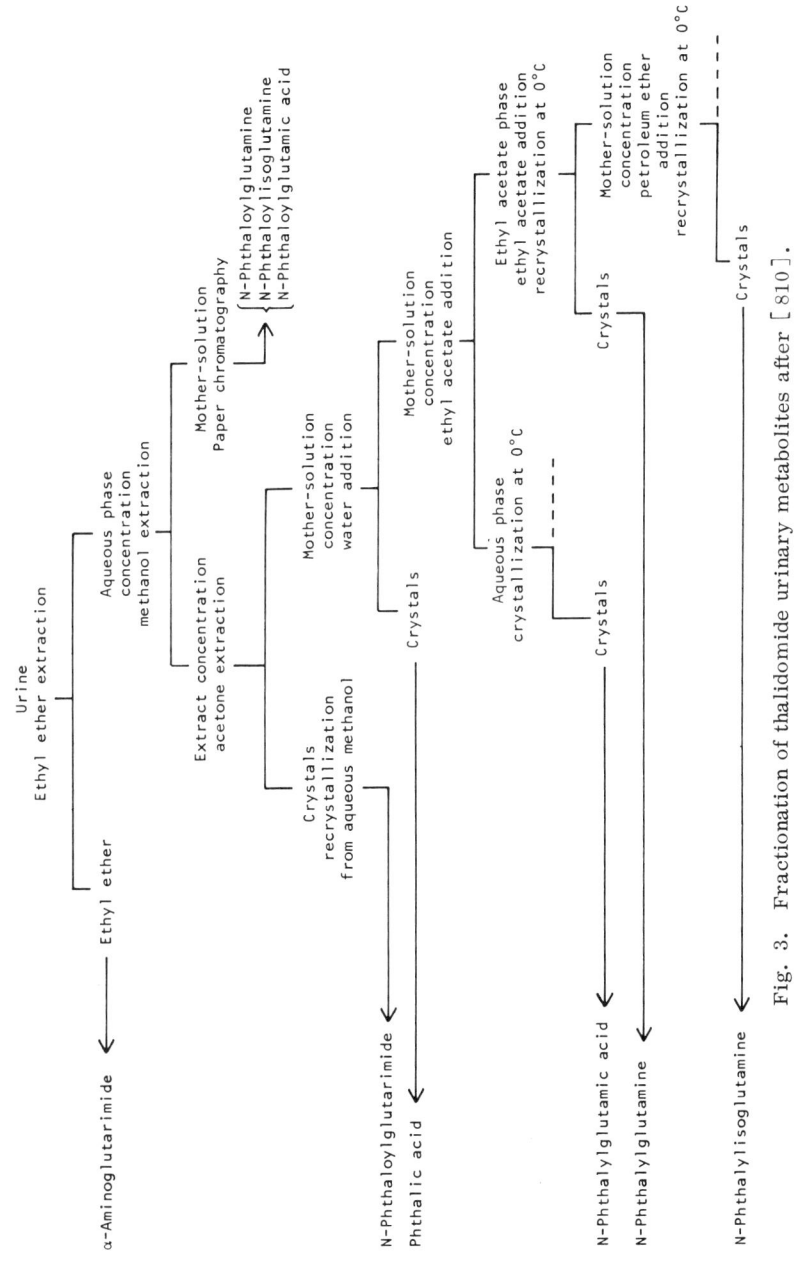

Fig. 3. Fractionation of thalidomide urinary metabolites after [810].

which represented 4% of the administered dose. Similar analysis was applied to other glutarimides.

Bernhard et al. [82] used a ^{14}C-labeled glutethimide and studied its distribution and excretion in the rat. The unchanged drug was determined in urine by isotopic dilution. An ethyl ether extract of urine to which inactive glutethimide was added was purified by chromatography on alumina and then chromatographed on paper (butanol : acetic acid : water 4 : 1 : 5). Two metabolites were thus detected with slight amounts of the unchanged drug. The same procedure was applied to the bile. Eighty percent of the administered radioactivity was recovered in urine.

Sheppard et al. [835] developed a colorimetric determination procedure for glutethimide and its metabolite, α-phenylglutarimide. It was based on the formation of an hydroxamic derivative, the ferric complex of which was purple. Urine was extracted with benzene to determine the unconjugated compounds and then was hydrolyzed for 6 hours with hydrochloric acid under reflux. The urine was extracted again with benzene to determine the conjugated compounds. Both benzene solutions were chromatographed on an alumina column. After the column had been washed with benzene, the compounds were eluted by the mixture chloroform : methanol 9 : 1. The eluates were also chromatographed on paper (petroleum ether : methanol 1 : 1):

- glutethimide, R_f 0.83
- α-phenylglutarimide, R_f 0.67

These assays showed that glutethimide and α-phenylglutarimide were both excreted in the conjugated forms.

Subsequently, the glucuronides in dog urine were separated [499] by the method of Kamil et al. [480]. The crude urine (6-12 l) was first extracted with ethyl ether in neutral medium. The extract was analyzed by chromatography on alumina under the same conditions as previously described. In the fraction eluted by the mixture benzene : chloroform 4 : 1, a product was separated that, on the basis of its elemental analysis and UV and IR spectra, was identified as α-phenyl-α-ethylglutaconimide. Urine was then adjusted to pH 4 and treated with neutral lead acetate. After the precipitate had been discarded, the urine was adjusted to pH 8-8.4 by ammonium hydroxide, and basic lead acetate was added. The precipitate was suspended in water and treated with hydrogen sulfide, and the resulting solution was evaporated.

The crude glucuronides mixture was further fractionated. A part was dissolved in ethyl acetate, treated with Norite, and concentrated. After dilution with methanol and addition of ethyl ether, an amorphous product separated. It was dried, redissolved in butanol, and analyzed by counter-current distribution (butanol:water — 100 transfers). A purified fraction was thus obtained, which was again precipitated from its methylethylketone solution by ethyl ether. Its UV spectrum was that of glutethimide. Its elemental analysis and its potentiometric equivalent weight corresponded to the combination of one molecule of glucuronic acid with one molecule of hydroxylated glutethimide.

Alternatively, the crude glucuronides were methylated by diazomethane in methanol and acetylated by acetic anhydride in pyridine. The mixture of methyl esters obtained was dissolved in cold methanol and an insoluble compound separated, which was recrystallized in methanol. On cooling to 0°C, a second compound crystallized from the concentrated methanol solution. After dilution of the residual solution, the crystals of a third component were obtained at 0°C. These three metabolites were examined by functional analysis and acetyl, C-methyl, N-methyl, and O-methyl groups and active hydrogens were determined. The IR spectra were also recorded. The aglycone structures were defined by degradation. Reflux with a 0.5N sodium methylate methanolic solution gave one neutral and one acid product and it was possible to methylate the latter with diazomethane. The determinations of C-methyl, N-methyl, and active hydrogens, as well as the UV and IR spectra, made it possible to establish the formula of these aglycones. The results were confirmed by synthesis (see also [427]). The three metabolites were the O-glucuronides of the 5- and 1'-hydroxylated derivatives of N-methylglutethimide and of the 5-hydroxylated derivative of glutethimide. However, it was not certain that the methylation of the NH was not due to the diazomethane treatment.

Bütikofer et al. [170] administered the ^{14}C-labeled drug to man. The unchanged drug, α-phenylglutarimide, and α-phenyl-α-ethylglutaconimide were characterized and determined by isotopic dilution. Two glucuronides were isolated by lead acetate, regenerated by hydrogen sulfide, and crystallized after methylation and acetylation. They were then identified with products already known in the animals by their UV and IR spectra, melting points, and R_f on thin layer.

Keberle et al. [497] further extended their investigations to the metabolic stereospecificity of glutethimide. They first reported that the two glucuronides corresponding to N-methylated glutethimide described in their preceding paper [499] actually originated from the separation procedure. They only obtained the two glutethimide 5- and 1'-O-glucuronides on mild methylation of the crude glucuronides mixture. Therefore there were four metabolites: the two glucuronides, α-phenyl-α-ethylglutaconimide, and α-phenylglutarimide.

The authors prepared the two enantiomers of glutethimide. The glucuronides were isolated as stated previously by the method of Kamil et al. [480]. They were thoroughly methylated and acetylated and chromatographed on a silica gel column (benzene : ethyl acetate 1 : 1). After (+) glutethimide administration a single metabolite was thus isolated, which corresponded to the 5-O-glucuronide. A single metabolite was obtained with the (-) isomer and corresponded to the 1'-glucuronide. These two glucuronides were identified by chromatographic comparison on silica gel thin layer prior to and after methylation (Table 42).

The same methods were applied by Keberle et al. [498] to the metabolism of the optical isomers of N-methylglutethimide.

Aminoglutethimide metabolism was the subject of a rough study by Douglas and Nicholls [274]. Urine at pH 7 was extracted with ethylene dichloride

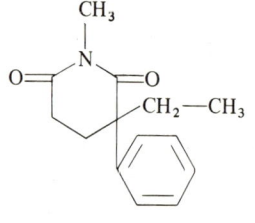

N-methylglutethimide

aminoglutethimide

TABLE 42

Solvent systems	R_f:	
	5-O-Glucuronide	1'-O-Glucuronide
Chloroform : ethyl ether 9 : 1	0.32	0.24
Cyclohexane : ethyl acetate 1 : 1	0.75	0.45
Chloroform : methanol 99 : 1	0.48	0.85

and chromatographed on paper (butanol : acetic acid : water 12 : 3 : 5). The p-dimethylaminobenzaldehyde reagent revealed only the unchanged drug which was colorimetrically determined in the extract.

Tsukamoto and Yoshimura [942] extended their investigations on cyclohexene drugs to two glutethimide derivatives: 2-methyl- and 2-phenyl-2-(1-cyclohexenyl)-glutarimide. The free metabolites were extracted from urine by benzene and chromatographed on an alumina column (chloroform; chloroform : methanol 1 : 1). The residual urine was hydrolyzed at pH 1 and extracted for 30 hours with ethyl ether or ethyl acetate. The extracts were chromatographed in the same way on alumina. These metabolites were compared to the chromic acid oxidation products of the drugs by paper chromatography (chloroform : methanol 1 : 1 — ascending), UV spectrophotometry, and formation of the dinitrophenylhydrazones. They were found to be similar, but the oxidation products were not identified. The author was nevertheless able to conclude that cyclohexene oxidation yielded a 3-oxo derivative.

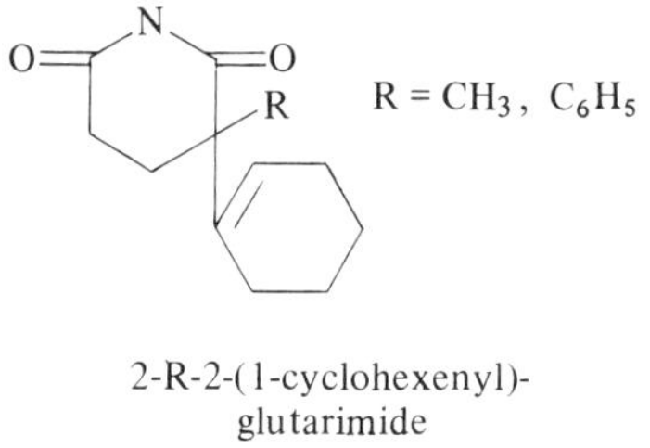

$R = CH_3, C_6H_5$

2-R-2-(1-cyclohexenyl)-glutarimide

A brief paper by Dill et al. [259] reported that ethosuximide could be determined after extraction into chloroform by gas-liquid chromatography (5% of XE-60 on Gas Chrom Z at 155°C). An internal standard α,α-dimethyl-β-methylsuccinimide was added to the sample before extraction. The same procedure, after β-glucuronidase hydrolysis of urine, revealed five metabolites.

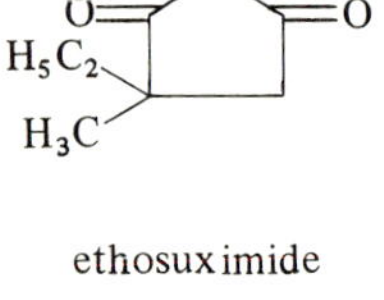

ethosuximide

Chapter XII

HYDRAZIDES AND HYDRAZINES

The largest part of this chapter will be devoted to the metabolism of isoniazid which has been extensively studied since 1952-1953. Since this drug shows variable rates among patients, it was given special attention and many chromatographic procedures for detecting isoniazid metabolites are found in the literature. Since most of them are only of clinical interest, major attention will be placed only on those that provide valid methods for the study of in vivo transformations.

I. ISONIAZID

Roth and Manthei [787] administered isoniazid ^{14}C-labeled on the carboxyl in 1952. Urine was chromatographed on paper (water saturated butanol) either "as is" or after extraction with isobutyl alcohol. The homogenized organs were also treated with the same solvent. These authors found isoniazid and five metabolites in urine. Isonicotinic acid represented 55% of the activity.

In 1953 Hughes [443] extracted the major metabolite of isoniazid from monkey or human urine by means of the mixture ethylene dichloride : isoamyl alcohol 7 : 3, after alkalinization and saturation with ammonium sulfate. The

---

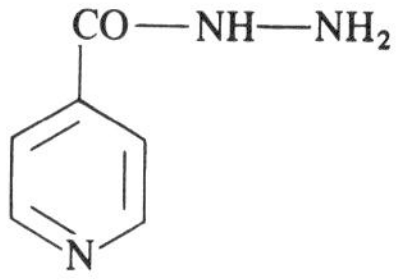

$CO-NH-NH_2$ — pyridine ring with N

isoniazid

metabolite was redissolved in water in alkaline medium, reextracted, and dissolved again in 0.1 N hydrochloric acid. It was crystallized from its ethanol solution by the addition of ethyl ether. Counter-current distribution (butanol + ethylene dichloride 9 + 1: 2M phosphate buffer, pH 5.1 — 8 transfers) showed it was pure. It was identified with 1-isonicotinoyl-2-acetylhydrazine (acetylisoniazid) by its elemental analysis, its UV spectrum, and the detection of the hydrazine function after acid hydrolysis. Confirmation was obtained by comparing the sample to the counter-current distribution of the authentic compound. The quantitative study of its excretion was performed by extraction and colorimetric determination with p-dimethylaminobenzaldehyde. The true yield was calculated on the basis of the known partition coefficients.

At the same time Ozawa and Kiyomoto [698] studied the fate of isoniazid in rabbit blood. Three conjugated derivatives were detected by paper chromatography. One was identified by its R_f and its UV spectrum.

Ceriotti et al. [198] investigated isoniazid metabolism in an attempt to explain the activity differences in vitro and in vivo. Urine was chromatographed directly on paper (isoamyl alcohol : water : acetic acid 50 : 50 : 1.5) and the separated products were detected by cyanogen bromide and determined microbiologically (Table 43).

The paper chromatography of human urine concentrated by lyophilization was also applied by Cuthbertson et al. [239] with the systems: water saturated butanol; methylethylketone : acetic acid : water 49 : 1 : 50 and propanol : water 4 : 1. They detected isonicotinoylglycine and isonicotinic acid.

Defranceschi and Zamboni [249] identified isonicotinoylglycine and 1-isonicotinoyl-2-acetylhydrazine in rat urine by paper chromatography (isopropanol : water 85 : 15). They confirmed the nature of these derivatives by their UV spectra, by their color reactions, and by identification of their hydrolysis products.

They further detected [1039] the hydrazones of pyruvic and α-ketoglutaric acids by paper chromatography (isopropanol : water 85 : 15) by the above same methods (Table 44).

The pyruvic and α-ketoglutaric acid hydrazones were later found by Iwainsky [448] by the same paper chromatography method. He identified them by comparison with the synthetic derivatives.

TABLE 43

Product	R_f
Isoniazid	0.80
Nicotinic acid	0.51
Nicotinamide	0.48
Isonicotinic acid	0.42

TABLE 44

Product	R_f
Acetylisoniazid	0.76
Isoniazid	0.66
Pyruvic acid isonicotinoylhydrazone two spots (a)	0.52
(b)	0.20
Isonicotinic acid	0.38
Isonicotinoylglycine	0.22
α-Ketoglutaric acid isonicotinoylhydrazone	0.06

Yamaguchi [1027] extracted ammonium sulfate saturated urine with alcohol and isolated glucose isonicotinoylhydrazone, acetylisoniazid, and isoniazid by counter-current distribution; he confirmed their presence by paper chromatography.

Abe [1] chromatographed tissues, plasma, and urine extracts in four solvent systems (butanol:ethanol:water 2:2:1; butanol:pyridine:water 16:4:3; ethanol:1.5N ammonium hydroxide:water 17:1:2; phenol:isopropanol:water 16:1:5). The isolated products were eluted to obtain their UV spectra. In this way isoniazid, isonicotinic acid, acetylisoniazid, and glucose and pyruvic acid hydrazones were separated.

Albert and Rees [16] studied isoniazid oxidation as catalyzed by hemin in vitro. The oxidation products were analyzed by paper chromatography (0.5M ammonium chloride) and determined by spectrophotometry after elution (Table 45).

Boone et al. [100, 101] also used paper chromatography [(I) water saturated butanol; (II) propanol:water 80:20]. They were thus able to identify isonicotinic acid and N-isonicotinoyl-N'-acetylhydrazine in rat urine. It was impossible to identify two other metabolites (Table 46).

The paper chromatography of metabolized labeled isoniazid was examined in detail by Diller et al. [260] who tested 300 solvent systems and retained the following three (Table 47):

(I) Isopropanol : 25% ammonium hydroxide, 85 : 15
(II) Isopropanol : water, 85 : 15
(III) Isopropanol : formic acid : water, 80 : 10 : 10

TABLE 45

Product	R_f	λ max	$Log_{10} \varepsilon$
Isoniazid	0.73	262 mµ	3.647
Isonicotinic acid	0.81	262 mµ	3.565
Diisonicotinoylhydrazine	0.68	266 mµ	3.880

TABLE 46

Product	R_f in system: I	R_f in system: II
Isoniazid	0.43	0.66
Isonicotinic acid	0.16	0.41
1-Isonicotinoyl-2-acetylhydrazine	0.63	0.76

Barreto published a series of articles devoted to the paper chromatography of isoniazid metabolites in the blood. He considered a large number of mono- and bidimensional paper chromatography systems, some of which were buffered or treated with EDTA. At first, blood was deproteinated by centrifugal ultrafiltration [50]. This procedure was later replaced by simple paper chromatography, where the blood was applied to a paper strip and, without

TABLE 47

Product	R_f in system:		
	I	II	III
Isoniazid	0.47	0.52	0.40
Ethanal isonicotinoylhydrazone	0.48	0.52	0.42
Acetone isonicotinoylhydrazone	0.48	0.52	0.40
Pyruvic acid isonicotinoylhydrazone	0.18	0.12	0.38
α-Ketoglutaric acid isonicotinoylhydrazone	0.02	0.04	0.36
Glucose isonicotinoylhydrazone	0.18	0.30	0.10
Pyridoxal isonicotinoylhydrazone	0.24	0.57	0.43
N,N'-Diisonicotinoylhydrazine	0.36	0.64	0.47
1-Isonicotinoyl-2-acetylhydrazine	0.34	0.65	0.54
1-Isonicotinoyl-2-glycylhydrazine	0.10	0.23	0.16
1-Methylpyridinium-4-carboxyhydrazine iodide	—	0.04	0.16
Isonicotinic acid	0.32	0.32	0.62
Isonicotinamide	0.63	0.65	0.47
N-Isonicotinoylglycine	0.17	0.21	0.55
Acetylhydrazine	0.40	0.42	0.41

drying, developed by the mixture pyridine:water 65:35. The metabolites were carried to the end of the strip and concentrated by evaporation [52]. A third procedure handled larger volumes. Plasma or serum were chromatographed on an anhydrous sodium sulfate column (chloroform:diethylamine 9:1) and the eluted metabolites were concentrated [53]. Bidimensional paper electrophoresis completed the chromatography of the acid metabolites [51]. Isoniazid, acetylisoniazid, pyruvic acid and ethanal hydrazones, isonicotinamide, isonicotinic acid, and diisonicotinoylhydrazine were separated and detected.

Abiko et al. [3] introduced ion-exchange chromatography. The sample at pH 8 was adsorbed on to a Dowex 1-X 10 (Cl⁻) column. Water displaced the isoniazid, a 0.15M acetate buffer pH 6 eluted acetylisoniazid and 1M sodium chloride eluted glucuronic hydrazone. Since the separation of the two last components was not complete, a second chromatography of the sample at pH 6 separated isoniazid and acetylisoniazid (water) from glucuronic hydrazone (1M sodium chloride). Elution with 0.1N hydrochloric acid yielded α-ketoglutaric hydrazone.

Heller et al. [411] described the separation of acetylisoniazid on a column of Dowex 1-X 8 (pyruvate) and tested the efficiency with labeled isoniazid.

Kakemi et al. [475] developed a method for the determination of isoniazid and its metabolites in urine on the basis of ion-exchange resins. The sample was percolated through a Dowex 1-X 8 (Cl$^-$) column and the effluent continuously transferred to the top of an Amberlite CG-50 (H$^+$) column. The first column retained isonicotinic acid and pyruvic acid isonicotinoylhydrazone. The second column retained the unchanged isoniazid. Isonicotinic acid was eluted by a sodium chloride solution and pyruvic hydrazone was eluted in its turn by 0.5N hydrochloric acid. Unchanged isoniazid was extracted from the second column by 0.5N hydrochloric acid.

Two Dowex-1 columns were used to separate 1-isonicotinoyl-2-acetylhydrazine and glucose isonicotinoylhydrazone. The urine sample was percolated through the first column to remove isonicotinic acid and pyruvic hydrazone. The acidified effluent was oxidized by potassium dichromate to convert isoniazid and glucose hydrazone into isonicotinic acid, which was further retained on the second Dowex column, whereas 1-isonicotinoyl-2-acetylhydrazine was eluted. Isonicotinic acid was extracted from the column by 0.5N hydrochloric acid. All metabolites were determined by cyanogen bromide. It was thus possible to calculate the individual amounts of each metabolite.

Peters et al. [709] studied the metabolic reaction, which could explain great differences in metabolic rates observed among individuals. They were critical of the previous determination methods and developed a technique for quantitative detection of all isoniazid metabolites [708] that separated isoniazid derivatives by ion-exclusion chromatography and isonicotinic acid derivatives by ion-exchange chromatography. The entire analytical procedure is summarized in Fig. 4 (page 197).

Neutral urine was chromatographed on a cation-exchange resin in the NH$_4^+$ form. The acid components flowed through the column and the neutral or ionized compounds were eluted in three groups. They were determined by a series of differential determinations without further separation. Isonicotinic acid and isonicotinoylglycine were determined [671] as the color

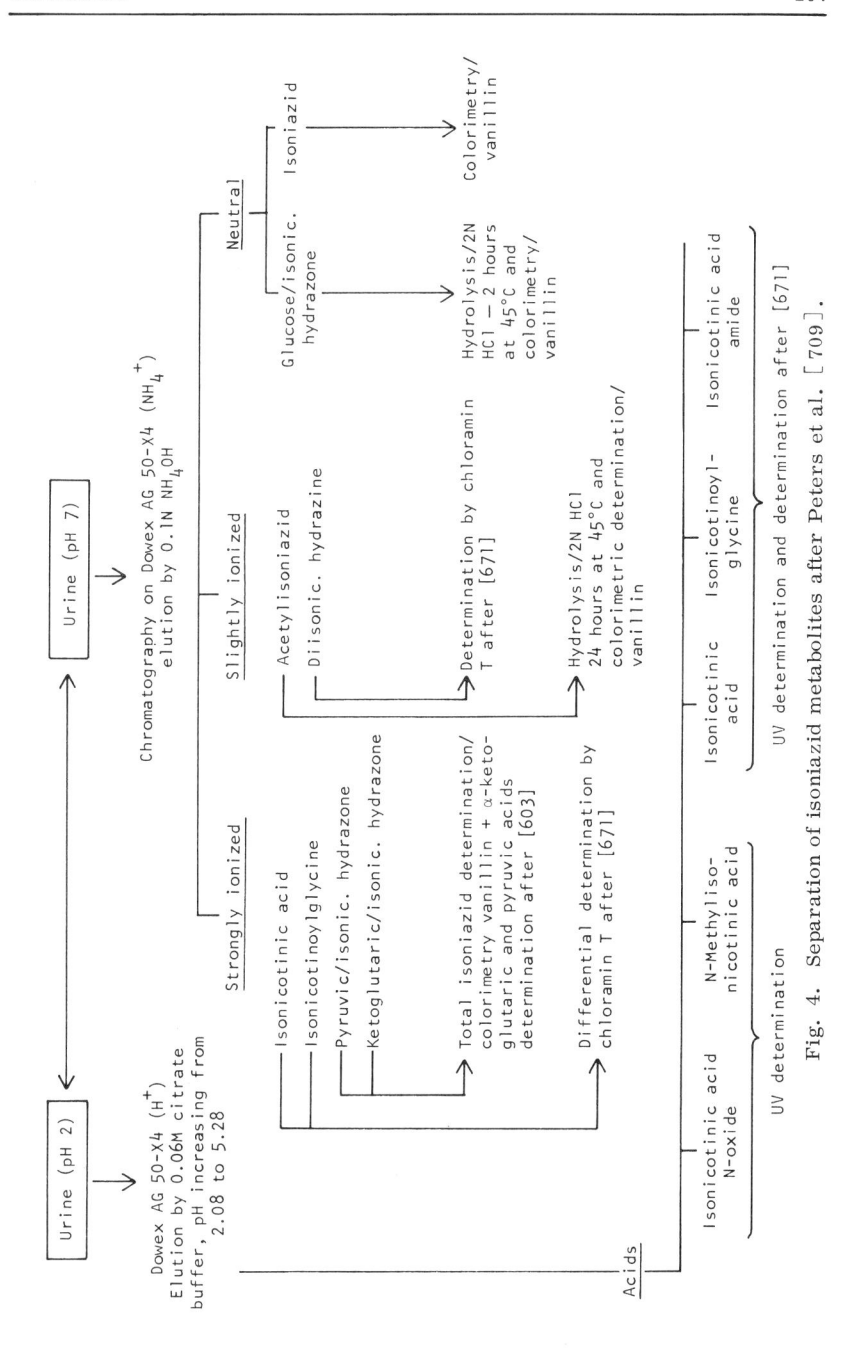

ISONIAZID 197

Fig. 4. Separation of isoniazid metabolites after Peters et al. [709].

derivatives of chloramin T, spectrophotometric absorbance at 600 and 620 mµ, respectively. The two hydrazones were hydrolyzed and the pyruvic and α-ketoglutaric acids sequentially determined by the enzymic procedure of Marks [603]. The method of Peters et al. (described in detail in [708]) is an excellent example of analytical procedure development when the nature of the various metabolites has been previously elucidated. It is most probably the best method published to date for isoniazid metabolites.

Nielsch and Giefer [671, 672] had also described a set of colorimetric procedures to determine free isoniazid, isonicotinic acid, N-isonicotinoyl-glycine, and the sum of the metabolites originating from isonicotinic acid.

Manthei [600] demonstrated the presence of diisonicotinoylhydrazine in the urine of mice and guinea pigs by isotopic dilution after synthesis of the metabolite.

A general review has been given by Venkataraman et al. [959].

II. OTHER HYDRAZIDES AND HYDRAZINES

Iproniazid is extensively metabolized and its study did not require sophisticated analytical procedures. Hess et al. [423] characterized the major metabolite, isonicotinic acid, by paper chromatography (propanol:water 4:1) and determined it colorimetrically after extraction.

Koechlin et al. [513, 516] used iproniazid ^{14}C-labeled on the CH. The greatest portion of the radioactivity was recovered in the expired carbon dioxide and acetone, which was retained as its dinitrophenylhydrazone. Urine was analyzed by paper electrophoresis (0.5M citrate-phosphate buffer, pH 3.7) which was followed by chromatography in the second dimension (water

---

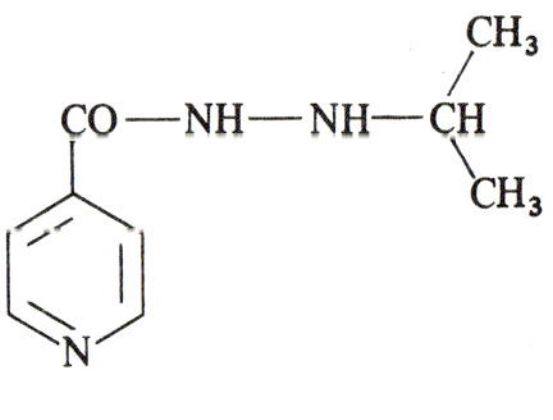

iproniazid

saturated butanol). Two basic metabolites and one acid metabolite were detected.

Nair [658], after administration of iproniazid ^{14}C-labeled on the carbonyl, extracted urine with chloroform. The extract and the aqueous phase were chromatographed on paper (propanol : water 4 : 1). He confirmed the presence of isonicotinic acid in addition to the unchanged drug.

Schwartz prepared isocarboxazid ^{14}C-labeled on the methylene of the benzyl. Almost all the activity was recovered in urinary sodium hippurate, detected by paper chromatography, identified by reverse isotopic dilution, and isolated as azlactone [816]. The in vitro transformations were studied by paper chromatography (t-butanol : methanol : water 5 : 4 : 1), which revealed benzylhydrazine [817].

Koechlin et al. [516] pursued this investigation in man but paper chromatography only revealed sodium hippurate.

Hydralazine was first examined by Douglass and Hogan [275] who showed that it underwent N-acetylation. In fact, hydralazine interfered with the biological acetylation of sulfanilamide and competition for the free acetyl groups was postulated. Assays in vitro indicated that the free hydralazine disappeared as soon as all the components of an acetylation system were present. Ethyl acetate extraction of urine made it possible to identify

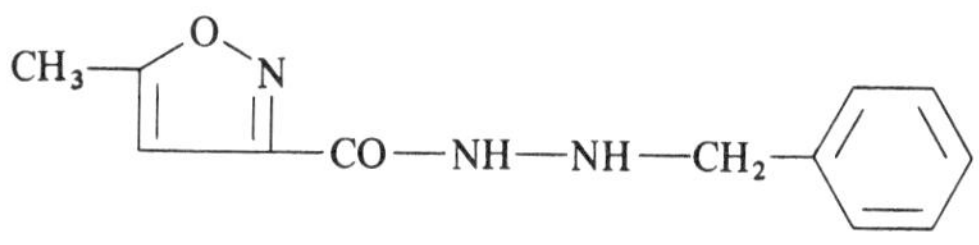

isocarboxazid

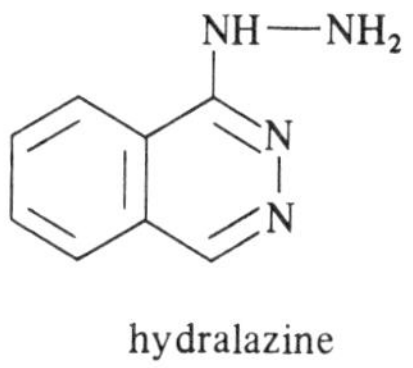

hydralazine

N-acetylhydrazine by paper chromatography; it was compared with the synthetic compound in seven solvent systems. In addition, the metabolite was isolated by paper chromatography and its UV spectrum determined.

A more extensive investigation was conducted by McIsaac and Kanda [576] after synthesis of the 1-^{14}C-labeled compound. The stability of the drug was studied as a function of pH by isotopic dilution in order to define the sampling conditions. Urine, urease treated and concentrated, was chromatographed on paper and a series of derivatives allowed the identification of metabolites (Table 48).

The metabolites were identified by their R_f values and the detection reactions. The glucuronides were hydrolyzed and the aglycones analyzed by paper chromatography. Three urinary metabolites were found beside the unchanged drug: hydralazine O-glucuronide, N-acetylhydralazine, and pyruvic acid hydralazine hydrazone. These metabolites were determined directly on the chromatograms and also by isotopic dilution. An almost complete metabolic balance was obtained.

Salicylohydrazide, although not a drug, exhibited an antitubercular activity. It essentially underwent glucuronoconjugation, as McIsaac and Williams [577] showed when they precipitated its glucuronide as the lead salt. Paper chromatography (ammonium hydroxide saturated propanol) revealed the salicylic acid metabolites: salicyluric and gentisic acids.

Procarbazine metabolism was briefly described by Raaflaub and Schwartz [741]. The unchanged drug was extracted with chloroform from the ethanol deproteinized plasma and then reextracted into diluted hydrochloric acid.

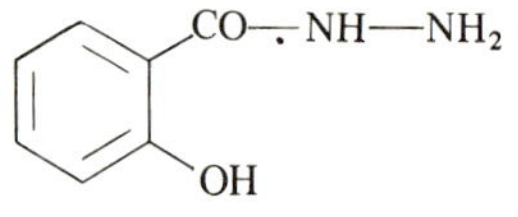

salicylohydrazide

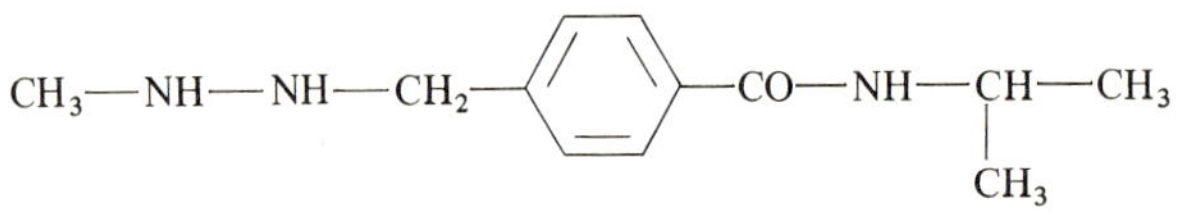

procarbazine

TABLE 48

Product	R_f in system:[a]		
	I	II	III
Hydralazine	0.90	1.00	0.90
N-Acetylhydralazine	0.87	0.85	0.90
Pyruvic acid hydralazine hydrazone	0.55	0.20	0.70
α-Ketoglutaric acid hydralazine hydrazone	0.30	0.00	0.55
1-Hydroxyhydralazine	0.65	0.15	—
7-Hydroxyhydralazine	0.95	0.95	0.78
1-Hydroxyhydralazine glucuronide	0.75	0.80	—
Phtalazine	0.85	1.00	0.82
1-Hydroxyphtalazine	0.42	0.14	0.70
1,4-Dihydroxyphtalazine	0.19	—	—
1,4-Dihydroxyphtalazine glucuronide	0.45	0.15	—

[a]Solvent systems: I, propanol:ammonium hydroxide 7:3; II, methylethylketone:2N ammonium hydroxide 2:1; III, butanol:acetic acid:water 4:1:5.

After neutralization, potassium ferricyanide was added and the resulting ferrocyanide determined by colorimetry. The chloroform phase also contained the azo derivative that corresponded to the procarbazine, which was detected after administration of the ^{14}C-labeled drug. Terephthalic acid isopropylamide was found in the aqueous phase by thin-layer chromatography. The conditions were not reported.

Schwartz [813] deproteinized the acidified urine with ethanol at 0°C and then converted the amines into their dinitrophenyl derivatives by fluorodinitrobenzene at pH 6.0 for 2 hours plus one hour at pH 8.5. The identity of the obtained derivatives was established by thin-layer chromatography in five solvent systems. Schwartz thus detected methylamine as a metabolite. He also showed $^{14}CO_2$ formation after 1-$^{14}CH_3$-procarbazine administration.

Similarly, Baggiolini et al. [47] collected $^{14}CO_2$ in experiments on normal and phenobarbital-treated rats, and on perfused isolated rat livers.

Kreis et al. [526] demonstrated that the 1-methyl group of procarbazine not only underwent an oxidation, but also a transposition on the purine bases of the organism to which procarbazine had been administered. In fact, after injection of the 1-^{14}C-labeled drug, it was possible to isolate the radioactive methylated derivatives of guanine, adenine, and hypoxanthine in a urine extract by paper chromatography.

Baggiolini and Bickel [46] amplified the N-demethylation mechanism. After the elimination of proteins by ethanol and chloroform, the aqueous solution was diluted to a saline concentration equivalent to 30mM and chromatographed on a column of carboxymethylcellulose equilibrated against a 30mM buffer, pH 7. Thus methylhydrazine was determined in the presence of procarbazine. The drug was separated from its monodemethylated metabolite by thin-layer chromatography on silica gel (toluene : acetone : methanol : ammonium hydroxide 50 : 30 : 10 : 1) of the extract obtained by the mixture ethanol : chloroform 1 : 6 at pH 7.5. N-Isopropylterephtalamic acid was characterized by silica gel thin-layer chromatography (toluene : acetone : methanol : acetic acid 50 : 30 : 10 : 2; ethylene dichloride : propanol : acetic acid 50 : 20 : 2) when the extraction was performed at pH 3.

Reed and Dost [749] demonstrated with ^{14}C- and ^{3}H-labeled drug the conversion into methane of the methylhydrazine that originated from procarbazine.

Methylglyoxalbisguanylhydrazone is a carcinostatic agent and the in vivo transformations were studied by Oliverio et al. [692] with the drug ^{14}C-labeled on the carbons of both guanyl groups. The likely metabolites were investigated in urine by paper chromatography, high voltage paper electrophoresis, and chromatography on Bio Rex 70 (H$^+$) column. Only the unchanged drug was found in urine.

$$\begin{array}{c} CH_3 \\ | \\ C=N-NH-C(=NH)-NH_2 \\ | \\ HC=N-NH-C(=NH)-NH_2 \end{array}$$

methylglyoxalbisguanylhydrazone

Chapter XIII

HETEROCYCLES WITH ONE NITROGEN

Heterocyclic compounds hold important positions in pharmaceutical chemistry. Actually, many of the drugs previously considered were heterocyclic compounds and this chapter and those following are reserved for those products that could not be readily otherwise classified. Thus, these remaining compounds may not exhibit much chemical or pharmacological similarity.

Heterocyclic compounds will be treated in four chapters, on the basis of containing one nitrogen, two nitrogens, more than two nitrogens, or other heteroatoms.

I. PYRROLIDINE DERIVATIVES

Prodilidine appears to be extensively metabolized in the rat since Weikel and Labudde [989] recovered only traces of the unchanged drug in urine. The methyl orange method of Brodie and Axelrod [130] was used after they had checked that the procedure determined only the intact drug. The product extracted from urine by the mixture ethylene dichloride : carbon tetrachloride : isoamyl alcohol 50 : 50 : 1.5 was analyzed by counter-current distribution (chloroform : 0.1M phosphate buffer, pH 4.4 and 7.4), by paper chromatography (isoamyl alcohol : ethanol : formic acid : water 100 : 15 : 10 : 100), and

----

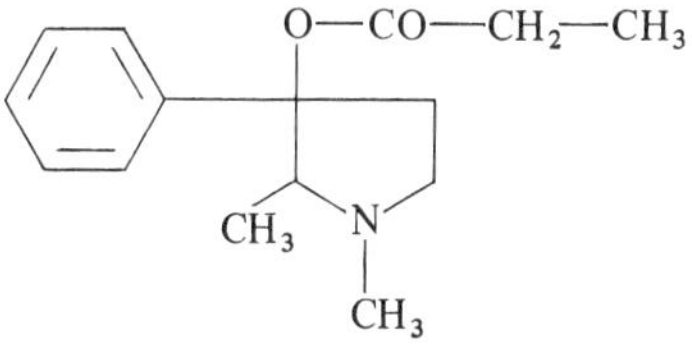

prodilidine

by paper electrophoresis (0.02M phosphate buffer, pH 5.5, on Whatman CM 50). Paper chromatography revealed the presence of the demethylated product after incubation of the drug with the liver microsomes of various animals. The dinitrophenyl derivative was crystallized.

Doxapram was determined by Bruce et al. [145] by oxidation into benzophenone. Chromatographies on silica gel thin layer (methanol; ethanol) revealed at least three metabolites. Urine extraction at various pH values before and after hydrolysis, followed by thin-layer chromatography of the various fractions, showed a large number of metabolites. The most abundant basic metabolite was precipitated by acidification, recrystallized, and identified with the product resulting from the opening of the cyclic morpholine by elemental analysis, melting point, and IR spectrum as compared to an authentic sample. Another amine metabolite was the 2-aminoethylpyrrolidinone corresponding to doxapram. In fact, its R_f value and that of its acetylated derivative on thin layer were identical to that for the reference compound. The IR spectrum was also identical. Five acid metabolites were present, but were not studied.

II. OXAZOLIDINE DERIVATIVES

The trimethadione demethylation product was isolated from dog urine by Butler et al. [183]. It was extracted into ethyl ether, purified by distribution between ethylene dichloride and water, and identified by its melting point and its UV spectrum.

Butler further described [173] an extraction and spectrophotometric determination procedure for this metabolite. He checked its specificity by its partition between a solvent and buffers of various pH.

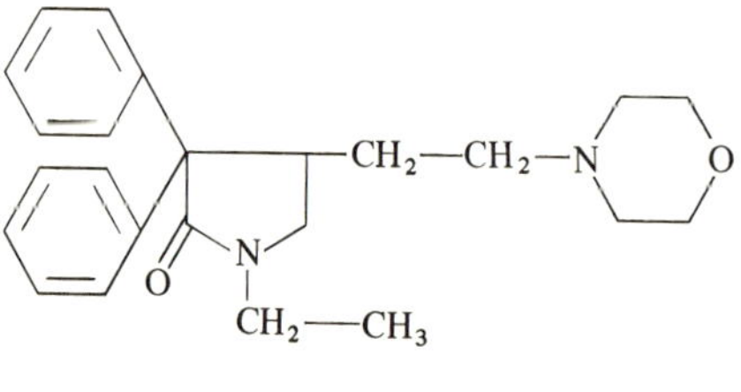

doxapram

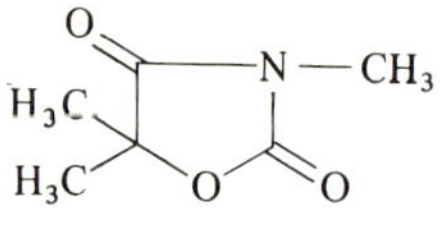

trimethadione

OXAZOLIDINE DERIVATIVES

Paramethadione was demethylated similarly in the dog. Butler [176] isolated the metabolite as its piperidine salt after a series of extractions and washings. He identified it by UV spectrum comparison with the authentic compound, as well as by its partition coefficient. As above, he developed a spectrophotometric method for the metabolite.

Mephenoxalone is almost quantitatively hydroxylated in a mixture of two isomeric phenols. Morrison [645] used ^{14}C- and ^{3}H-labeled molecules. He isolated two metabolites by counter-current distribution in the four systems of the below-mentioned paper chromatography (220 transfers). The UV and fluorescence spectra and various reactions suggested phenolic derivatives. Paper chromatography (butanol : acetic acid : water 4 : 1 : 5; butanol : ammonium hydroxide : water 4 : 1 : 5; ethylene dichloride : acetic acid : water 8 : 1 : 5; ethylene dichloride : ammonium hydroxide : water 8 : 1 : 5) confirmed these results and showed that these hydroxylated derivatives were partly conjugated as glucuronides.

The major metabolite of metaxalone results from the oxidation of one of its methyl groups. Bruce et al. [146] saturated dog urine with ammonium sulfate and extracted the resulting precipitate with hot ethanol. The extracted product, after recrystallization, was identical with 5-(3-methyl-5-carboxyphenoxymethyl)-2-oxazolidinone on UV and IR spectrophotometry and thin-layer

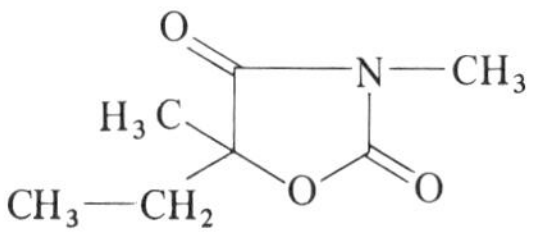

paramethadione

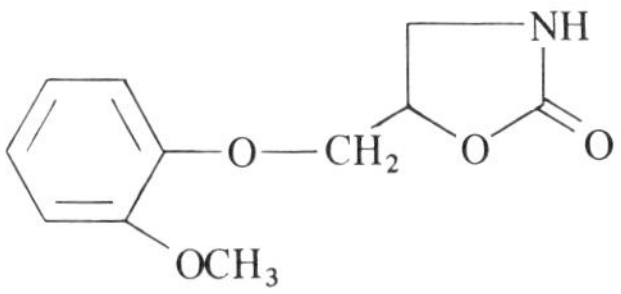

mephenoxalone

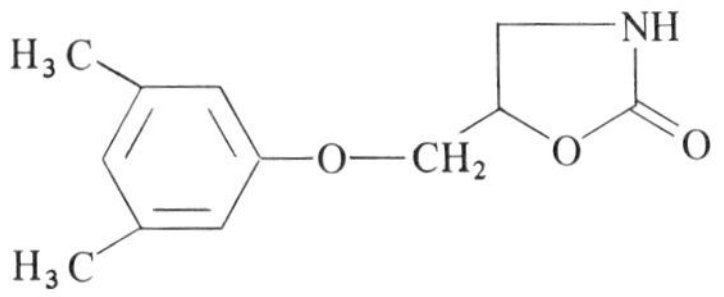

metaxalone

chromatography. This metabolite was also present as a glucuronide. Isolation was effected by acetylation and methylation of the ammonium sulfate precipitate after ethyl ether extraction to eliminate the free metabolite. A fraction containing this glucuronide was isolated by chromatography on a Florisil column (acetone : benzene mixtures). This fraction corresponded to the major component detected by thin-layer chromatography (benzene : acetone 1 : 1). This glucuronide was identified by comparison with the synthetic compound prepared by condensing the metabolite with methylbromacetyl-glucuronate. 3,5-Xylenol was detected by extraction of acidified urine with ethyl ether and gas-liquid chromatography (10% of Apiezon on Chromosorb W at 185° C).

III. THIAZOLINE DERIVATIVES

Thin-layer chromatography on silica gel (chloroform : methanol 95 : 5) established the presence of five metabolites of antazonite. Allewijn and Demoen [20] extracted them from chicken feces with ethyl ether and applied chromatography on a silica gel column (methanol, 1% of ammonium hydroxide in methanol). The major metabolite was thus isolated. The structure elucidated by UV and IR spectrophotometry and elemental analysis was the cyclic 5,6-dihydro-6-(2-thienyl)imidazo-[2,1-b]-thiazole. The synthesized compound was four times more active than the administered drug. Two other metabolites were identified by chromatographic comparison.

IV. PYRIDINE DERIVATIVES

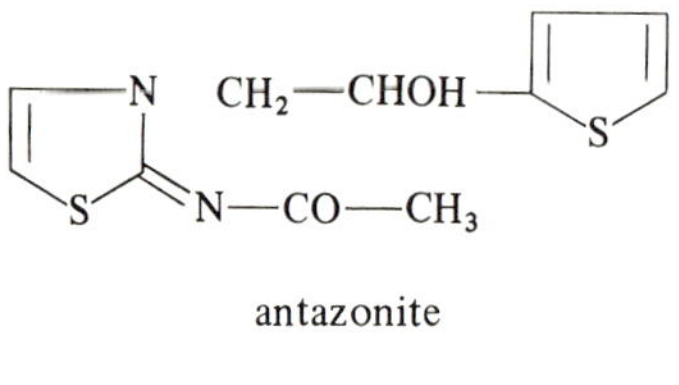

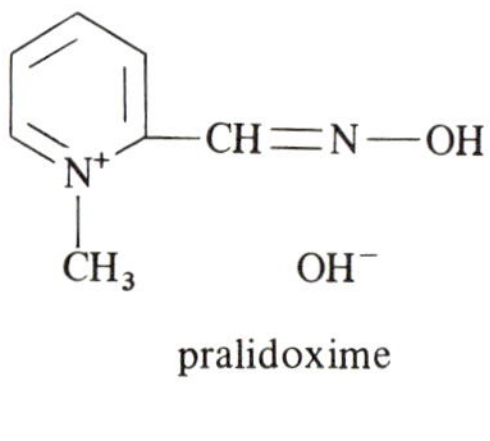

Way et al. [987] briefly reported in 1960 that the perfusion of ^{14}C-labeled pralidoxime iodide in isolated rat livers resulted in the isolation of three radioactive fractions by chromatography on Dowex-50 (Na$^+$) of the alcohol soluble fraction of the perfusion medium. The column was eluted by a sodium chloride solution of increasing concentration. Each fraction was purified by adsorption on Darco S-51 and elution by 50% alcohol. One of the separated components was related by its UV spectrum and the presence of a CN function to 2-cyano-N-methylpyridinium.

Ellin and Easterday [304] studied the paper chromatographic systems that could separate the various degradation products of pralidoxime iodide. The best results were obtained with the system butanol:acetic acid:water 5:1:3. However, two spots were always obtained and one corresponded to hydrogen iodide (Table 49).

Kramer [522] determined the nature of one of the pralidoxime iodide metabolites contained in human urine. It was separated by paper chromatography from the unchanged drug, which was identified by its R_f and UV spectrum. The metabolite was analyzed by UV, IR, and fluorescence spectrophotometry and assigned the structure of N-methylpicolinic acid.

Enander et al. [310] investigated pralidoxime methanesulfonate and chemically determined the excreted thiocyanide. They also [311] detected 2-cyano-N-methylpyridinium by paper chromatography and determined it by the Zinke-Königs reaction. They further [312] used $^{14}CH_3$-labeled pralidoxime iodide. The absence of $^{14}CO_2$ in the expired air revealed the lack of demethylation. On paper chromatography (butanol:acetic acid:water 4:1:1) five radioactive fractions were observed. The unchanged drug and three metabolites were identified by their R_f, compared to authentic compounds. For some, reverse isotopic dilution was also used.

TABLE 49

Product	R_f
Pralidoxime iodide	0.51
N-Methyl-α-pyridone	0.81
2-Cyano-N-methylpyridinium iodide	0.42
2-Formyl-N-methylpyridinium iodide	0.62
2-Carboxy-N-methylpyridinium iodide	0.38
2-Carboxamido-N-methylpyridinium iodide	0.37
Hydrogen iodide	0.20

Way described [982] a spectrophotometric determination procedure for pralidoxime iodide. He isolated and identified two metabolites in the perfusate medium of isolated rat livers on the use of a $^{14}CH_3$-labeled drug. These metabolites were separated by first adding cold alcohol and centrifuging and then by chromatographing on a column of Celite : Darco S-51 1 : 1. The column was washed subsequently with water and 0.01M EDTA, pH 4.75, and eluted by 47.5% alcohol. The concentrated eluate was chromatographed on Dowex-50 (Na$^+$). The first metabolite was eluted by 0.2M sodium chloride [983], the second one, in a separate experiment [986], by a 0.25M sodium chloride solution in 0.0001N hydrochloric acid. These metabolites were purified by paper electrophoresis or chromatography. The first metabolite [983] appeared to be an O-conjugated derivative of pyridinium ion, on the basis of its UV spectrum and the study of its alkaline degradation. This hypothesis was confirmed by synthesis of 2-ethoxy-N-methylpyridinium iodide. Nevertheless, this conjugate was hydrolyzed neither by glucuronidase nor by sulfatase. The second metabolite [986] was assigned the structure of 2-cyano-N-methylpyridinium on comparison of its spectral, chromatographic, and electrophoretic properties with the authentic compound.

Way has more recently described [984] the "multiple spot" phenomenon that appears in the paper chromatography of the pralidoxime iodide metabolite, as well as the influence of the anion on the R_f value.

Miranda et al. [634] briefly reported that they had separated several metabolites of pralidoxime iodide by gas-liquid chromatography. One of these was N-methyl-α-pyridone which was characterized by UV spectrophotometry, spectrofluorometry, paper chromatography, and electrophoresis.

Trimedoxime is also a cholinesterase reactivator. Way et al. [985] separated its metabolites by chromatographing urine on activated carbon. The ethanol-eluted fraction was purified by chromatography on a cation-exchange resin, and by paper chromatography and electrophoresis (conditions not reported). One of the metabolites thus isolated, in addition to the unchanged drug, was found to be similar to synthetic 1-(4-aldoximinopyridinium)-3-(4-cyanopyridinium) propane on paper chromatography (propanol : water 7 : 3; propanol : acetic acid : water 6 : 1 : 3). Confirmation was achieved by paper electrophoresis, spectrophotometry, and isotopic dilution.

Klicka et al. [509] reported in a short communication that they had isolated a metabolite of metyrapone by methylene chloride extraction and column chromatography. The UV, IR, and mass spectra did not permit definition of its structure.

Ethionamide was easily titrated in biological media by polarography. Kane [482] observed a secondary wave that he was able to ascribe to ethionamide sulfoxide by comparison with the authentic compound. The determination of the partition coefficient confirmed this identification.

Bieder and Mazeau [91] extracted urine at pH 6.5 with chloroform. The concentrated chloroform solution was purified on an alumina + calcium carbonate column, then chromatographed on silica coated paper (water saturated butanol; water saturated butanol + 1.6% of acetic acid; butanol : heptane 10 : 3, water saturated). Several metabolites were found in human urine in addition to ethionamide. One of these was probably 2-ethyl-4-carbamoylpyridine, as shown by comparison of its R_f value with the authentic compound. Three other metabolites fluoresced.

The same authors [89] determined both ethionamide and its sulfoxide by spectrophotometry or polarography with comparable results.

Bieder and Mazeau [92] further applied thin-layer chromatography. The chloroform extract from urine, chromatographed on silica gel thin layer

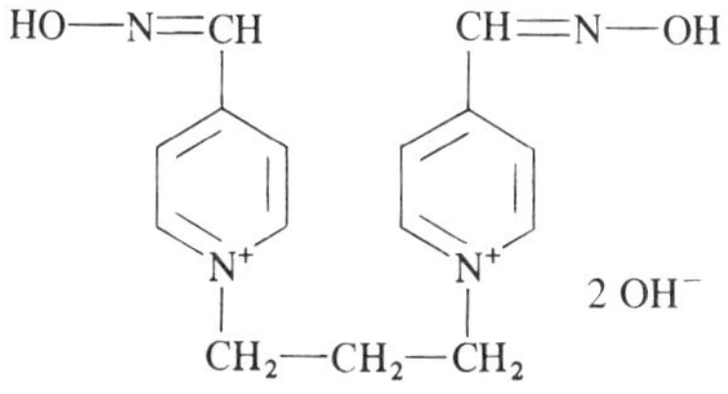

---

trimedoxime

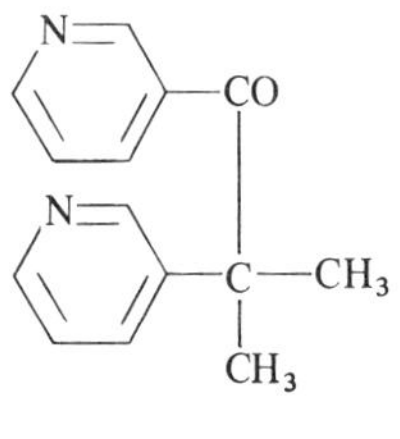

metyrapone

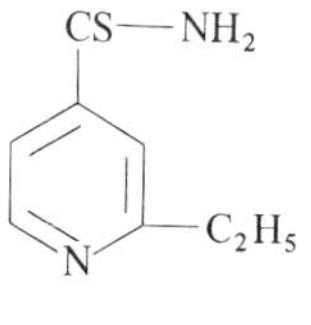

ethionamide

with the mixture chloroform : methanol 100 : 5 showed five fluorescent components. The area corresponding to the first three was eluted into methanol and chromatographed again on silica gel (propanol). Four compounds were thus detected and two corresponded to ethionamide and its sulfoxide. The ethyl acetate extract of urine was chromatographed on cellulose powder with the system 3% ammonium hydroxide saturated butanol and 2-ethylnicotinic acid was isolated. Its identity was confirmed by the UV spectrum. Urine was percolated through an Amberlite IRC-50 (Na^+) column, the basic components eluted by 0.1N hydrochloric acid and then chromatographed on alumina (water saturated butanol). A compound was detected with an R_f value that was the same as 1-methyl-2-ethyl-4-thiocarbamoylpyridinium. Its identity was confirmed by the UV spectrum.

The same authors [90] further elucidated the nature of three other metabolites, by comparison with synthesized compounds. The amyl alcohol extract of concentrated urine was analyzed by preparative thin-layer chromatography on silica gel (chloroform : methanol 10 : 1). The R_f 0.41 fraction was eluted into methanol and chromatographed on silica gel (ethyl acetate). In this way ethionamide was separated from the metabolite 1-methyl-2-ethyl-4-thiocarbamoyl-6-oxodihydropyridine which was chromatographed again (ethyl acetate). A second fraction from preparative chromatography — R_f 0.24-0.30 — was chromatographed on silica gel (propanol). Two metabolites were thus separated; they were purified by one or two other chromatographies. They were identified, as the preceding one, on the basis of their R_f and their spectral and polarographic characteristics as 1-methyl-2-ethyl-4-carbamoyl-6-oxodihydropyridine and 1-methyl-2-ethyl-S-oxo-4-thiocarbamoyl-6-oxodihydropyridine.

Iwainsky et al. [449] extracted urine with chloroform and the resulting solution was chromatographed on paper (propanol : buffer, pH 6.5 85 : 15). Six to eight metabolites were detected, including the sulfoxide, which was located by its R_f.

Weinman and Geissman [993] studied tripelennamine using the drug ^{14}C-labeled on the benzyl methylene. By a series of extractions urine was fractionated in hydrophilic, lipophilic, and intermediary components. Paper chromatography (butanol : acetic acid : water 100 : 14 : 38) revealed various metabolites. These included one or several conjugates, which

PYRIDINE DERIVATIVES

were hydrolyzed by β-glucuronidase. Paper electrophoresis allowed tripelennamine to be distinguished from a metabolite of close R_f.

Bisacodyl is completely deacetylated in vivo. Ferlemann and Vogt [318] demonstrated this by determining the phenolic compounds before and after hydrolysis. Diphesatine is less hydrolyzed. The diphenols are the active compounds. Vogt et al. [961] further showed the presence of a glucuronide of the bisacodyl deacetylation product. This glucuronide was isolated from bile freed from biliary acids after ethyl ether extraction at pH 1 by chromatography on Dowex-1 (OH$^-$). The product, eluted by 2N formic acid, was chromatographed again using formic acid of increasing concentration. A fraction was obtained in which the respective amounts of diphenol and glucuronic acid corresponded to a monoglucuronide.

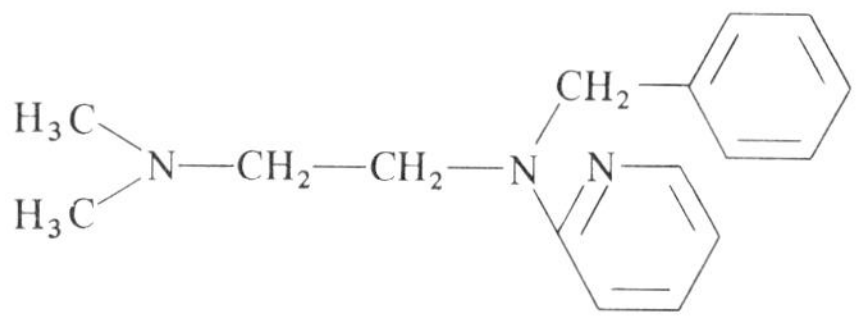

tripelennamine

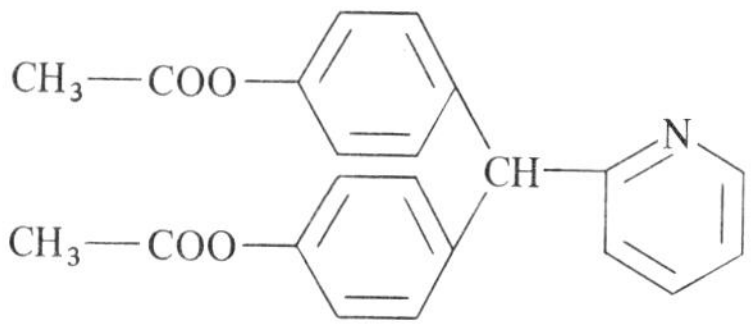

bisacodyl

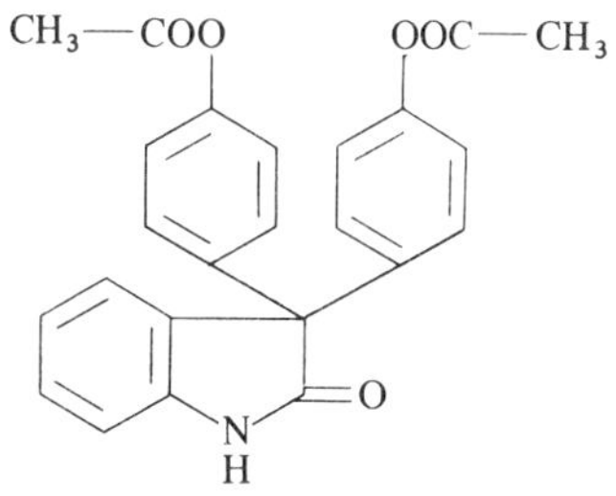

diphesatine

V. PIPERIDINE DERIVATIVES

After $^{14}CH_3$-labeled <u>mepivacaine</u> administration, a significant $^{14}CO_2$ excretion was recorded. Hansson et al. [401] applied silica gel thin-layer chromatography (chloroform : methanol 2 : 1; ethanol : acetone : benzene : ammonium hydroxide 5 : 40 : 50 : 5) to chloroform extracts of urine and bile. Several radioactive metabolites were found. One was comparable to synthetic mepivacaine hydroxylated in position 4 on the phenyl ring. The major metabolite was conjugated and β-glucuronidase hydrolyzed it.

<u>Pethidine</u> was also extensively metabolized. Way et al. [977] determined the excreted drug by the methyl orange method of Brodie [136] and checked its specificity by counter-current distribution (ethylene dichloride : 2M phosphate buffer, pH 5.45 — 8 transfers).

Plotnikoff et al. [712] prepared the N-$^{14}CH_3$-pethidine and analyzed the metabolites by counter-current distribution as above. Demethylated pethidine was identified in this way after N-demethylation had been indicated by $^{14}CO_2$ formation. On addition, 1-methyl-4-phenylisonipecotic acid was crystallized as the picrate and identified by its melting point.

Burns et al. [155] continued Plotnikoff's investigations by the same methods. Demethylated pethidine was isolated and identified by its melting point and IR spectrum compared to the authentic compound. In addition, 4-phenylisonipecotic acid was evidenced, as the authors showed that the amount of demethylated pethidine increased by esterification.

Plotnikoff et al. [713] attempted to account for all of the administered radioactivity by the same methods of counter-current distribution and methyl orange determinations. In addition, conjugates were hydrolyzed for 15

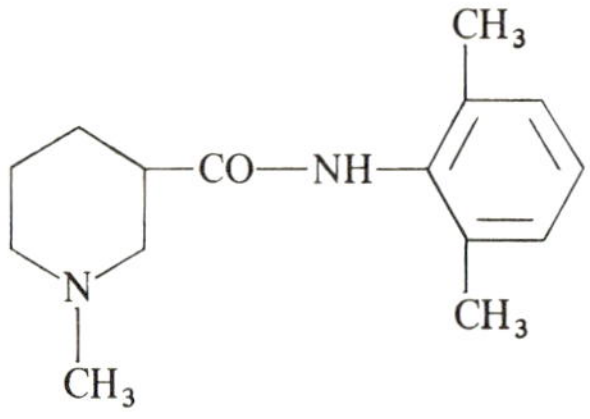

mepivacaine

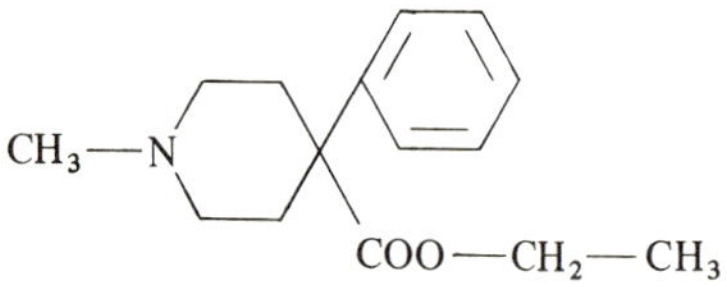

pethidine

minutes at 100° C by sulfuric acid. The comprised compounds determined represented 30 to 100% of the dose and varied with the patient.

Anileridine is structurally close to pethidine. Porter [715] studied its in vivo transformations. He applied the diazotization determination of Bratton and Marshall [108] combined with extractions and hydrolyses, to determine anileridine, acetylanileridine, the two corresponding acids, as well as a conjugate.

Lin and Way [553] investigated the same problem. Alkalinized rat urine was extracted with benzene and the concentrated extract was chromatographed on paper buffered at pH 6 (t-amyl alcohol : butyl ether : water 10 : 1 : 5) and on silica gel thin layer (benzene : dioxane : ethanol : ammonium hydroxide 50 : 40 : 5 : 5; benzene : dioxane : ethanol : acetic acid 50 : 40 : 5 : 5). A component with the same R_f value as demethylated pethidine was observed in addition to the unchanged drug and acetylanileridine. Demethylated pethidine was eluted and identified by its distribution curve (benzene : 0.2M phosphate buffer, pH 7.4 — 9 transfers), its UV and IR spectra, and by formation of the picrate. By extraction of acid urine with ethyl acetate and counter-current distribution or thin-layer chromatography, the presence of p-acetylaminophenylacetic acid was detected. It was identified by its R_f, its distribution curve, and by hydrolysis into p-aminophenylacetic acid.

Another derivative closely related to pethidine is ethoheptazine which is included here, although it is not a piperidine derivative. Walkenstein et al.

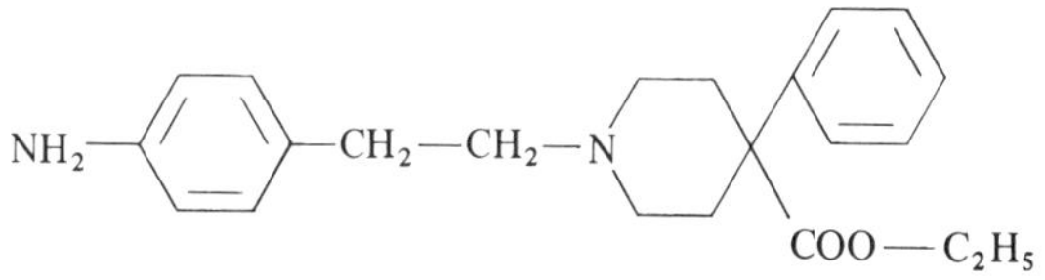

anileridine

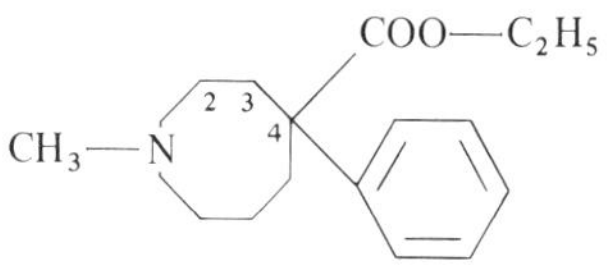

ethoheptazine

[968] studied the 4-^{14}C-labeled drug. Paper chromatography (water saturated butanol; 1M ammonium acetate saturated butanol pH 7.0; ascending) showed five radioactive metabolites in addition to unchanged drug. This was extracted from urine at pH 11 into toluene and characterized by its R_f, IR spectrum, and distribution curve (phosphate buffer, pH 6:toluene). One of the metabolites was a hydroxylated derivative which was isolated from toluene-treated urine by ethyl ether and chloroform at pH 5.5. Its elemental analysis and its IR spectrum indicated the presence of an OH group. A second metabolite was the deesterified drug, which was converted back to ethoheptazine by reacting sulfurated ethanol with urine that contained neither the unchanged drug nor the hydroxylated metabolite. A third excretion product was a hydrolyzed derivative that corresponded to hydroxyethoheptazine as shown by its IR spectrum.

The 3-methyl derivative of ethoheptazine, proheptazine, was also examined by Walkenstein et al. [966], who used the drug ^{14}C labeled on the carboxyl. Paper chromatography of urine was performed directly (methanol: ethyl acetate:1M ammonium acetate, pH 7, 26:2:2). The N-demethylated metabolite was detected when its hydrolysis was prevented by collecting the urine in a Dry Ice-cooled flask. Two products resulting from the ester function hydrolysis were also characterized by comparison with reference compounds.

In the rat, methyprylone underwent slight transformations. Bernhard et al. [80] prepared the 6-^{14}C-labeled drug and determined the unchanged compound by isotopic dilution. Several metabolites were detected in the dog [81] by paper chromatography [(I) butyl ether : isopropyl ether : methylcellosolve : 5% acetic acid 10:1:1:1]. These metabolites are marked (+) in Table 50.

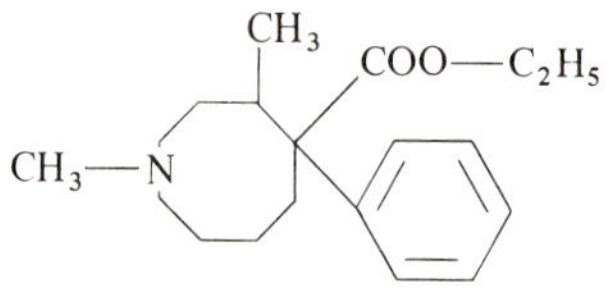

---

proheptazine

TABLE 50

Product	R_f in system: I	II
Methyprylone	0.50	0.90
2,4-Dioxo-3,3-diethyl-5-methyl tetrahydropyridine (+)	0.72	0.94
2,4-Dioxo-3,3-diethyl-5-hydroxymethyl tetrahydropyridine (+)	0.11	0.84
4,6-Dioxo-5,5-diethyl tetrahydronicotinaldehyde	0.63	—
4,6-Dioxo-5,5-diethyl tetrahydronicotinic acid (+)	0.34	0.30
2,4-Dioxo-3,3-diethyl tetrahydropyridine	0.54	—
2,4,6-Trioxo-3,3-diethyl-5-methyl-piperidine (++)	—	0.50

The authors did not succeed in purifying the glucuronides by the method of Kamil et al. [480] but they were able to characterize the aglycones after hydrolysis by paper chromatography. The chromatographic identifications were completed by elemental analyses and by UV and IR spectra.

Bösche and Schmidt [103] chromatographed the ethyl ether and chloroform extract of human urine on paper with system II: 5N ammonium hydroxide saturated butanol. They separated the three previously characterized metabolites and found a fourth component (Table 50). It was isolated on 12 paper sheets and identified by its elemental analysis and its UV, IR, and mass spectra as the 2,4,6-trioxo derivative (++).

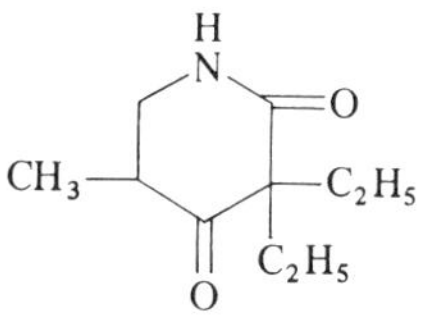

methyprylone

Piribenzil gives a single metabolite in man as well as in the rat. Beckmann [70] used a tritiated drug in the animal and the unlabeled drug in man. He separated the unchanged compound by paper chromatography [(I) pyridine : amyl alcohol : water 7 : 7 : 6; (II) butanol : dioxane : 2N ammonium hydroxide 4 : 1 : 5]; by thin-layer chromatography on cellulose powder [(III) butanol : acetic acid : water 5 : 1 : 4]; and by paper electrophoresis (barbital buffer, pH 8.6, 100 V, 16 hours) (Table 51).

TABLE 51

Product	R_f in system: I	II	III	Electrophoretic migration (cm)
Piribenzil	0.67	decomp.	0.15	9
N,N-Dimethyl-2-hydroxy-methylpiperidinium	0.43	0.31	0.67	15

VI. INDOLE DERIVATIVES

Harman et al. [404] found two metabolites of ^{14}C-labeled indomethacin. They were the N-deschlorobenzoyl derivative (A) and the O-desmethyl derivative (B), which were partly free and partly glucuronidated. These metabolites were separated by paper chromatography (I, II, III) or by silica gel thin-layer chromatography (IV, V) (Table 52).

The metabolites were identified by IR or UV spectrophotometry and by comparison with specially prepared authentic compounds [876]. It was

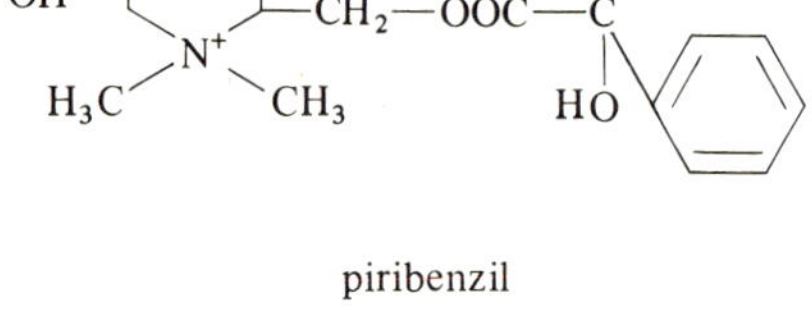

piribenzil

TABLE 52

System[a]	R_f for:					
	Indomethacin		Derivative A		Derivative B	
	Free	Glucuronide	Free	Glucuronide	Free	Glucuronide
I	0.75	—	0.40	—	0.66	—
II	0.95	0.75	0.88	0.55	0.92	0.69
III	0.95	0.00	0.95	0.00	0.95	0.00
IV	0.50	—	0.55	—	0.65	—
V	0.57	—	0.37	—	0.15	—

[a]Solvent systems: I, isopropyl alcohol : 15N ammonium hydroxide : water 8:1:1; II, methanol : water : butanol : benzene 2:1:1:1; III, acetic acid : isopropyl alcohol 5:95; IV, ethyl acetate : isopropyl alcohol : 10% ammonium hydroxide 5:4:3; V, acetic acid : chloroform 5:95.

impossible to purify indomethacin glucuronide. It was characterized by the Dische reaction, β-glucuronidase hydrolysis, and inhibition of this hydrolysis by saccharolactone. The glucuronide was prepared also by administering labeled indomethacin to rats. The rat urine was maintained at pH 7 by intraperitoneal injections of ammonium chloride, extracted by ethyl acetate at pH 2, and purified by counter-current distribution (0.5M phosphate buffer, pH 6.6: ethyl acetate + secondary butyl alcohol 328 + 72 − 60 transfers). This glucuronide was identified by the formation of its triacetylmethyl ester

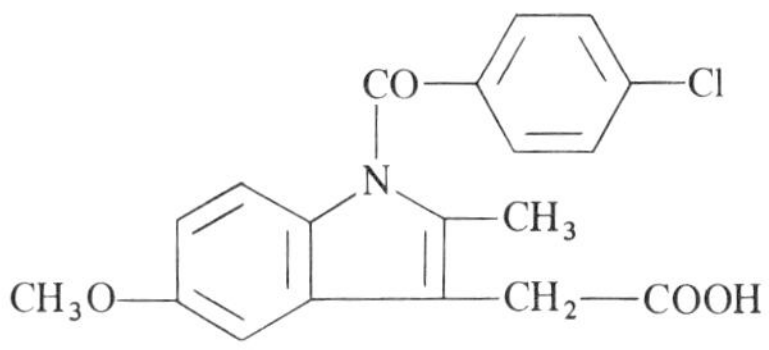

indomethacin

and compared with the synthetic product. The glucuronide of B, also isolated by the counter-current distribution, was purified by chromatography on a carbon:Supercel 1:1 column (water; methanol:water 1:1; phenol:water 7:93). Finally, the glucuronide of A was obtained when synthetic A was administered to ammonium chloride-treated rats and the urine was extracted at pH 2 with ethyl acetate.

The metabolites of 1-^{14}C- and 3-^{14}C- labeled chlortalidone were separated by chromatography on paper (benzene:ammonium hydroxide:methanol:ethanol 5:1:3:2) and on silica gel thin layer (methanol). Beisenherz et al. [73] characterized the metabolites by R_f comparison.

Benzoxazoles can be linked to indoles. Conney et al. investigated two closely related compounds. Zoxazolamine [218] was essentially metabolized by 6-hydroxylation and glucuronoconjugation. Urine, after β-glucuronidase hydrolysis (5 hours at 37°C, pH 5), was extracted by ethyl ether for 24 hours. The concentrated extract was chromatographed on a silicic acid column with ethyl ether. Isolated 6-hydroxyzoxazolamine was identified by its UV and IR spectra, its elemental analysis, and its partition coefficient and was compared to the synthetic product. This metabolite was detected by paper chromatography (chloroform:acetic acid:water 2:1:1) in in vitro experiments. Acidified urine was extracted with petroleum ether containing 1.5% of isoamyl alcohol and gave a minor metabolite that UV spectrum and partition coefficient identified as chlorzoxazone.

Chlorzoxazone itself was also studied [217]. Its major metabolite is 6-hydroxychlorzoxazone glucuronide, which was isolated and identified as above.

These hydroxylated metabolites were synthesized by Plampin and Cain [711].

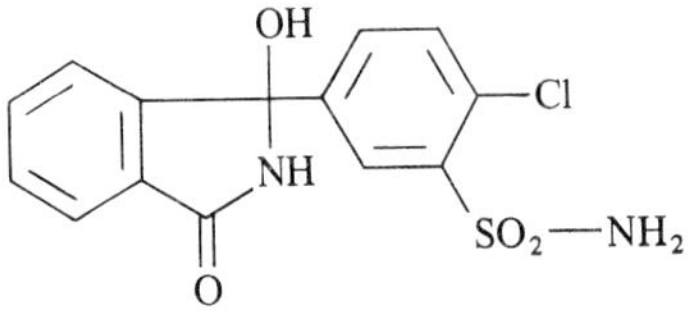

chlorthalidone

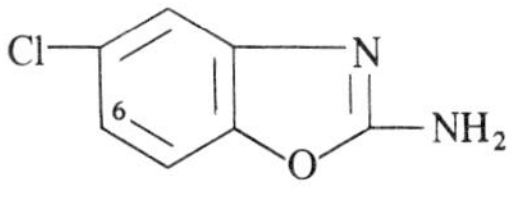

zoxazolamine

VII. QUINOLINE DERIVATIVES

Chloroquine and oxychloroquine were the subjects of a very lengthy paper by Titus et al. [907]. The chloroform extract of urine was fractionated by counter-current distribution (2M phosphate buffer, pH 6.36: chloroform — 25 transfers). Fifty percent of the administered dose of chloroquine was recovered unchanged and 25% as the corresponding N-monodesethyl derivative. This was purified by a second distribution, molecular distillation, and crystallization as the picrate. The same procedure led to similar results with oxychloroquine. In addition, the residual urine after chloroform extraction was treated with butanol, and the concentrated extract was purified by successive distributions (3% ammonium hydroxide : butanol — 16 transfers). The component thus isolated appeared to be oxychloroquine-N-oxide glucuronide. The corresponding aglycone was not identified, but it gave a mixture of picrates of which one was that of oxychloroquine.

McChesney et al. [575] confirmed these results in man by the same counter-current distribution procedure and extended them to the rat.

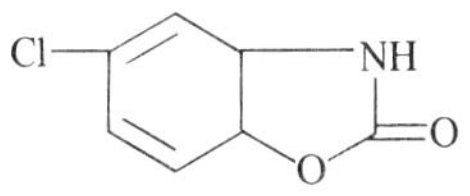

chlorzoxazone

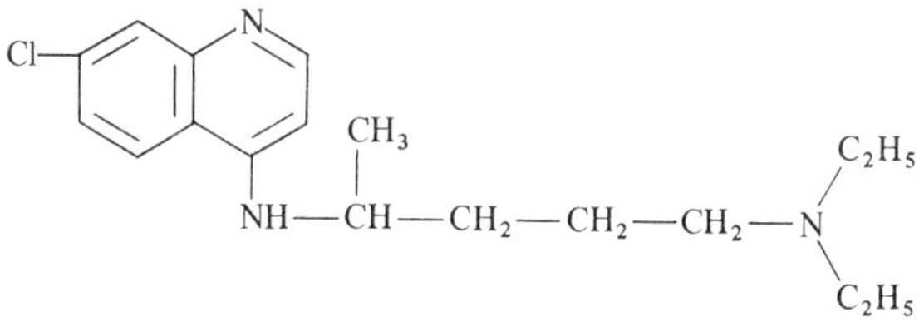

chloroquine

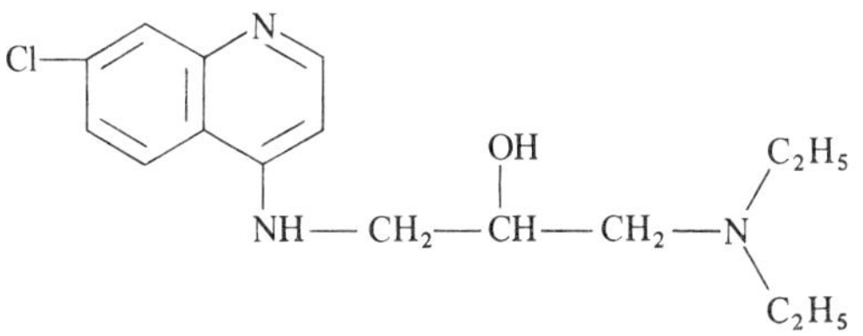

oxychloroquine

Kuroda [532] undertook the study of chloroquine metabolism by paper chromatography (butanol : acetic acid : water 10 : 1 : 5). Four metabolites were detected in urine and one was identified as 4-amino-7-chloroquinoline by R_f and UV spectrum comparison with the synthetic compound.

McChesney et al. [574] further extended their investigations to chloroquine, hydroxychloroquine and several of their metabolites. They extracted urine at pH 10-11 by heptane or ethylene dichloride containing 2 to 10% of isoamyl alcohol and fractionated the metabolites by chromatography on silica gel thin layer (20% ammonium hydroxide : isopropanol : ethyl acetate 6 : 15 : 79) (Table 53). The various fluorescent components were eluted into 0.1N hydrochloric acid in methanol and determined by fluorometry or spectrophotometry.

Glaphenine is excreted in the urine of man as the corresponding acid, 4-(O-carboxyphenylamino)-7-chloroquinoline. Mallein et al. [593] detected it by paper chromatography (butanol : 0.1N sodium hydroxide : water 90 : 5 : 5; isoamyl alcohol : ethanol : 0.1N sodium hydroxide 50 : 10 : 5, ascending).

Brodie and Udenfriend [137] isolated two metabolites of pamaquine that they were unable to identify. Josephson et al. [471] studied one of these

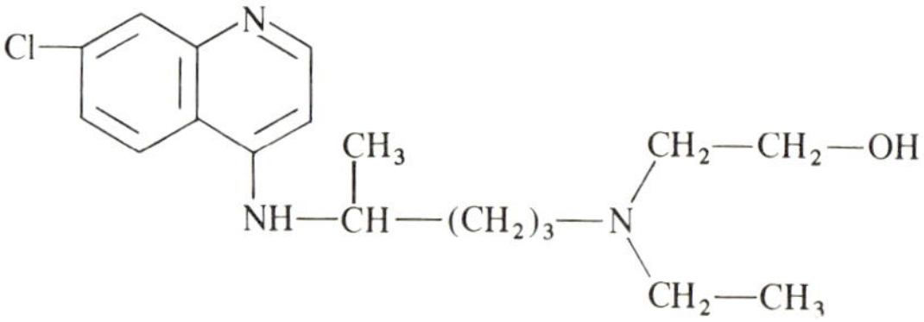

hydroxychloroquine

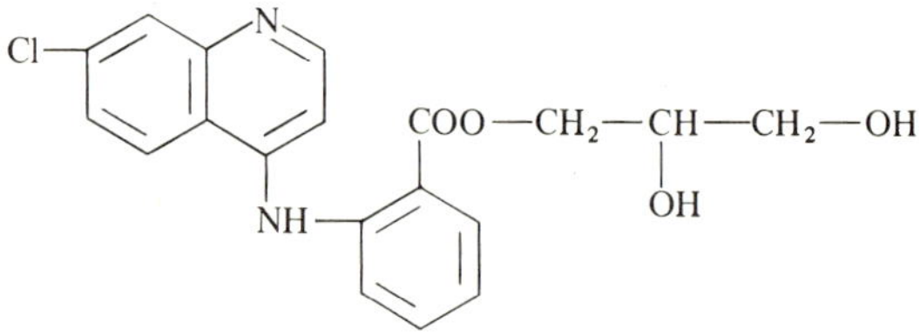

glafenine

TABLE 53

Product	R_f
Chloroquine	0.44
Hydroxychloroquine	0.31
Chloroquinemonodesethyl derivative	0.17
Chloroquinedidesethyl derivative	0.04
Hydroxychloroquinedesethyl derivative	0.11
4-Amino-7-chloroquinoline	0.75
4-(7-chloro-4-quinolylamino)-4-methyl-1-butanol	0.63

by UV spectrophotometry, then purified it [470] by chromatography on a cellulose column and identified it with the corresponding 5,6-quinolinequinone by comparison with the reference product.

Smith [851] examined the fate of pentaquine with two compounds ^{14}C-labeled on the methoxyl and on the terminal isopropyl. In addition to the measurement of the expired $^{14}CO_2$, the urinary metabolites were analyzed.

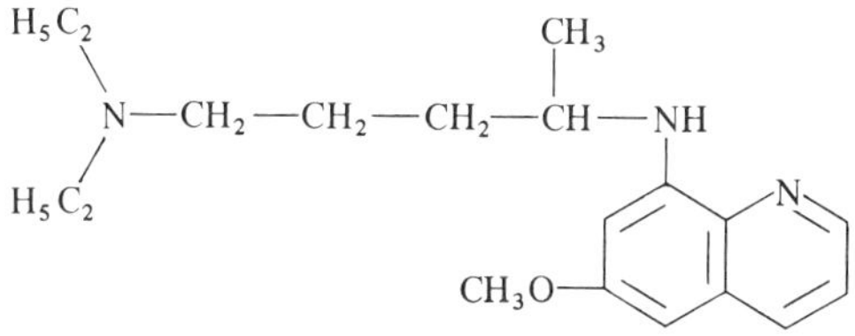

pamaquine

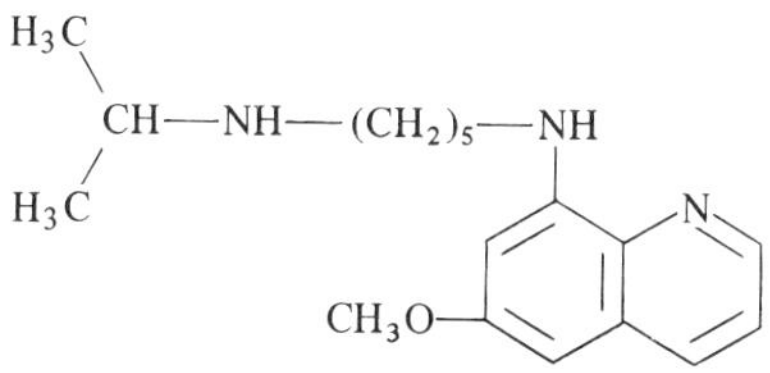

pentaquine

Urine at pH 7.5, to which ammonium sulfate had been added, was extracted with butanol and the extract chromatographed on a cellulose column. The elution was performed with butanol containing decreasing proportions of benzene (from 30% to 0%). Three radioactive metabolites were observed with the first compound and six with the second one. None were identified.

Clioquinol and diiodohydroxyquinoline evidently give a glucuronide and a sulfate in vivo. Haskins and Luttermoser [408] extracted them from urine into butanol. They precipitated these by cooling the glucuronides and recovered the sulfates by evaporating the solvent and redissolving in methylcellosolve.

Ritter and Jermann [769] compared clioquinol, 5-chloro-8-hydroxyquinoline, and 5-chloro-8-acetoxyquinoline. They determined the free quinolinols by ethyl ether extraction and colorimetry of their ferric complex. The sulfates were hydrolyzed in boiling hydrochloric acid for 30 seconds, which did not modify the glucuronides. These were hydrolyzed by β-glucuronidase.

Benzquinamide was studied by Wiseman et al. [1013] and by Koe and Pinson [511]. The former authors mainly studied distribution and excretion of the tritiated drug, whereas the latter isolated and characterized 11 metabolites in dog and human urine.

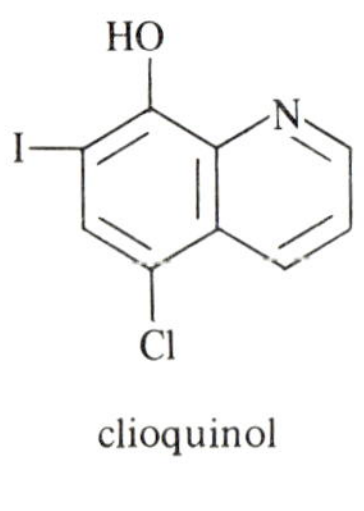

clioquinol

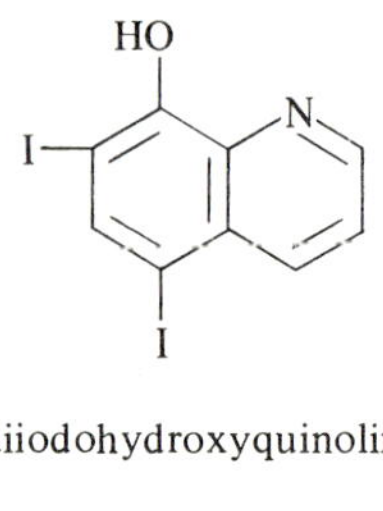

diiodohydroxyquinoline

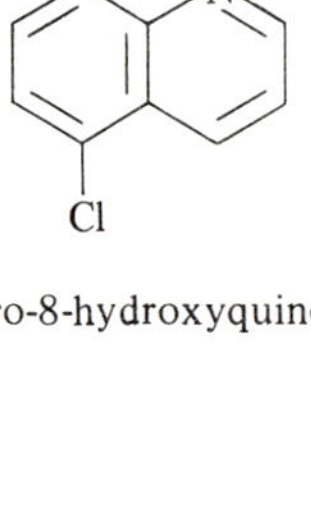

5-chloro-8-hydroxyquinoline

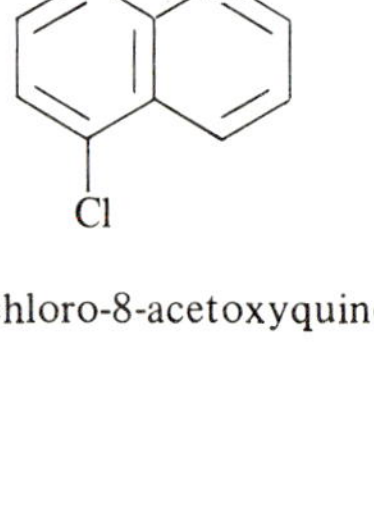

5-chloro-8-acetoxyquinoline

Crude urine was chromatographed on paper that had been previously treated with the mixture formamide : acetone 2 : 3 in the three solvent systems [(I) hexane : benzene : diethylamine 27 : 9 : 4; (II) benzene : diethylamine 9 : 1, formamide saturated; (III) benzene : chloroform : diethylamine 13 : 6 : 1, formamide saturated] and showed a series of fluorescent components. These components and intermediates in benzquinamide synthesis revealed that the 2-acetoxy derivatives exhibited yellow fluorescence and the 2-hydroxy derivatives blue fluorescence. Deacetylation by methanolysis and acetylation, followed by further chromatographies related the two types of compounds, the corresponding alcohols and acetates of various benzoquinolizines. Diazomethane methylation showed the presence of several pairs of metabolites that differed in the degree of O-methylation.

Chromatography on alumina column (benzene : chloroform, then chloroform : methanol of variable concentrations), followed by preparative paper chromatography, led to the isolation of six crystallized metabolites in addition to the unchanged drug. These products were identified by elemental analysis, UV and IR spectra, and R_f comparison with authentic compounds. Three O-demethylated metabolites were not isolated in the pure state but were characterized on the chromatograms by methylation assays. Three other fluorescent metabolites were detected. The conjugated metabolites were studied either after enzymic hydrolysis or after separation by paper chromatography (butanol : acetic acid : water 5 : 1 : 4).

This very complete work is a remarkable application of paper chromatography combined with chemical reactions. It was facilitated by the fluorescence of the studied compounds.

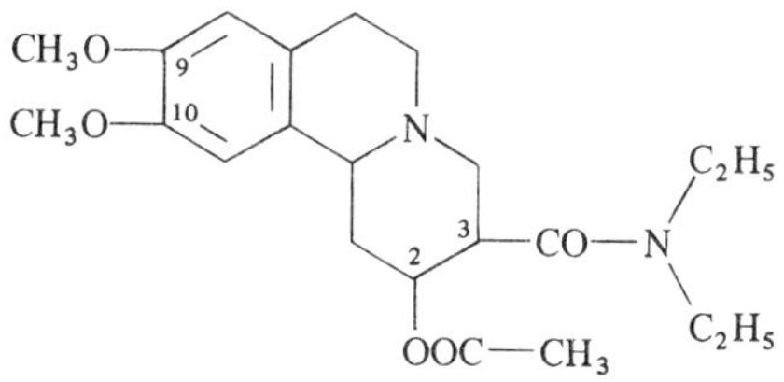

benzquinamide

Tetrabenazine was roughly examined by Quinn et al. [740]. They applied the counter-current distribution method of Brodie and Udenfriend [136] to test their extraction procedure of the unchanged drug before its fluorometric determination.

Schwartz et al. [815] almost completely elucidated tetrabenazine metabolism. Five nonconjugated metabolites were revealed by benzene extraction of the concentrated urine, followed by silica gel thin-layer chromatography in the four solvent systems mentioned in Table 54. These were identified either by R_f comparison with synthetic compounds or after isolation by silica gel column chromatography. The fractionation was achieved by a series of solvents of increasing polarity (isopropyl ether : chloroform 2 : 1, then 1 : 2, chloroform, chloroform : butanol 98 : 2, 90 : 10, and 50 : 50). The metabolites thus isolated were characterized by their elemental analysis and their high resolution mass spectra.

The conjugated metabolites were precipitated and purified by the method of Kamil et al. [480], then hydrolyzed by Glusulase (β-glucuronidase + sulfatase) for 24 hours at 37°C, pH 4.5. The aglycones were analyzed as above and two of them were identified.

Metofoline is an isoquinoline whose biotransformations were elucidated by Schwartz et al. [814]. The metabolites were separated, after acid or enzymic hydrolysis of urine, by silica gel thin-layer chromatography (t-butanol : butyl ether : 2% ammonium hydroxide 70 : 20 : 7.5; chloroform : butanol : 2% ammonium hydroxide 70 : 30 : 0.6; water saturated cyclohexanol; butanol : formic acid : water 84 : 8 : 8) and by electrophoresis on silica gel thin layer. The glucuronides were isolated by the method of Kamil et al. [480].

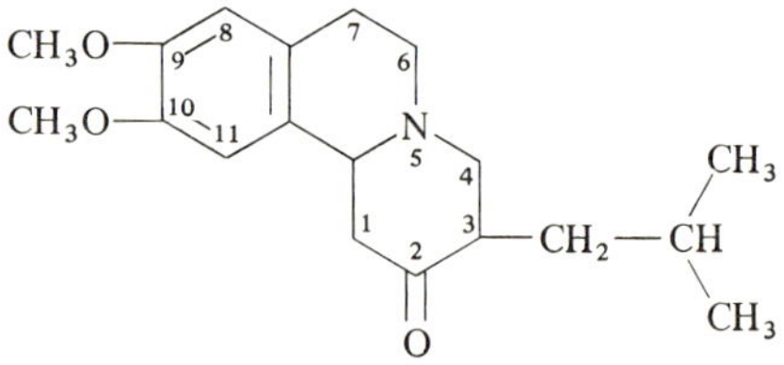

tetrabenazine

TABLE 54

Product	Mobility as referred to tetrabenazine in system:[a]			
	I	II	III	IV
Tetrabenazine	1.00	1.00	1.00	1.00
2-Hydroxylated metabolite (α-isomer)	0.83	0.69	0.77	0.38
2-Hydroxylated metabolite (β-isomer)	0.69	0.33	0.85	0.30
2'-Hydroxylated metabolite	0.53	0.44	0.57	0.23
2(α),2'-Dihydroxylated metabolite	0.41	0.20	0.38	0.12
2(β),2'-Dihydroxylated metabolite	0.20	0.10	0.32	0.06
Aglycones:				
9-Hydroxylated metabolite	1.00	0.97	0.85	1.20
2(α),9-Dihydroxylated metabolite	0.83	0.34	0.55	0.54

[a] Solvent systems: I, t-amyl alcohol : butyl ether : 0.25% ammonium hydroxide 80 : 7 : 13; II, chloroform : butanol : 2.5% ammonium hydroxide 80 : 20 : 0.6; III, toluene : acetone : 25% ammonium hydroxide 50 : 50 : 1; IV, water saturated cyclohexanol.

The metabolites that resulted from N-demethylation and O-demethylation were identified by comparison with the synthetic compounds. The unchanged drug and some metabolites were determined by fluorimetry, after separation according to their solubilities.

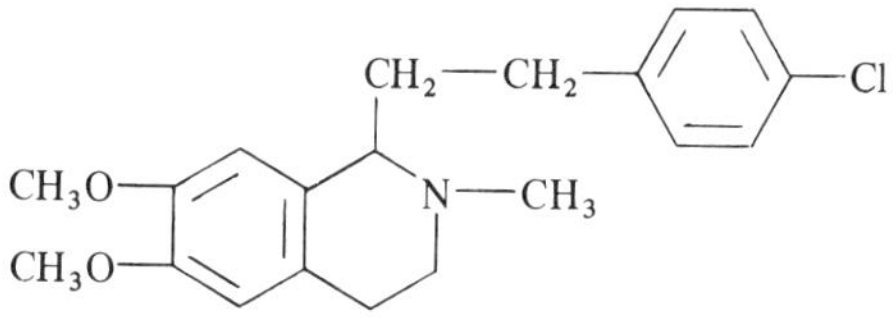

metofoline

The major metabolite of quindonium is its glucuronide. Dubnick et al. [281] isolated it by chromatographing urine at pH 5 on IRC-50 (acetate) followed by Dowex-1 X10 (acetate). The product was purified by alternate passages through these two columns and showed a molar ratio quindonium/glucuronic acid close to one. Quindonium was also characterized by thin-layer chromatography after β-glucuronidase hydrolysis. After simultaneous administration of quindonium and ^{14}C-labeled glucose, counter-current distribution separated a radioactive fraction that did not contain free quindonium but that liberated quindonium and glucuronic acid on hydrolysis.

Tacrine, according to Kaul [490], gave several metabolites which can be separated by paper chromatography of the chloroform extract of urine (butanol:acetic acid:water 4:1:5).

2-Quinolyl-1-piperazine is unstable in chloroform solution when it is extracted from biological media. This problem was studied by Leeling et al. [548] who emphasized the difficulties due to the use of chlorinated hydrocarbons in the extraction of secondary amines.

Cinchophene gave a series of phenolic metabolites. One of these was isolated in the crystalline state as early as 1912 by Dohrn [264] who established its structure by elemental analysis. Sipal and Jindra [847] identified a series of free hydroxylated derivatives, which were extracted by ethyl ether at

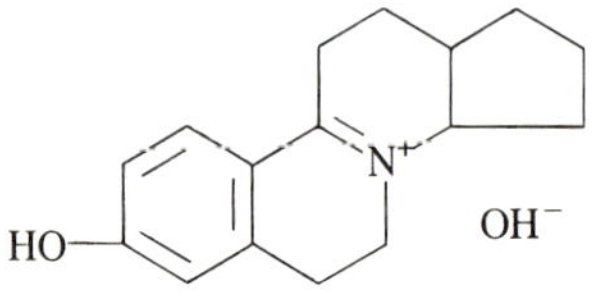

quindonium

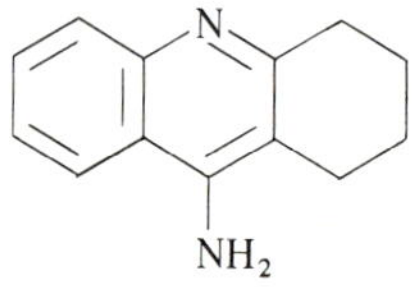

tacrine

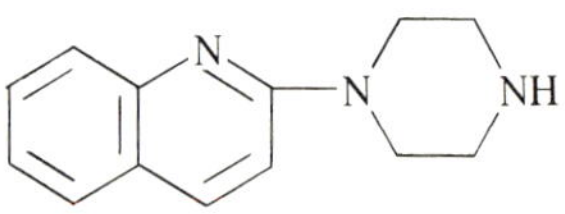

2-quinolyl-1-piperazine

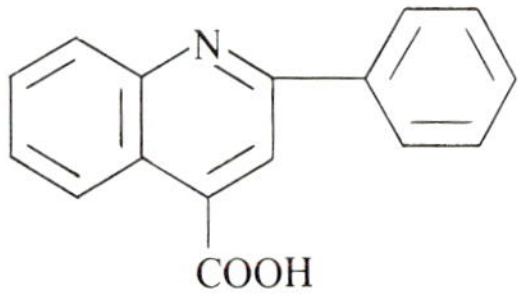

cinchophene

pH 3-4. They were redissolved into water, adsorbed on the ion exchanger Wofatit KPS 200 (H^+), and eluted by 5N ammonium hydroxide. They were then also chromatographed on paper with three solvent systems (Table 55):

(I) Methylethylketone : 5N ammonium hydroxide 1 : 1
(II) Butanol : 5N ammonium hydroxide 1 : 1, ascending
(III) Benzene : propanol : 5N ammonium hydroxide 2 : 5 : 2, ascending

TABLE 55

Product	R_f in system:		
	I	II	III
Cinchophene	0.25	0.64	0.66
6-Hydroxycinchophene	0.07	0.31	0.35
8-Hydroxycinchophene	0.31	0.52	0.58
4'-Hydroxycinchophene	0.06	0.25	0.29
4',6-Dihydroxycinchophene	< 0.02	0.05	0.06
4',8-Dihydroxycinchophene	< 0.02	0.18	0.19

Chapter XIV

HETEROCYCLES WITH TWO NITROGENS

I. PYRAZOLE DERIVATIVES

The hypoglycemic activity of 3,5-dimethylpyrazole appeared more slowly after intravenous injection than after oral administration. This drug was inactive in animals that were treated with a liver enzyme inhibitor. This evidence implies that the activity is due to a metabolite produced in the liver. Smith et al. [852], using gas-liquid and thin-layer chromatography, showed that rat urine did not contain the unchanged drug. The hypoglycemically active metabolite was extracted with ethyl acetate at pH 1, purified on a column of Dowex-1 (H$^+$), and eluted by 0.2N hydrochloric acid. After ethyl acetate extraction, it was precipitated by benzene, then submitted to a sublimation. Its potentiometric microtitration revealed a single acid function. Infrared spectrum and elemental analysis led to the formula of 5-methylpyrazole-3-carboxylic acid which was confirmed by synthesis. This metabolite was 200 times more active than tolbutamide, and was responsible for the total urinary hypoglycemic activity, although it only represented 15% of the oral dose.

The same authors [853] studied the fate of the other 85% with the drug ^{14}C-labeled in the 3 or 5 position. Urine and its ethyl acetate extract were

---

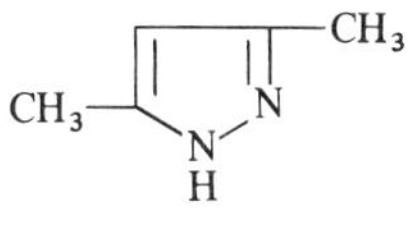

3,5-dimethylpyrazole

chromatographed on paper (butanol : acetic acid : water 2 : 1 : 1; butanol : piperidine : water 81 : 2 : 17; potassium chloride aqueous solution 20 g/100 ml) and on a thin layer of fluorescent silica gel (ethyl acetate : water : formic acid 95 : 5 : 1; ethyl acetate : dimethylformamide : water : formic acid 85 : 15 : 5 : 2). No unchanged drug was found in urine, but four radioactive derivatives were. One of these was 5-methylpyrazole-3-carboxylic acid. The second and the third metabolites were conjugated and had to be hydrolyzed and the former gave 5-methylpyrazole-3-carboxylic acid. The latter was extracted and purified by chromatography on a silica gel column (chloroform : methanol in variable proportions). The resulting product, studied by UV, IR, and NMR spectrometry and by elemental analysis, was consistent with 3,5-dimethyl-4-hydroxypyrazole. The fourth metabolite represented only 1% of the administered dose and it was not possible to isolate it.

The biological transformations of phenazone (antipyrine) and aminophenazone (amidopyrine) had been studied since the beginning of this century. Jaffe [452] isolated and described some aminophenazone excretion products in 1901. They were a glucuronide and rubazonic acid with the latter a result of air oxidation of one of the metabolites. Later Jaffe [453] reported the presence of 4-ureidoaminophenazone. Lawrow found the glucuronide of an hydroxylated derivative of phenazone [541] and aminophenazone [540] in the urine. All these investigations lacked definitive analytical evidence.

Halberkann and Fretwurst [391, 392], in 1940, considered aminophenazone metabolism as a demethylation followed by an acetylation. They synthesized 4-amino- and 4-acetylamino-1-phenyl-2,3-dimethyl-5-pyrazolone and compared them to the urinary metabolites.

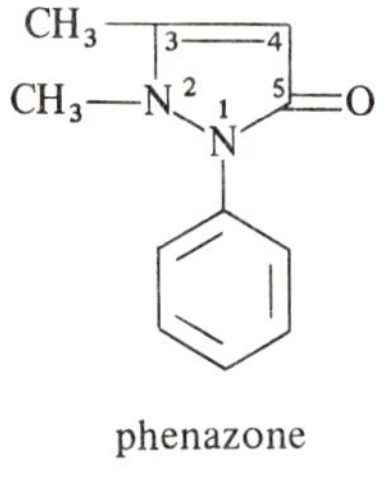

phenazone

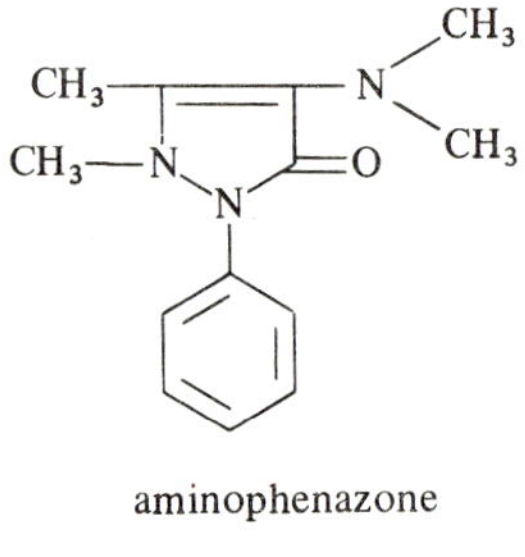

aminophenazone

Brodie and Axelrod [128], in 1950, isolated the 4-hydroxylated metabolite of phenazone from acid hydrolyzed urine by counter-current distribution (30% petroleum ether in chloroform : 1M phosphate buffer, pH 9.3 — 8 transfers). Its melting point and UV spectrum corresponded to those of the authentic compound. It is likely that this metabolite was excreted as the glucuronide because the urinary levels of glucuronic acid determined according to the method of Maughan et al. [614] were increased on administration of the drug. Brodie and Axelrod [129] extracted from the urine of an aminophenazone-dosed man two chloroform-soluble metabolites, one of which was removed by washing with 1N hydrochloric acid. They were identified by comparing their partition coefficients in the system chloroform : water at various pH values and were 4-amino- and 4-acetylamino-1-phenyl-2,3-dimethyl-5-pyrazolone. These two metabolites represented 50% of the administered dose.

Hennig and Weiler [414] proposed an analytical procedure for three aminophenazone metabolites. Rubazonic acid was retained on an alumina column, then eluted by ammoniated ethanol at pH 8 and determined by colorimetry. 4-Amino-1-phenyl-2,3-dimethyl-5-pyrazolone was directly determined from urine percolated on alumina by colorimetry with phenol and potassium ferricyanide. The corresponding acetylamino derivative was titrated in the same way after sulfuric acid hydrolysis.

Tanaka [893] found five phenazone metabolites by paper chromatography. He also synthesized five possible metabolites of aminopropylone [894] and was able to show that this drug was excreted unchanged.

Pechtold [703] examined aminophenazone urinary metabolites after studying its in vitro oxidation products. Acidified sodium chloride saturated urine was extracted with methylene chloride and the extract chromatographed on silica gel (chloroform : ethyl ether : methanol 7 : 3 : 1; chloroform containing

---

aminopropylone

1% of ethanol). The separated components were identified by comparison with synthetic derivatives. It was thus found that the major metabolite in man and the dog was the mono-N-demethylated derivative, which accompanied the unchanged drug. In addition the previously described methylrubazonic acid in the dog and the two 4-amino and 4-acetylamino metabolites in man were characterized.

At the same time, Preuss and Voigt [731] showed that some of the aminophenazone excretion products that had not yet been identified were comprised of methylrubazonic acid and 1-phenyl-3-methyl-4,5-pyrazolinedione. They applied chromatography on paper treated by a 30% formamide solution in methanol (chloroform : toluene : water 10 : 45 : 15). Methylrubazonic acid was also isolated by partition between solvents and by column and thin-layer chromatography (conditions not reported). It was identified after crystallization by its melting point and IR spectrum.

Schüppel and Soehring [812] extracted aminophenazone excretion products from urine at pH 4 into chloroform, after elimination of rubazonic acids by ethyl ether at pH 1. They separated the metabolites by bidimensional thin-layer chromatography on silica gel with the solvent systems:

Solvents:
  (I)   Chloroform : ethanol : acetone 148.5 : 1.5 : 50
  (II)   Chloroform : ethanol 9 : 1
  (III) Butanol : acetic acid : water 4 : 1 : 5

Systems:
  I + III, II + III, I + II

In this way they characterized by comparison four excretion products: the 4-amino and 4-acetylamino derivatives, rubazonic acid, and the unchanged drug. Two other metabolites were identified by their chromatographic behavior and their UV spectrum as methylrubazonic acid and the aminophenazone monodemethylated derivative (Table 56).

More recently Schüppel [811] briefly reported that he had characterized 1-phenyl-3-methyl-5-pyrazolone as a metabolite of phenazone. The metabolite was excreted as a N-glucuronide precipitable by lead acetate. This metabolite was identified by spectrophotometry, thin-layer chromatography, and chemical reactions.

Richarz et al. [760] studied the N-isopropyl homolog of aminophenazone: isopyrine. They generally applied the methods which had been previously

TABLE 56

Product	R_f in system:		
	I	II	III
Aminophenazone	0.22	0.66	0.19
4-Amino-1-phenyl-2,3-dimethyl-5-pyrazolone	0.10	0.36	0.39
4-Acetylamino-1-phenyl-2,3-dimethyl-5-pyrazolone	0.04	0.29	0.35
4-Methylamino-1-phenyl-2,3-dimethyl-5-pyrazolone	0.17	0.49	0.30
Rubazonic acid	0.78	0.87	0.78
Methylrubazonic acid	0.61	0.80	0.65

used for determining aminophenazone metabolites. However, only 3% of the administered dose was recovered in rat urine, as compared with 30% for aminophenazone.

Nifenazone is mostly excreted unchanged, with a small amount of 4-aminophenazone. Hoffmann and Roller [425] separated these two compounds by paper chromatography (butanol : isoamyl alcohol : acetic acid : water : ethanol 18 : 54 : 12 : 15 : 4) or on silica gel thin layer (water).

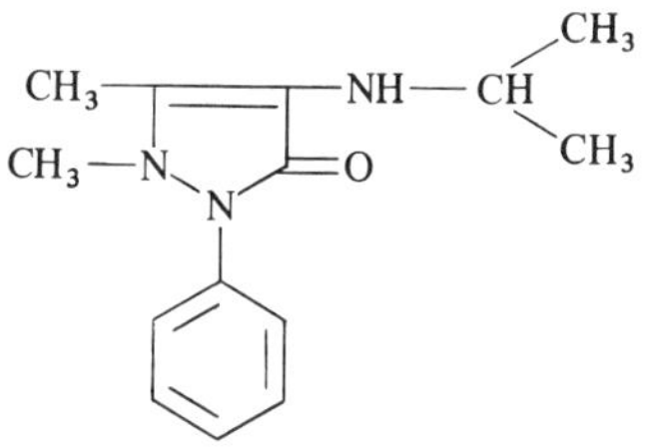

isopyrine

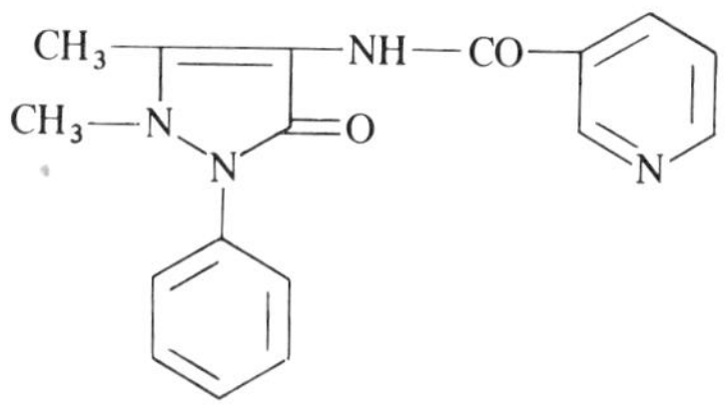

nifenazone

^{131}I-Labeled 4-iodophenazone is used to measure blood flow. Its metabolism was studied by Straub et al. [877] who only reported the presence of several radioactive metabolites on paper chromatograms (butanol : acetic acid : water 20 : 3 : 7.5).

Anghileri et al. [27] showed that 4-iodophenazone was rapidly converted into phenazone and iodide by paper chromatography (chloroform : ethanol : water 45 : 45 : 10).

Burns et al. [156] observed that phenylbutazone was almost completely metabolized. They determined the drug by spectrophotometry after an extraction the specificity of which had been checked by counter-current distribution. They further [157] isolated two metabolites in human urine by counter-current distribution (ethylene dichloride : 0.3M phosphate buffer, pH 7.2 — 8 transfers). These metabolites were identified by their elemental analysis and UV spectra. One was the monohydroxylated derivative in para of one of the two phenyl rings and the other compound had a β-hydroxyl group on the butyl chain. These two metabolites were synthesized and compared to the urinary excretion products by their melting point and UV and IR spectra.

In the course of their investigations on an aminophenazone homolog used in association with phenylbutazone, Richarz et al. [760] roughly examined phenylbutazone metabolism and they applied the procedures of Burns [156].

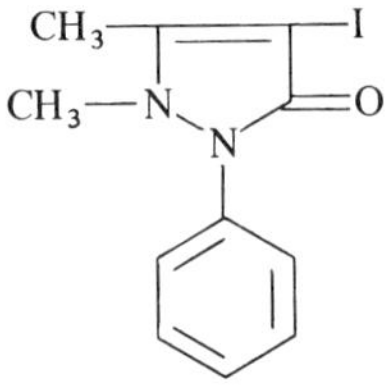

4-iodophenazone

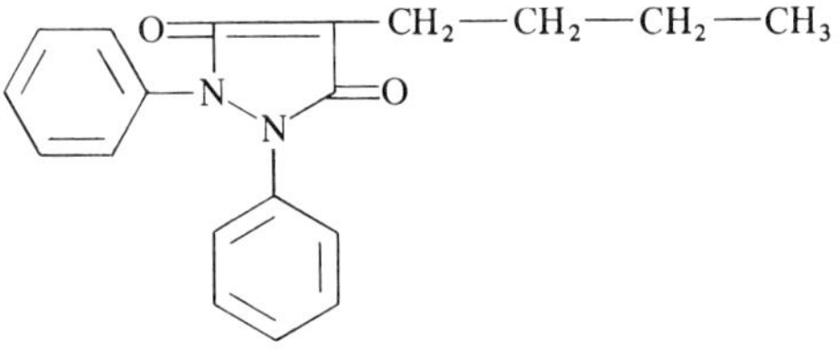

phenylbutazone

Burns et al. also studied the phenylbutazone analog, 4-phenylthioethyl-1,2-diphenyl-3,5-pyrazolidinedione [160]. They isolated its metabolite from urine by extraction and counter-current distribution as above and found it to be identical to the synthetically prepared corresponding sulfoxide by UV and IR spectrophotometry. This metabolite was sulfinpyrazone; it is responsible for the activity of the drug and has been introduced into therapeutics. The metabolism of sulfinpyrazone was investigated by Dayton et al. [245] using the same analytical methods. The metabolite was isolated by successive counter-current distributions and was found similar to synthetic hydroxysulfinpyrazone by UV and IR spectrophotometry.

Siblikova et al. [843] detected two metabolites of ketophenylbutazone by paper chromatography (formamide : benzene 25 : 75).

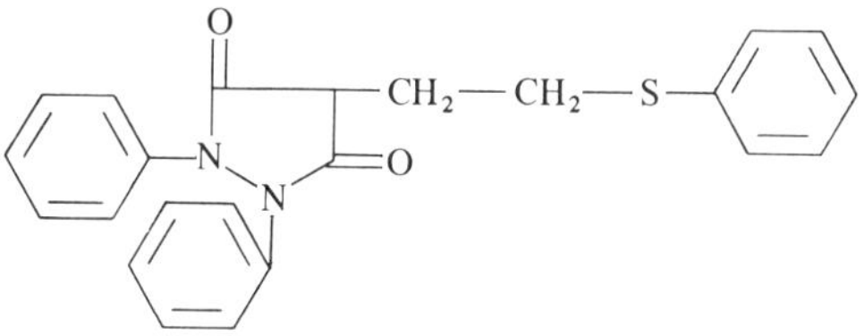

4-phenylthioethyl-1,2-diphenyl-
3,5-pyrazolidinedione

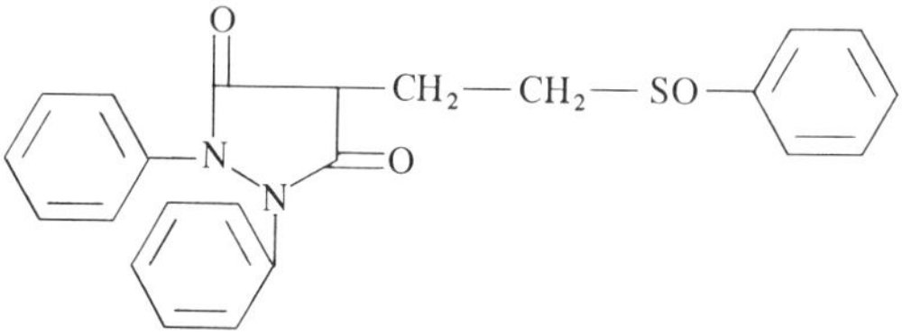

sulfinpyrazone

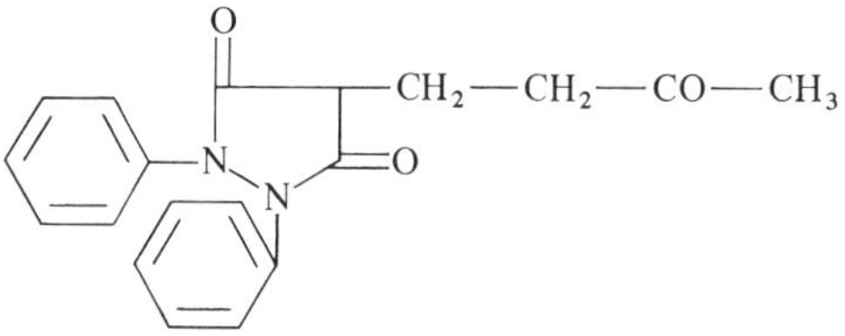

ketophenylbutazone

Nemecek et al. [668] extracted urine at pH 5.5, then at pH 2 by ethylene dichloride, to isolate ketophenylbutazone metabolites. Paper chromatography revealed two derivatives and the chromatogram was not modified by hydrolysis. The two metabolites were purified on an alumina column and the unchanged drug was eluted by chloroform. The first metabolite was eluted by methanol and the second by the mixture methanol : 25% ammonium hydroxide 5 : 1. Both excretion products were identified by their chromatographic behavior on paper or silica gel thin layer (Table 57).

Finger and Zicha [320] separated several metabolites of oxyphenbutazone on silica gel thin layer (chloroform : acetone : acetic acid 7 : 2 : 1).

II. OTHER DERIVATIVES

Antiseptic drugs with the nitrofuran group are easily determined by microbiological methods and it has been observed that they are mostly excreted unchanged. Buzard et al. [187] investigated nitrofurantoin in the rat and determined it by colorimetry as nitrofuraldehyde. Jones et al. [469] applied polarography and Meythaler et al. [631] thin-layer chromatography on silica gel (isopropanol : ethyl acetate : ammonium hydroxide : water 45 : 30 : 17 : 8).

Ings et al. [445] administered metronidazole to the dog and its urine was chromatographed on paper after concentration (propanol : ammonium hydroxide

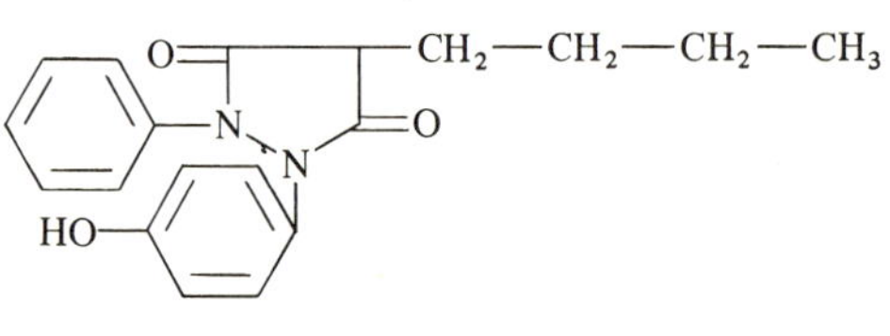

oxyphenbutazone

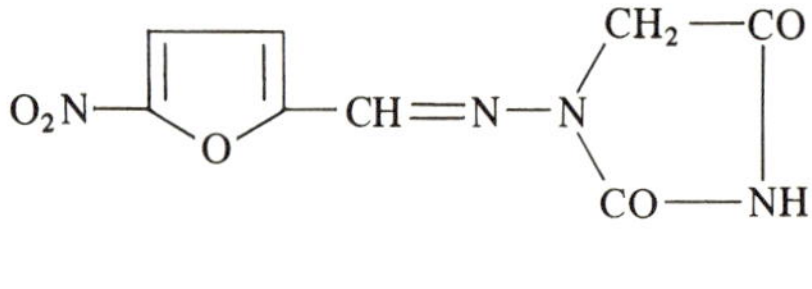

nitrofurantoin

TABLE 57

	R_f		
Product	On paper[a]		On thin layer[b]
	I	II	III
Ketophenylbutazone	0.35	0.85	0.19
Phenylbutazone	0.93	0.95	0.70
1-(p-Hydroxyphenyl)-2-phenyl-3,5-dioxo-4-(3-hydroxybutyl)-pyrazolidine	0.00	0.00	0.06
1-(p-Hydroxyphenyl)-2-phenyl-3,5-dioxo-4-(3-oxobutyl)-pyrazolidine	0.00	0.11	0.09

[a]Paper treated by a 40% formamide ethanolic solution containing 5% of ammonium formate: solvent I, benzene; solvent II, chloroform.

[b]Thin layer: solvent III, chloroform : ethanol 10 : 1.

7 : 3). Three metabolites appeared. They were eluted and characterized by chromatographic comparison with reference compounds. They were also isolated on Amberlite IR-45 (OH$^-$) from which they were eluted by 0.1N hydrochloric acid and water. They were further fractionated by chromatography on a dry column of microcrystalline cellulose (Avicel) with the system pyridine : butanol : water 1 : 1 : 1. Each fraction was eluted after extrusion and sectioning of the column. Two of the recrystallized metabolites

---

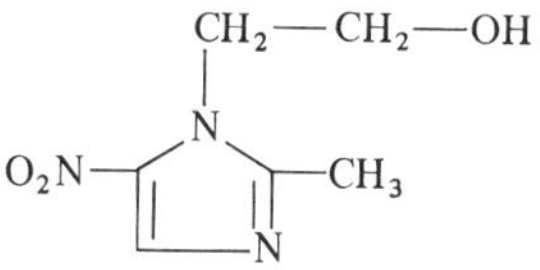

metronidazole

were identified by their melting point with the unchanged drug and 2-methyl-5-nitro-1-imidazolylacetic acid; the third one was most likely the glucuronide of this acid.

Brodie et al. [123] studied the fate of tolazoline. The unchanged drug was extracted from alkaline urine by chloroform containing 1.5% of isoamyl alcohol and identified by counter-current distribution (chloroform with 1.5% of isoamyl alcohol : 0.05M borate buffer, pH 8.6 — 25 transfers). It was determined by the methyl orange method. Almost all of the drug was excreted unchanged in man.

Silvestrini et al. [846] determined unchanged oxolamine by UV spectrophotometry either after selective ethyl ether extraction or after paper chromatography (butanol : acetic acid : water 65 : 15 : 25). They detected diethylamine, which was titrated after distillation, and a neutral compound the UV spectrum of which corresponded to 5-phenyl-1,2,4-oxadiazole.

Meclozine loses its 3-methylbenzyl group and is converted into norcyclizine. Narrod et al. [662] extracted this metabolite into ethylene dichloride

---

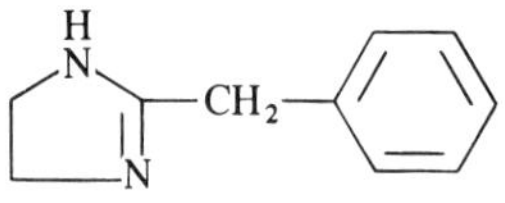

tolazoline

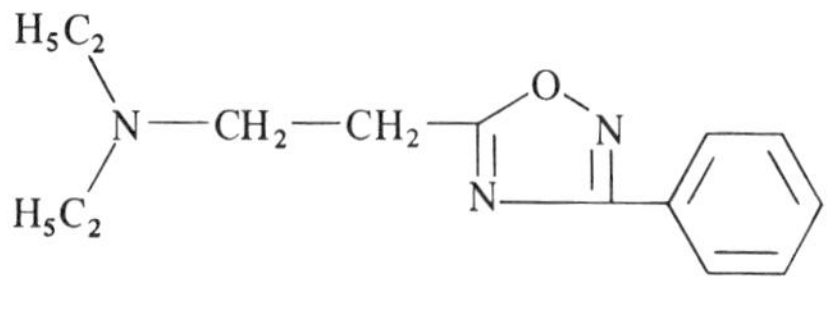

oxolamine

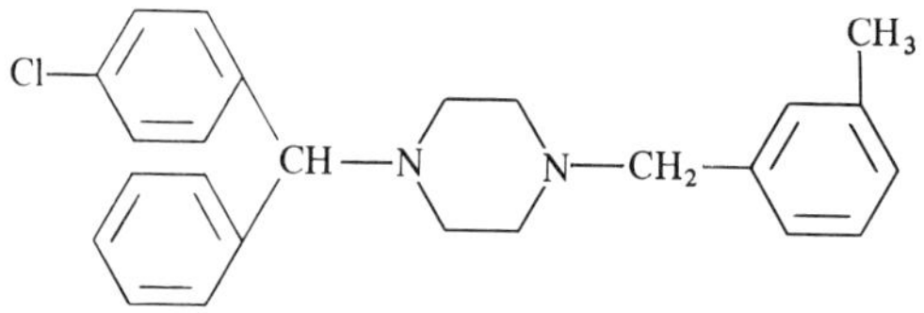

meclozine

at pH 10 and identified it by its UV spectrum and its R_f on paper buffered at pH 3 (isoamyl alcohol : 5% citric acid 9 : 1; benzene : ethylene dichloride 1 : 1) or on unbuffered paper (benzene : methanol : 10% acetic acid 50 : 25 : 25).

Similarly, chlorcyclizine and cyclizine are demethylated in vivo: Kuntzman et al. [531] demonstrated this fact by thin-layer chromatography on alumina (toluene : ethanol : water : acetic acid 20 : 30 : 10 : 1) and on silica gel (benzene : ethanol : water : acetic acid 20 : 20 : 10 : 1). They determined the unchanged drug and its metabolite by the methyl orange method after separation between solvents.

Idoxuridine metabolism was studied in 1960 by Prusoff et al. [738]. It hydrolyzed into iodouracil and was dehalogenated into uracil and iodide. The authors administered the ^{131}I- or ^{3}H-labeled drug and investigated the metabolites in urine and tissues extracted by cold 0.5N perchloric acid. These metabolites were separated on a Dowex-1 (formate) column with a formic acid solution of increasing concentration (0.1N → 6N).

Subsequently, Fox and Prusoff [335] used thin-layer chromatography on silica gel (isopropanol : chloroform 2 : 8) to analyze these metabolites.

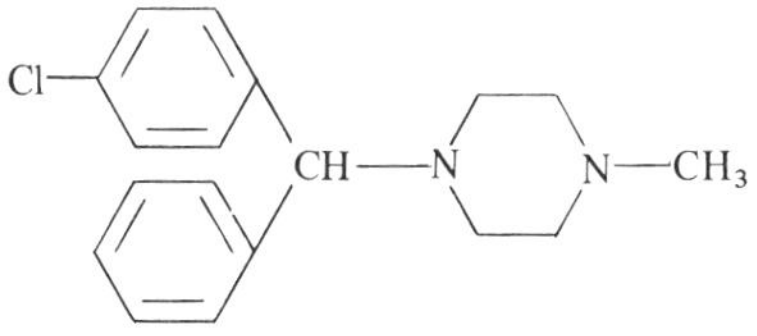

chlorcyclizine

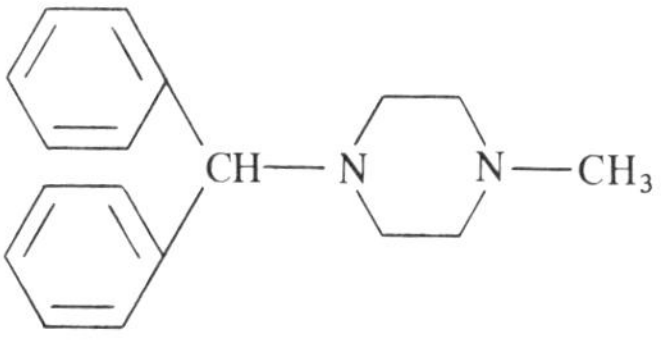

cyclizine

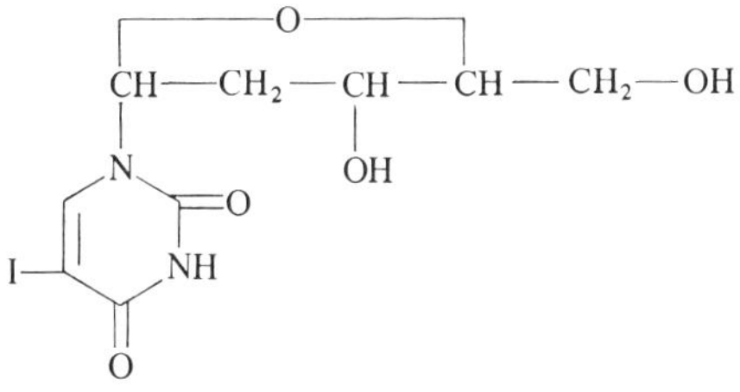

idoxuridine

5-Fluorocytosine, a fungistatic agent, was studied by Koechlin et al. [514]. After administration of the 2-^{14}C-labeled drug the unchanged compound and its metabolites were separated by bidimensional thin-layer chromatography on a support (not mentioned) [(I) ethyl acetate : formic acid : water 60 : 5 : 35; (II) isopropanol : ammonium hydroxide 20 : 1; (III) methanol : water 85 : 15 — I + II or III + II].

Tiabendazole metabolism was studied by Tocco et al. [910] who used material ^{14}C-labeled on the benzene ring and ^{35}S-labeled. Sheep urine was chromatographed on paper (butanol : acetic acid : water 4 : 1 : 1). In addition to the unchanged drug, five derivatives appeared. 5-Hydroxytiabendazole, its sulfate, and its glucuronide represented the major metabolites and were recognized after the synthesis of the reference compounds (Table 58).

5-Hydroxytiabendazole was extracted from urine hydrolyzed by Glusulase (β-glucuronidase + sulfatase) into methylene chloride, subsequent to the elimination of the unconjugated compounds by treating crude urine with ethyl acetate and methylene chloride. The metabolite was purified by paper chromatography, crystallized, and identified by its UV and IR spectra. The corresponding glucuronide was separated on a column of cation-exchange

TABLE 58

Product	R_f
Tiabendazole	0.73
5-Hydroxytiabendazole	0.59
5-Hydroxytiabendazole sulfate	0.27
5-Hydroxytiabendazole glucuronide	0.12

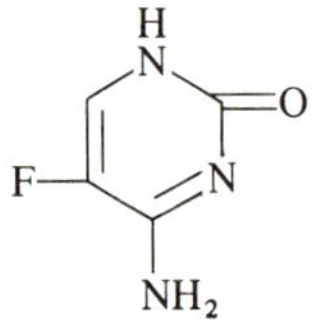

5-fluorocytosine

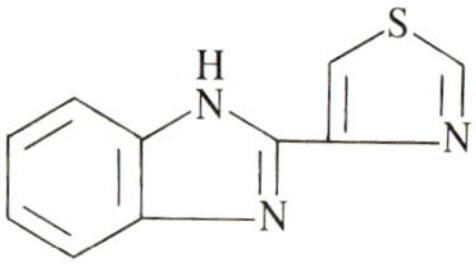

tiabendazole

resin CG-120 (NH_4^+) and eluted at increasing pH. The radioactive fraction was analyzed by UV, IR, and NMR spectrometry. Enzymic hydrolysis to 5-hydroxytiabendazole confirmed that glucuronic acid was attached in the 5 position and not the alternative 6 position. The sulfate was isolated by preparative paper electrophoresis. R_f values allowed its characterization after hydrolysis to SO_4^{2-} and 5-hydroxytiabendazole. These three metabolites represented 95% of the administered activity. Two minor metabolites were detected by paper chromatography.

Tocco et al. [911] extended their radiolabeled analytical procedures by chemical methods to plasma, feces, and tissues. The original drug and its hydroxylated metabolite, after ethyl acetate extraction, were fluorometrically determined at different excitation wavelengths. Similarly, the glucuronide and the sulfate were determined by spectrofluorometry before and after enzymic hydrolysis.

Tocco et al. [912] applied the same methods to man and the dog. An important fraction of the radioactivity was excreted in the urine in an unknown form.

Benzydamine was mostly excreted without modification. Thin-layer chromatography on silica gel (benzene:acetic acid:methanol 4:1:3) revealed to Catanese et al. [197] the presence of a metabolite which seemed to be hydroxylated.

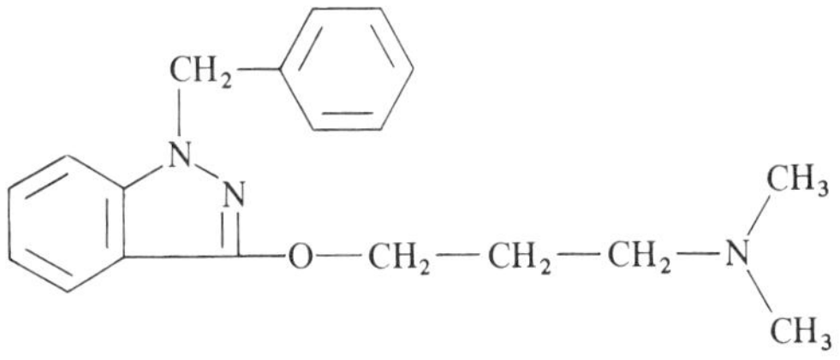

benzydamine

Akagi et al. [13] showed that methaqualone was practically not excreted unchanged in man on spectrophotometric determination. An increase of total glucuronic acid in urine was observed after administration of the drug. A metabolite was further isolated [14] by ethyl ether extraction of urine and chromatography on silica gel (conditions not reported). The UV, IR, and NMR spectra, the study of alkaline degradation products by paper chromatography, and various reactions led to the assignment of the structure 2-methyl-3-(2'-hydroxymethylphenyl)-(3H)-4-quinazolinone.

Cohen et al. [213] used ^{14}C-labeled methaqualone and noted in the mouse the presence of four urinary metabolites beside the unchanged drug. They applied paper chromatography (butanol : acetic acid : water 4 : 1 : 5, ascending).

Beyer [83] studied the hydrolysis of methaqualone conjugated metabolites. The yield on enzymic hydrolysis was always less than that obtained by acid hydrolysis. However a mixture of β-glucuronidase and sulfatase liberated more methaqualone than β-glucuronidase alone. These facts suggested the existence of two conjugated metabolites.

More detailed experiments were undertaken in various animal species by Nowak et al. [678]. They first synthesized the ten possible monohydroxylated derivatives of methaqualone and then compared these products with the metabolites obtained from urine. Three kinds of extractions were applied: (a) acid hydrolysis for 30 minutes at 80°C and ethyl acetate extraction; (b) two successive β-glucuronidase hydrolyses for 12 hours at 37°C and ethyl acetate treatment; (c) percolation of urine at pH 0.5 - 1.0 through a Dowex-50W X8 (Na$^+$) column, elution by 2% ammonium hydroxide, and then enzymic hydrolysis as in (b). The metabolites were chromatographed on silica gel thin layer [(I) chloroform : acetone : 25% ammonium hydroxide 50 : 50 : 2; (II)

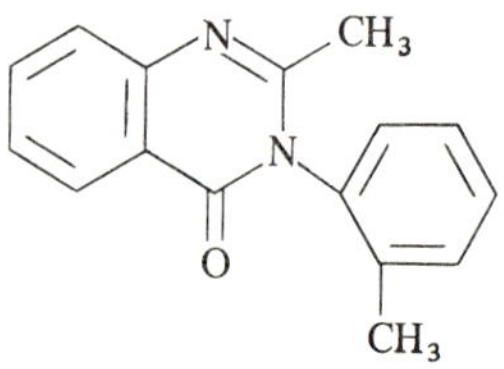

methaqualone

cyclohexane : chloroform : diethylamine 7 : 2 : 1; (III) propanol : ethyl acetate : 25% ammonium hydroxide 6 : 2 : 2] together with the reference compounds. Six metabolites (marked (*) in Table 59) were thus localized in addition to the unchanged drug.

These metabolites were separated in more sizable amounts on a 0.75 mm thick layer plate of silica gel, which was developed with system II, then system I. The metabolites were eluted by the mixture chloroform : methanol and were examined by UV spectrophotometry and thin-layer chromatography. The two hydroxylated derivatives that could not be separated by chromatography were analyzed by high voltage electrophoresis.

Almost all these metabolites gave glucuronides.

TABLE 59

Product	Migration as referred to methaqualone in system:		
	I	II	III
Methaqualone	1.00	1.00	1.00
*2-Hydroxymethyl-3-o-tolyl-(3H)-4-quinazolinone	0.93	0.59	0.97
*2-Methyl-3-(2'-hydroxymethylphenyl-(3H)-4-quinazolinone	0.70	0.36	0.98
2-Methyl-3-o-tolyl-5-hydroxy-(3H)-4-quinazolinone	0.98	0.80	0.97
*2-Methyl-3-o-tolyl-6-hydroxy-(3H)-4-quinazolinone	0.42	0.00	0.74
2-Methyl-3-o-tolyl-7-hydroxy-(3H)-4-quinazolinone	0.15	0.00	0.68
*2-Methyl-3-o-tolyl-8-hydroxy-(3H)-4-quinazolinone	0.67	0.00	0.76
*2-Methyl-3-(3'-hydroxy-2'-methylphenyl)-(3H)-4-quinazolinone	0.64	0.00	0.93
*2-Methyl-3-(4'-hydroxy-2'-methylphenyl)-(3H)-4-quinazolinone	0.55	0.00	0.85
2-Methyl-3-(5'-hydroxy-2'-methylphenyl)-(3H)-4-quinazolinone	0.52	0.00	0.87
2-Methyl-3-(6'-hydroxy-2'-methylphenyl)-(3H)-4-quinazolinone	0.42	0.00	0.87

At the same time Preuss et al. published two papers on methaqualone metabolism in man. In the first [723] they described their analytical procedure. Urine at pH 8-9 was extracted for 48 hours with ethyl ether. The residual aqueous solution was hydrolyzed for 12 hours with boiling hydrochloric acid (5%) and then extracted with ethyl ether. The first extract was chromatographed on a silica gel column with isopropyl ether, and the second extract on a similar column with petroleum ether containing increasing amounts of ethyl ether. Thirteen compounds that did not appear on the direct thin-layer chromatography of a urinary extract were thus detected. One of these compounds was identified as the metabolite described by Akagi et al. [13] by comparing its IR and UV spectra with those of the authentic sample. Two others corresponded to unchanged methaqualone which was not completely eliminated from urine by the first extraction.

Preuss et al. [724] also reported the identification of three other phenolic metabolites. These were compared with authentic compounds by IR spectrophotometry, chromatography, and melting point determination.

Chapter XV

HETEROCYCLES WITH MORE THAN TWO NITROGENS

The short-lived action of the hypnotic 5-ethyl-6-azauracil was investigated by Rupp and Handschumacher [790]. Concentrated urine was chromatographed on paper (ethyl acetate : formic acid : water 65 : 5 : 25). The two metabolites and the unchanged drug were eluted. The major metabolite was purified by chromatography on a column of Dowex-1 X 4 (formate) using formic acid of increasing concentration. The pure recrystallized product was examined by elemental analysis and UV and IR spectrophotometry and was assigned the structure of 5-(1-hydroxyethyl)-6-azauracil. The second metabolite, eluted from its paper chromatogram, was purified by passage through a Dowex-1 X 4 column after conversion into its borate complex. Elution by a solution of 0.1M ammonium formate : 0.007M potassium borate, pH 7, permitted its separation from the major metabolite. The UV spectra at various pH values suggested an azauracil ribonucleoside and confirmation was obtained by the orcinol reaction. Hydrochloric acid hydrolysis yielded 5-ethyl-6-azauracil.

The vasodilator action of diallylmelamine is delayed in its onset, which suggests that the action is mediated by a metabolite. Zins [1044] studied the transformations in the rat with a ^{14}C-labeled drug. Confirmation was

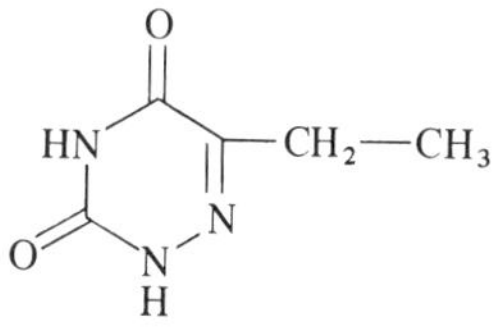

5-ethyl-6-azauracil

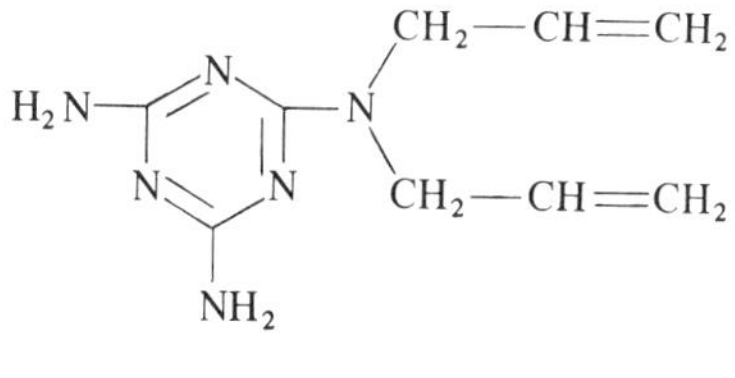

diallylmelamine

obtained by the fact that spectral titration of the unchanged drug after selective extraction showed that the change in arterial pressure did not correspond to the drug blood levels. Also, the administration of SKF 525 A inhibited diallylmelamine metabolism and lengthened hypotensive action. The chloroform extraction of blood, diluted with a buffer, pH 9, gave a solution which was concentrated and chromatographed on formamide-coated paper in two dimensions [(I) formamide and water saturated butyl acetate; (II) butanol: acetic acid:water 2:1:1]. Four metabolites appeared, one of which was only a trace. A good relationship was shown between the blood levels of one of these chromatographically separated four metabolites and the arterial pressure. The three main metabolites were isolated from the urine of 100 rats, lyophilized, and chloroform extracted by counter-current distribution in a 1,000 tube apparatus (ethanol:water:ethyl acetate:cyclohexane 1:1:1:1). Two metabolites were obtained after 2,000 transfers, a third one after 4,000 transfers. They were tested for their hypotensive action and only one was found to be active. The three metabolites were identified by comparison of UV, IR, and NMR spectra, melting points, elemental analysis, ionization constants, R_f, and biological activities with synthesized compounds. The active metabolite was a consequence of N-oxidation and became a drug (oxonazine). The other metabolites resulted from deallylation or dihydroxylation on one allyl chain.

Angelucci et al. [26] used ^{14}C-labeled 4-acetamido-2-morpholino-1,3,5-triazine. Successive paper chromatographies in various solvent systems and comparison with reference compounds showed the presence in urine or plasma of the deacetylated drug and of the deacetylated 6-hydroxylated derivative.

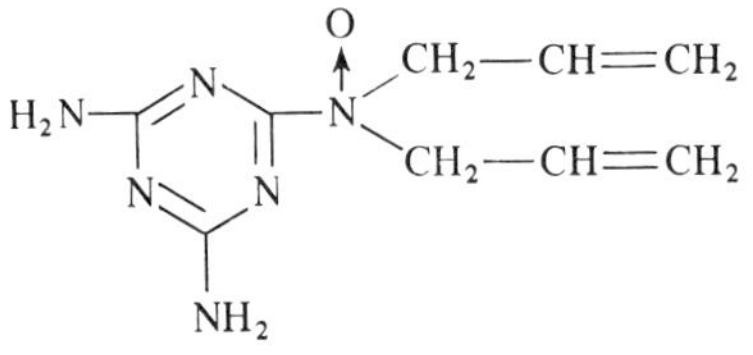

oxonazine

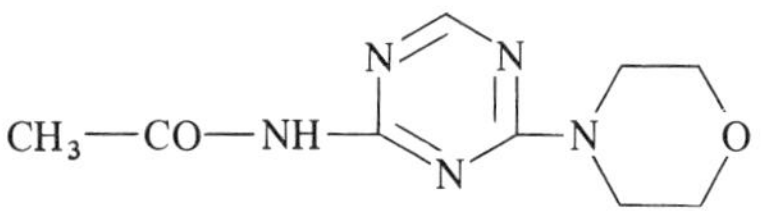

4-acetamido-2-morpholino-1,3,5-triazine

XV. HETEROCYCLES WITH MORE THAN TWO NITROGENS

The antiviral triazine, 2-[(o-phenoxyphenyl)amino]-4-amino-1,3,5-triazine was studied by Cresseri et al. [231]. The main fraction of the radioactivity was excreted in bile after the administration of the 6-^{14}C-labeled drug. The bile was treated with cold 4N sulfuric acid to precipitate glycocholic acid. It was then saturated with ammonium sulfate at pH 6.8 and extracted by butanol. The extracted products were redissolved into water and reextracted into ethyl acetate. Chromatography on silica gel thin layer (acetone : acetic acid 4 : 1) produced a spot with an R_f value which differed from that of the original product. Elemental analysis and synthesis of the supposed derivative assigned it to 2-[(o-phenoxyphenyl) amino]-4-amino-6-hydroxy-1,3,5-triazine.

5-Azacytidine is metabolized into various products, which include 5-azauracil and its degradation products, guanidine, and guanylurea ribonucleoside. Raska et al. [743] chromatographed urine on paper [(I) butanol : acetic acid : water 10 : 1 : 3] after administration of the labeled drug. The spots were eluted and chromatographed again [(II) isobutyric acid : water : ammonium hydroxide 66 : 33 : 1.5] or fractionated by paper electrophoresis (0.001M phosphate buffer, pH 7.4 — 600 V — 2 hours).

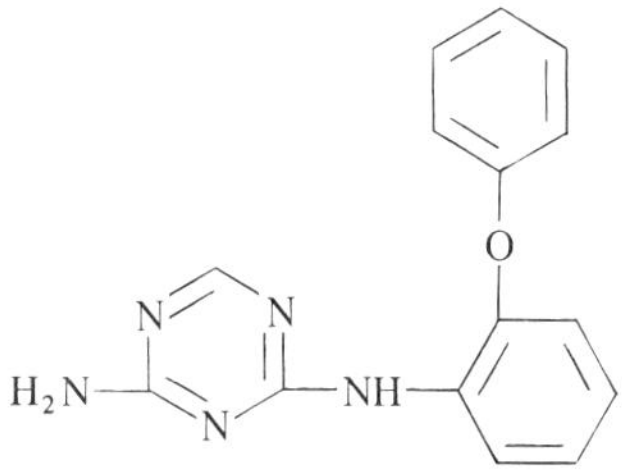

2-[(o-phenoxyphenyl)amino]-4-amino-1,3,5-triazine

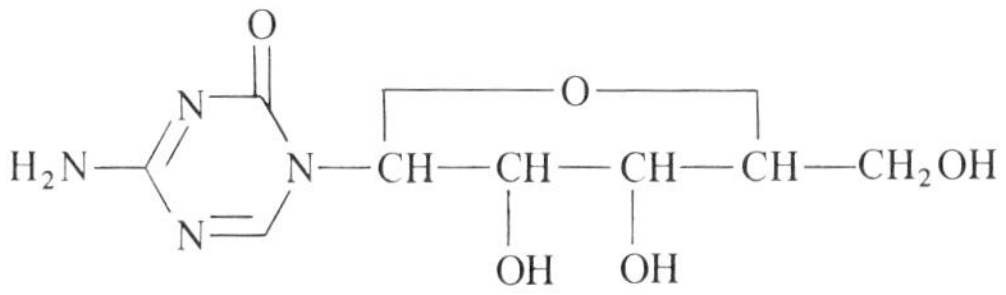

5-azacytidine

The same authors [472], in their in vitro experiments, separated 5-azacytidine from its nucleotide by three successive paper chromatographies (system I above, water saturated butanol, system II) followed by electrophoresis (citrate buffer pH 3.8). This nucleotide was further hydrolyzed by snake venom and chromatographed. The mono-, di-, and triphosphates of 5-azacytidine were separated in system II.

Raska et al. [744] also applied these analytical methods to the study of 5-azacytidine transformations in mouse brain.

Novotny et al. [677] showed that 6-azacytidine is deaminated into 6-azauridine by mono- and bidimensional paper chromatography (isopropanol: ammonium hydroxide: water 70:15:15; water saturated butanol; butanol: acetic acid: water 10:2:5).

Pentetrazole is metabolized into a biologically inactive compound as reported by Esplin and Woodbury [313]. After administration of the ^{14}C-labeled drug, only one radioactive compound was recovered in urine and feces and it behaved exactly as pentetrazole on paper chromatography (water). The microdistillation of a urine extract also gave the drug. This information can be reconciled with pentetrazole metabolism only if it is postulated that the metabolite is converted back to the drug either in urine or in the course of the analysis.

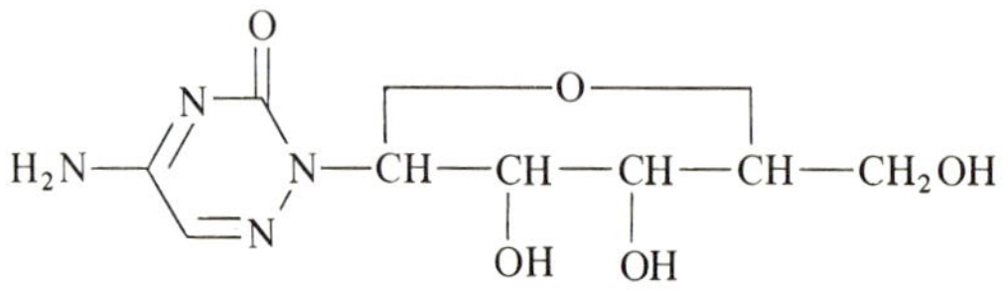

6-azacytidine

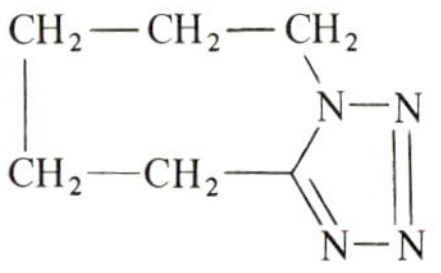

pentetrazole

Lehmann [549] separated urinary metabolites of triamterene by paper chromatography (isopropanol : water : 25% ammonium hydroxide 14 : 5 : 1; butanol : acetic acid : water 4 : 1 : 1; 3% ammonium chloride; 4% sodium citrate). Five fluorescent substances were detected. They were separated on a Dowex-1 X 2 column (0.5N acetic acid). Their chemical reactions and UV spectra, their methylation followed by paper chromatography, and the synthesis of reference products allowed the identification of the unchanged drug and three metabolites. The two main metabolites resulted from phenyl ring p-hydroxylation and further esterification by sulfuric acid.

The only metabolite of dipyridamole that Beisenherz et al. [74] were able to detect was the glucuronide conjugate. Fluorometry determined that it was essentially excreted fecally and that it underwent an enterohepatic cycle. The metabolite was isolated from bile by butanol extraction at pH 10. The biliary acids and pigments were precipitated by barium hydroxide, the excess of which was removed by sulfuric acid. After concentration of the aqueous solution, the crude metabolite was precipitated by ethyl ether. It was purified

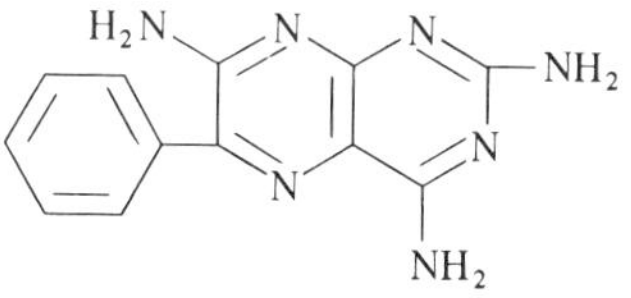

triamterene

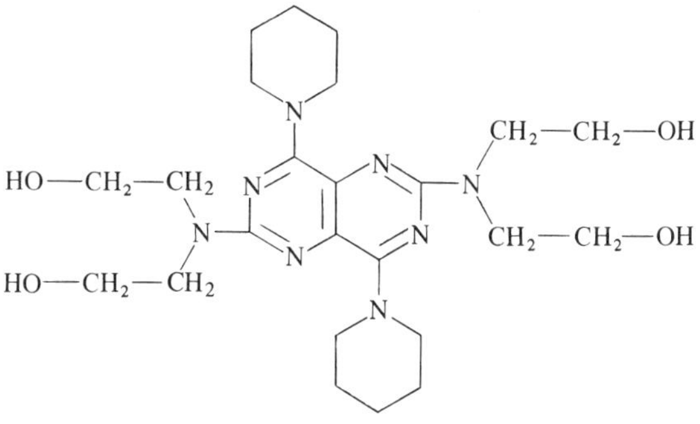

dipyridamole

by chromatography on a silica gel column at a temperature of 3°C (increasing concentrations of methanol in ethyl ether starting from 25%). The purity of the fractions was tested by thin-layer chromatography. The purified fraction was chromatographed on a column of Amberlite IRC-50 (H$^+$) at 3°C. After water washing of the column for 12 hours, the metabolite was eluted by 0.1N ammonium hydroxide. The obtained product was homogeneous on thin-layer chromatography in four different solvent systems. This metabolite was identified by its elemental analysis, its UV and IR spectra, its optical rotation, and its hydrolysis followed by paper chromatography. Glucuronic acid was detected by the carbazol reaction and determined by the naphtoresorcinol colorimetry of Fishman and Green [329], and by the decarboxylation procedure of Ogston and Stanier [686].

Zak et al. [1038] developed a more sensitive fluorometric method and found a second metabolite, dipyridamole diglucuronide. Diluted bile was introduced onto a Dowex-21 K (OH$^-$) column. After water washing, the unchanged drug and its two metabolites were eluted by 2N acetic acid. The eluate was concentrated, washed with ethyl acetate, and chromatographed on a column of aminoethylcellulose (formate). The diglucuronide was separated from the other compounds by elution with diluted formic acid. This separation was followed by a thin-layer chromatography. The crude metabolite was purified by ethyl acetate precipitations followed by a new chromatography on aminoethylcellulose. Dipyridamole resulted from hydrolysis of the metabolite as shown by IR spectra. The metabolite exhibited glucuronic acid reactions and its elemental analysis was consistent with the diglucuronide.

Beisenherz et al. [72] confirmed that dipyridamole was essentially excreted as a glucuronide. Detection was effected in bile by chromatography on silica gel thin layer (butanol : methylethylketone 8 : 2). The metabolite was extracted from the dried bile with absolute ethanol and then precipitated by adding ethyl ether. It was compared to the synthetic glucuronide. The authors did not specify whether the metabolite was a monoglucuronide in this paper.

Henderson et al. [412] prepared tritiated methotrexate. They chromatographed the urine or bile of treated animals on a DEAE-cellulose column with an ammonium bicarbonate, pH 8.3, solution with the concentration increasing from 0.1M to 0.4M. Almost all of the radioactivity was recovered in the unchanged drug peak. The same procedure was applied to human urine [413].

Redetzki et al. [748] isolated a metabolite from the urine of methotrexate-dosed rabbits. The metabolite precipitated on simple centrifugation of frozen and thawed urine. After recrystallization, its elemental analysis suggested a dihydroxylated derivative of methotrexate. Its UV spectrum was identical to that of the metabolite isolated after in vitro experiments using DEAE-cellulose column chromatography by Johns et al. [459]. These authors had assigned the metabolite to 7-hydroxymethotrexate. The metabolite isolated by Redetzki et al. exhibited a UV spectrum different from that of the synthetic 7-hydroxymethotrexate.

Oliverio and Davidson [693] studied the fate of ^{36}Cl-labeled dichloromethotrexate. Urine and bile were chromatographed on DEAE-cellulose column with a phosphate buffer, pH 8, with the concentration increasing from 0.01M to 0.4M in accordance with the method of Oliverio [691]. The fractions were identified by their elution time, UV spectrum, R_f on paper, and high voltage electrophoresis. Only one metabolite was characterized by comparison to a known compound: 4-deamino-4,7-dihydroxydichloromethotrexate.

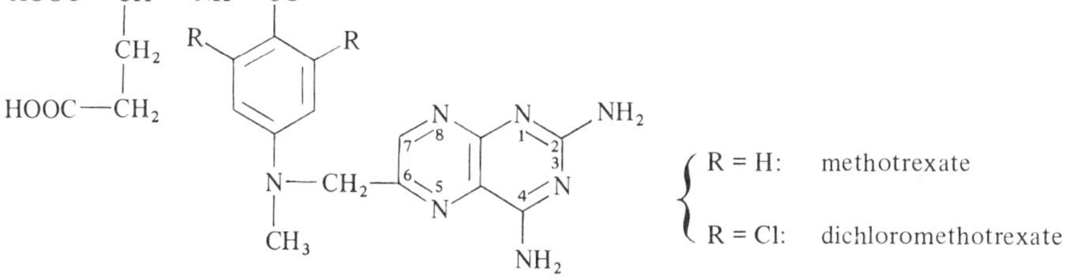

Loo and Adamson [559], after in vitro incubation, believed they had found the same metabolite on the basis of paper chromatography, high voltage electrophoresis, and UV spectrophotometry assay and stated [560] that the structure previously attributed to the metabolite was erroneous. This derivative was isolated by DEAE-cellulose column chromatography using an ammonium bicarbonate buffer with the concentration increasing from 0.01M to 0.4M. The fraction containing the metabolite was concentrated and redissolved in sodium hydroxide, and the pure metabolite was precipitated by acetic acid. Its elemental analysis seemed to indicate only one more oxygen than in the hydrated drug. Its UV spectrum in alkaline medium showed an hypsochromic shift similar to that of the 2,4-diamino-7-hydroxy-6-methylpteridines. Mild potassium permanganate oxidation in alkaline medium gave a pteridine which was identified as 2,4-diamino-7-hydroxy-6-pteridinecarboxylic acid by elemental analysis, UV spectrum, electrophoretic mobility, and behavior on DEAE-cellulose. Zinc reduction of the metabolite in phosphoric acid gave two products, separable on DEAE-cellulose. One was similar, on the basis of the above criteria, to 3,5-dichloro-4-methylaminobenzoylglutamic acid. These results showed that the metabolite was 7-hydroxydichloromethotrexate.

Numerous purine base analogs have been considered as anticancer agents. Their in vivo transformations have been studied since the mode of action of these compounds is by interference with the normal nucleic acids metabolism. Only those compounds whose therapeutic activity has been recognized will be examined.

Mercaptopurine was investigated in the mouse by Elion et al. [299] and in man by Hamilton and Elion [394] using 8-^{14}C- and ^{35}S-labeled derivatives. Paper chromatography of mouse urine (5% ammonium sulfate aqueous solution containing 5% of isopropanol) revealed the unchanged drug (R_f 0.44), 6-thiouric acid (R_f 0.27), and one unidentified metabolite. The mercaptopurine incorporation into the nucleic acids of various tissues was followed. Nucleic acids were extracted with trichloroacetic acid according to the procedure of Schneider [807]. Mercaptopurine partition between DNA and RNA was studied by separating the two types of nucleic acids by alcohol precipitation of the extract obtained with a hot 10% sodium chloride solution. In man an undefined combination of chromatography on paper and ion-exchange resin identified 6-thiouric acid, the unchanged drug, a sulfate conjugate, and one unknown metabolite.

These preliminary investigations were extended by Loo et al. [561] who synthesized 2-, 6-, and 8-thiouric acids. Human urine was chromatographed on a column of Dowex-1 X 8 (Cl$^-$). After washing with 1% acetic acid, thiouric acid was eluted by 2% hydrochloric acid. This eluate was chromatographed on Dowex-1 X 8 (Cl$^-$) with hydrochloric acid of concentrations increasing from 0.001N to 0.01N. 6-Thiouric acid was thus isolated, crystallized, and identified by elemental analysis and UV spectrophotometry. It was further found in mouse urine by paper chromatography, after Elion et al. [299], in addition to two or three undefined metabolites.

At the same time, Elion et al. [302] described the isolation and the synthesis of 6-thiouric acid. Human urine was filtered on Dowex-50 (H^+) to decrease the pH to 4 and then chromatographed on Dowex-1 (formate). After water washing, the column was eluted with 0.1N, then 0.5N, formic acid and the resultant thiouric acid was identified by its UV spectrum.

Elion et al. [300] applied similar methods to separate mercaptopurine and 6-thiouric acid while studying the influence of a xanthine-oxidase inhibitor.

Sarcione and Stutzman [794] characterized 6-methylmercaptopurine as an additional metabolite of mercaptopurine. An ethanolic extract of lyophilized urine was chromatographed on paper (butanol:acetic acid:water 35.5:9.6:16; 2N ammonium hydroxide saturated butanol). The metabolite was identified by its R_f and UV spectrum.

Scannell and Hitchings [795] administered 8-^{14}C-labeled mercaptopurine to tumor-bearing mice. Tumor tissue DNA was isolated and degraded to the deoxyribonucleosides by pancreatic deoxyribonuclease and snake venom. The digestion product was analyzed by bidimensional paper chromatography [(I) butanol:water 86:14; (II) 5% disodium phosphate solution or 5% ammonium sulfate solution + 5% of isopropanol]. Three radioactive components were thus detected and the most important corresponded to the R_f of thioguanosine deoxyriboside.

Remy [750] studied the in vitro metabolism of a series of thiopurines. He developed a paper chromatographic technique to follow 6-methylthiopurine in the rat and mouse. Lyophilized urine was extracted with ethanol and chromatographed on paper in two dimensions [(I) 86% butanol in ammonia atmosphere; (II) butanol:acetic acid:water 4:1:5]. Four major metabolites and eight to ten minor metabolites were revealed.

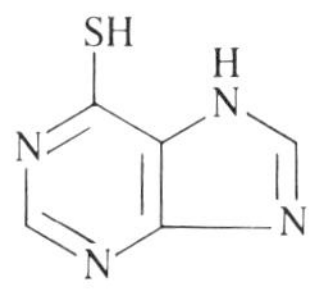

---

mercaptopurine

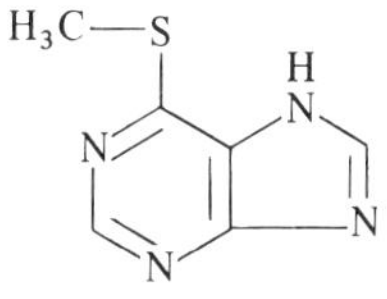

6-methylthiopurine

Hansen et al. investigated the in vivo transformations of 9-ethyl- and 9-butylmercaptopurine. They used the ^{35}S-labeled compounds and applied a double chromatography on Dowex-50 and -1 according to the methods of Elion et al. [302]. The only metabolite identified by enzymic hydrolysis was the glucuronide, although a second metabolite was present [398]. Subsequently, 9-ethylmercaptopurine ^{14}C-labeled on the ethyl group permitted demonstration of dealkylation by the same methods [399].

Duggan and Titus [282] administered 8-^{14}C-labeled 6-chloropurine to the rat. After alcohol precipitation, the plasma was chromatographed on paper. The first solvent system (I) was the diphasic system isoamyl alcohol: 5% disodium phosphate solution after Carter [194]. The second (II) was the mixture butanol:benzene:water 2:1:1 to which enough methanol was added to get a single phase. DNA and RNA were separated as in [299]. Mono- and bidimensional chromatography revealed five metabolites and the unchanged drug (Table 60).

Mouse tumor cells convert 6-chloropurine into its ribonucleotide. Hampton and Paterson [396] chromatographed the acid-soluble fraction of these cells (extracted by 0.4M, then 0.2M cold perchloric acid) on Dowex-1 X 8 (formate). The eluent was a formic acid solution, the concentration of which was increased from 0 to 11.7M. In this way 6-chloropurine and its ribonucleotide were separated. The metabolite was characterized by its R_f, paper electrophoretic migration, and hydrochloric acid hydrolysis.

Thioguanine metabolism by tumor cells in vitro was studied by Bieber and Sartorelli [88]. The acid-soluble fraction resulting from the cold extraction by 0.5M perchloric acid was neutralized, then chromatographed on a Dowex-1 (formate) column. One molar formic acid eluted thioguanine and thioguanosine;

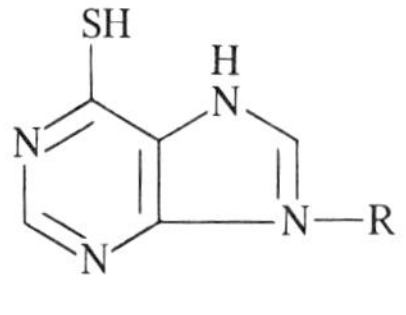

{ R = C$_2$H$_5$: 9-ethylmercaptopurine
{ R = C$_4$H$_9$: 9-butylmercaptopurine

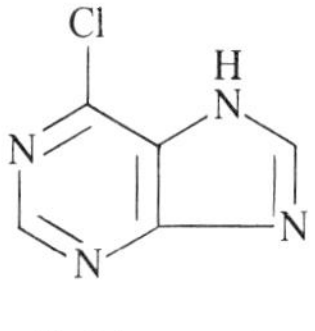

6-chloropurine

TABLE 60

Product	R_f in system:	
	I	II
Allantoin	0.75	0.09
Uric acid	0.40	0.21
Urea	0.73	0.40
Hypoxanthine	0.55	0.39
6-Chlorouric acid	0.40	0.53
6-Chloropurine	0.38	0.74

4M acid eluted 6-thiouric acid; and the mixture 4M formic acid + 1M ammonium formate all eluted thioguanine mononucleotide.

[35]S-Labeled 6-(4-carboxybutyl)-thiopurine was administered by Francova et al. [337] to various animals and man. The butanol extract from urine was chromatographed on paper (butanol : acetic acid : water 4 : 1 : 5; butanol : 2.5% ammonium hydroxide 86 : 14). Five excretion products were characterized by their R_f. The most important was 6-(2-carboxyethyl)-thiopurine.

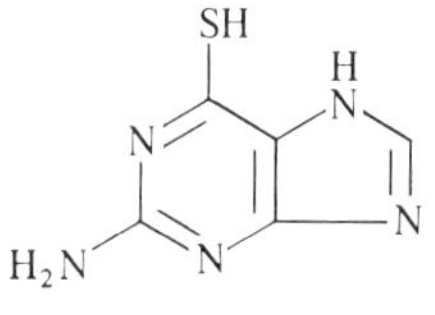

thioguanine

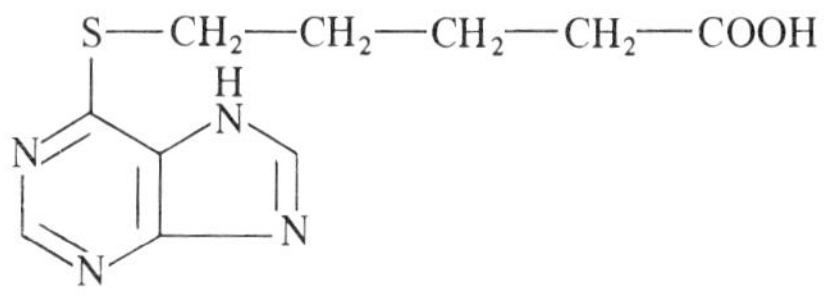

6-(4-carboxybutyl)-thiopurine

6-^{14}C-labeled allopurinol used by Elion et al. [301] showed that the urine of dosed mice contained only the unchanged drug and its 6-oxidation derivative, alloxanthine. The urine was chromatographed on Dowex-1 (formate) to remove the salts, and the metabolites were eluted by 0.1N formic acid. They were further separated on paper (water saturated butanol, ammonia atmosphere, ascending). Human urine was deionized and analyzed on a Dowex-50 (H$^+$) column. Alloxanthine was eluted by the water wash. The specific radioactivity of this alloxanthine decreased on purification by paper chromatography and a new component appeared with the same R$_f$ as allopurinol. It was a hydrolyzable allopurinol derivative that gave the ribose reaction with orcinol.

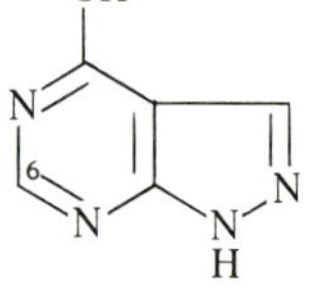

allopurinol

Chapter XVI

HETEROCYCLES NOT CONTAINING NITROGEN

McMahon studied the metabolism of a series of 1,4-benzodioxanes. The administration of butamoxane [581] ^{14}C-labeled on the butyl allowed the recovery of the greatest fraction of radioactivity in urine. After acid or β-glucuronidase hydrolysis, the radioactive compounds were extractable into ethyl ether. Chromatography on paper treated with 0.1M phosphate buffer, pH 6.1 (tertioamyl alcohol : dibutyl ether : water 80 : 7 : 13), revealed a single radioactive compound which reacted as a phenol with diazotized sulfanilamide. Comparison of the metabolite with an incomplete series of known derivatives and UV spectrophotometric study did not differentiate between 6- or 7-hydroxylation.

Ethomoxane is metabolized differently in the rat and the dog. McMahon et al. [588] demonstrated this by comparing the amounts of expired $^{14}CO_2$ after administration of the drug labeled either on the α- carbon of the O-ethyl or on the α-carbon of the N-butyl. Paper chromatography of an ethyl ether extract of rat urine (butanol : 1.5N ammonium hydroxide 1 : 1) revealed two metabolites in addition to unchanged drug. One of these was, most probably, 8-hydroxybutamoxane. The amount of this derivative was less than that of the ethomoxane which underwent cleavage of its ethoxy groups, and an important proportion of the radioactivity was not extracted.

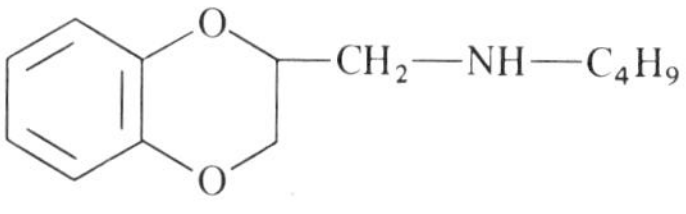

butamoxane

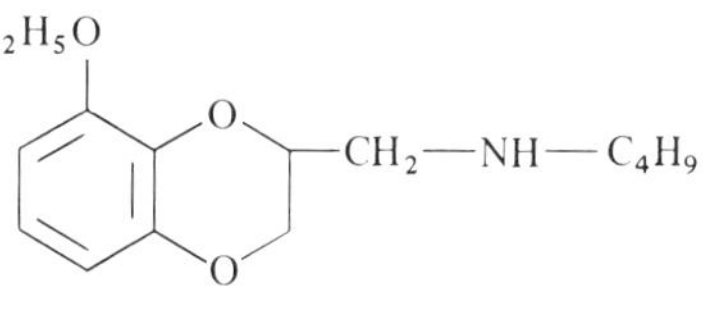

ethomoxane

258   XVI. HETEROCYCLES NOT CONTAINING NITROGEN

In the dog, paper chromatography identified 2-carboxy-8-ethoxy-1,4-benzodioxane as the only excretion product. This metabolite was isolated by silica gel column chromatography of the ethyl ether extract (benzene : ethyl acetate in various proportions). Its formula was confirmed by its melting point, its UV and IR spectra, and its X-ray diffraction diagram.

Chlorethomoxane was examined [583] by the same methods and the results were similar to those obtained for ethomoxane in the rat and dog.

After administration of guanoxan ^{14}C-labeled on the guanidine, Canas-Rodriguez [191] extracted the metabolites from urine concentrated in vacuo with methanol. They were then fractionated either by chromatography on Amberlite IRC-50, buffered at pH 6.5 (water), or by preparative paper chromatography (propanol : 1M ammonium carbonate : ammonium hydroxide 70 : 30 : 2). The metabolites were characterized by their R_f, their electrophoretic mobility, and the preparation of derivatives. They resulted from hydroxylation of guanoxan and hydrolysis of its guanidine group.

Dicoumarol has been the subject of many studies devoted to its distribution and excretion. Lee et al. [546], with the drug ^{14}C-labeled on the methylene bridge, showed by isotopic dilution that the product stored in rabbit liver was unchanged dicoumarol.

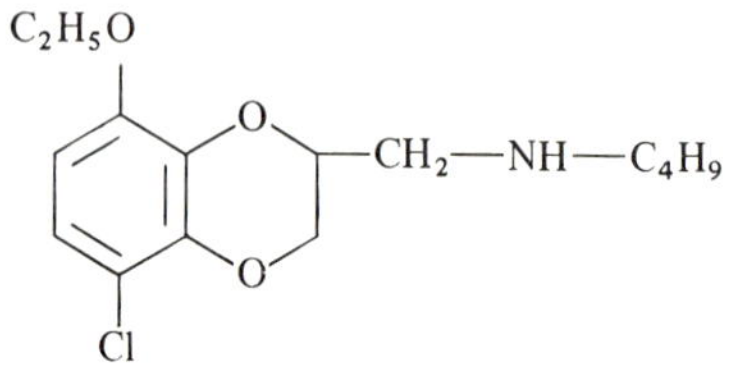

chlorethomoxane

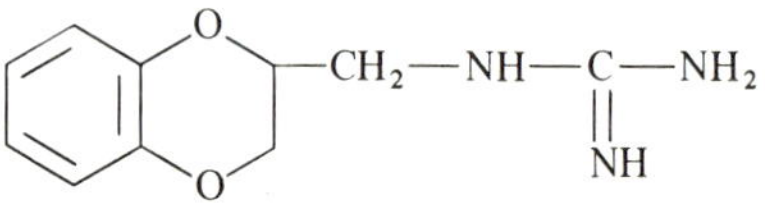

guanoxan

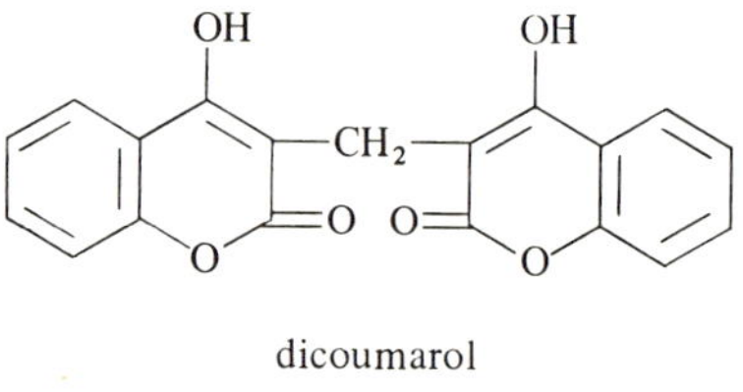

dicoumarol

Similarly, Hausner et al. [409] detected seven radioactive metabolites in rat urine and four in plasma by paper chromatography (methanol : butanol : water).

Christensen [202] developed a specific analysis for dicoumarol in biological media after paper chromatography (butanol : 3N ammonium hydroxide 1 : 1) and observed [203] the presence of several metabolites in the excreta. The greatest fraction had a UV spectrum similar to dicoumarol.

After administration of the drug ^{14}C-labeled on the methylene bridge, he was able to detect several metabolites [204]. Urine was acidified, sodium chloride saturated, and extracted with ethyl acetate. The feces were ground in 1N hydrochloric acid and extracted for 3 hours with acetone. After concentration, the acetone solution was extracted with ethyl acetate. The concentrated extracts were chromatographed on paper in one dimension [(I) butanol : 3N ammonium hydroxide 1 : 1; (II) benzene : acetic acid : water 2 : 2 : 1; (III) ethyl acetate : 3N ammonium hydroxide 1 : 1] or in two dimensions (I + II). None of the metabolites was identified.

The most important of these excretion products was isolated and crystallized by Christensen [205]. The feces were extracted with acetone and the concentrated sodium chloride saturated acetone solution was extracted with chloroform. The obtained product was chromatographed on paper (butanol : 3N ammonium hydroxide 1 : 1) twice successively, then chromatographed again with the system ethyl acetate : 3N ammonium hydroxide 1 : 1. The chemical reactions and the UV spectrum, as well as comparison with 7-hydroxy- and 7-methoxydicoumarol did not define this metabolite.

As early as 1949, ethylbiscoumacetate was the subject of investigations by Pulver and von Kaulla [739]. They showed that two metabolites were excreted in the rabbit in addition to the unchanged drug. One of these metabolites was the acid resulting from drug deesterification.

---

ethylbiscoumacetate

Burns et al. [158] demonstrated that the metabolism in man essentially involved hydroxylation of one of the two benzene rings. Counter-current distribution of the heptane extract of urine (40% isoamyl alcohol in heptane : 0.1M phosphate buffer, pH 7.6 — 9 transfers) revealed a single compound which was not the drug. This metabolite was isolated and its structure established by a degradation study [159]. Its elemental analysis indicated one more oxygen than in the original drug. Its UV spectrum was different from that of ethylbiscoumacetate, but its potentiometric determination curve was identical. This suggested the presence of an hydroxyl group. Its position in 4 was determined by fusion with potassium hydroxide and identification of the resulting hydroxyacids. In the rabbit, the nature of the acid found by Pulver and von Kaulla was confirmed by counter-current distribution and the UV spectra.

A single short communication by Link et al. [555] reported the identification of the 6-, 7-, and 8-hydroxylated metabolites of Warfarin by isotopic dilution and by paper and thin-layer chromatography.

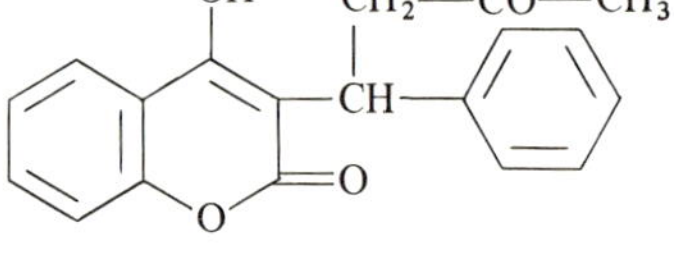

Warfarin

Chapter XVII

ALKALOIDS

The alkaloids are drugs which have been known for a long time. Studies on the metabolism of several were attempted before the development of the modern elegant analytical methods necessary to complete them. The importance and antiquity of alkaloids were counterbalanced by their molecular complexity, which discouraged metabolic research.

I. PHENANTHRENE DERIVATIVE ALKALOIDS

The metabolisms and analyses of morphine, codeine, their derivatives, and the synthetic analgesics with morphine-like activity have been reviewed in detail by Way and Adler in 1960-1962 [973, 974, 975, 976].

It had been suggested as early as 1938 that the organism could convert codeine into morphine. The toxicological and forensic importance of this hypothesis gave rise to research on codeine metabolism, in parallel to that on morphine which we shall consider later.

Oberst, who had already demonstrated that morphine was partly excreted as a conjugate [682] applied the same techniques (see page 264) to codeine-dosed patients. He found that the titratable amount of codeine increased

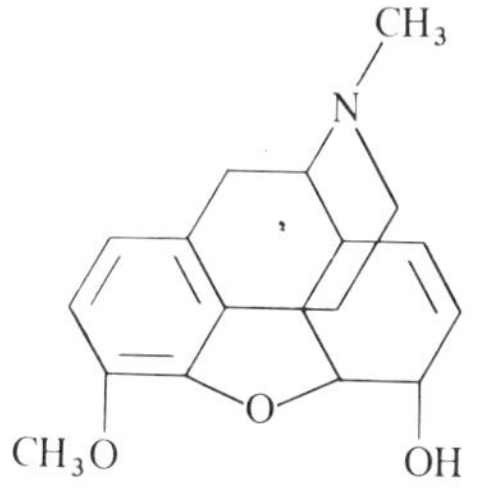

codeine

after acid hydrolysis. In order to determine codeine in the presence of morphine, he capitalized on the solubility differences of the silicotungstates of these two alkaloids [683]. Glucuronic acid determinations suggested that codeine was partly excreted as a glucuronide, although the attempted isolation of the conjugate by solvent extraction was unsuccessful [684].

Later, Latham and Elliott [537] studied the metabolism of codeine, ^{14}C-labeled on the methoxyl, in the rat. Paper chromatography indicated one or several radioactive metabolites. This study was completed by Brodie's techniques: liquid-liquid partition and methyl orange determinations. The only definite conclusion was to establish the fact of in vivo demethylation, since radioactive carbonic dioxide was collected in the expired air.

Adler and Shaw [11] incubated codeine with rat liver slices. After protein precipitation, the culture media were first extracted with chloroform at alkaline pH values to remove codeine and then at acidic pH values to remove acidic impurities. A metabolite was obtained from the residual solution after concentration and extraction. It was identified as morphine by its UV spectrum and by X-ray diffraction of its dinitrophenyl ether crystals.

This result was confirmed in vivo in man by Mannering et al. [597] who used a paper chromatographic procedure capable of detecting small amounts of morphine in the presence of related alkaloids [598]. Urine with 10% hydrochloric acid was hydrolyzed by autoclaving. The alkaloids were extracted into chloroform and chromatographed on paper (isoamyl alcohol: acetic acid:water 10:1:5, ascending). The iodoplatinate reagent revealed codeine (R_f 0.23) and morphine (R_f 0.16). The morphine spot was eluted with ammonium hydroxide and chromatographed again on paper buffered at pH 7 (water saturated isoamyl alcohol); this second separation allowed morphine to be distinguished from dihydromorphine and dihydromorphinone. The morphine spot was eluted, purified by microsublimation and converted into its dinitrophenyl ether. Its crystalline melting point and refractive index were the same as the pure compound.

Woods et al. [1024] determined morphine and codeine in the dog and monkey by the methyl orange method. Separation was achieved by partition between alkaline aqueous solution and ethylene dichloride. The isolated morphine was identified by its melting point, its R_f on paper, and the formation of its diacetylated derivative. The analytical details were described. The various metabolites comprised 38 to 53% of the administered dose.

Adler briefly reported that N-methyl oxidation of codeine resulted in norcodeine. This metabolite was identified in man on counter-current distribution [8]. Confirmation was achieved with codeine ^{14}C-labeled on the N-methyl [9]. A portion of the norcodeine could be determined only after acid hydrolysis with autoclaving.

Adler [10] further isolated codeine and norcodeine by counter-current distribution in 1955. The codeine metabolites characterized were conjugated codeine, free and conjugated morphine, and free and conjugated norcodeine. These metabolites and unchanged codeine represented 50 to 90% of the administered dose.

Anders and Mannering [25] attempted gas-liquid chromatographic analysis of codeine and its metabolites in 1962. As these alkaloids cannot be directly separated on nonpolar stationary phases, it was necessary to convert them into their acetates or propionates. The conversion was achieved in situ by injecting acetic or propionic anhydride concomitantly with the alkaloid. This procedure had its drawbacks as interfering peaks appeared in the blank injections.

Schmerzler et al. [805] avoided this difficulty by realizing that the acetylation rates of codeine and norcodeine were widely different under mild conditions. Norcodeine was thus removed although it has the same retention time as codeine on SE 30. Urine was extracted at pH 9 with chloroform containing 10% of butanol. Codeine and its metabolites were transferred to an aqueous acid phase and reextracted into chloroform at pH 13. An exception was morphine, which was recovered at pH 9 in chloroform containing 1% ethanol. Codeine and norcodeine were determined with two injections, with the second injection being made after 30 seconds contact with acetic anhydride. Morphine was injected after complete acetylation (2 hours at 60°C). The conjugated metabolites were hydrolyzed by 10% of hydrochloric acid for 1 hour at 100°C.

Morphine is one of the oldest examples of a drug metabolism study. Marme [604, 605] reported in 1883 the presence of pseudomorphine (2,2'-bimorphine) in the tissues of morphine-addicted dogs and suggested that this pseudomorphine could play an important part in the addiction phenomenon. In the same year, Stolnikow [875] noticed the presence of a morphine sulfuric ester in the dog, and later he synthesized this metabolite.

Oberst in 1940 [682] detected free and conjugated morphine in urine of morphine addicts. The alkaloid was determined by extraction, adsorption on Permutit and colorimetry with the Folin-Denis reagent [681]. The amount of morphine increased after boiling with diluted hydrochloric acid. Glucuronic acid determination showed that morphine was most likely present as its glucuronide. However, it was not possible to isolate this metabolite by solvent extraction [684].

A related technique enabled Gross and Thompson [387] to get the same results in the dog. These authors [903] noticed two types of conjugated morphine metabolites. One was hydrolyzed at pH 1 in 1 hour at 100°C, whereas the other required concentrated acid and heating under pressure.

Zauder [1041] heated acidified rat urine in the autoclave for 30 minutes at 125°C and determined morphine with the silicomolybdate reagent after purification by adsorption on Florisil and methanol elution.

Wood in 1954 [1020, 1019] undertook the first extensive study on conjugated morphine metabolites in the bile and urine of the dog and rat. He applied mono- and bidimensional paper chromatography with various mixtures of butanol:acetic acid:water to detect the metabolites. Column chromatography on Whatman paper No. 1 powder (butanol:acetic acid:water 4:1:2)

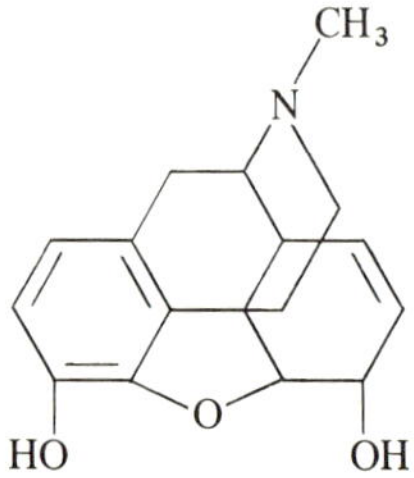

morphine

isolated a fraction containing conjugated morphine which was recrystallized several times from aqueous methanol. Morphine was released from its conjugate on hydrolysis, and the latter was identified by the formation of its diacetylated derivative. Elemental analysis and morphine and glucuronic acid determinations indicated a morphine glucuronide with two water molecules. The negative test with alkaline ferricyanide showed that the conjugation was on the phenol group. Woods reported that important amounts of another morphine metabolite were present in dog urine, but he was unable to obtain and purify it.

Simultaneously, Seibert et al. [823] separated free morphine from its conjugated derivative in dog urine by paper chromatography (butanol:acetic acid:water 10:3:4.5). The metabolite was hydrolyzed either by autoclaving in acid medium or by β-glucuronidase. The liberated morphine was characterized by paper chromatography.

Fujimoto and Way [346] isolated from the urine of a suicide a morphine glucuronide that corresponded to 95.8% of the total dose. This conjugated derivative was precipitated after the urine had been saturated with potassium carbonate. In 1956 they [345] briefly reported the purification of this glucuronide and indicated that it differed from the metabolite isolated in the dog by Woods [1020].

Details followed [347] when the urine of morphine addicts was saturated with potassium carbonate to isolate quantities of crude glucuronides. Free morphine was extracted by ethyl acetate from an acidified solution of these glucuronides. The conjugated metabolites were adsorbed on suspended activated carbon and were eluted from this carbon by acetic acid. Subsequent purification was effected by continuous paper electrophoresis, with the purified fraction fractionated by counter-current distribution (20% acetic acid: butanol, lower aqueous layer as the mobile phase). Even then absolute purity was not obtained, since several components were visualized with ninhydrin on paper chromatography. Further purification was effected by paper chromatography after reaction with fluorodinitrobenzene. The presence of morphine in the pure conjugate was established by X-ray diffraction. The UV spectrum in acid and alkaline media was consistent with the absence of a free phenolic group. The IR spectrum was identical to that of the Woods' monoglucuronide. Potentiometer titration gave two pK values. Thus, the authors assigned the structure of the isolated metabolite to a zwitterionic 3-morphine glucuronide dihydrate.

The glucuronides extracted from human and canine urines [348] were shown to be the same and the analytical data confirmed glucuronide attachment on the phenolic hydroxyl in position 3.

In addition to conjugation, morphine is metabolized by N-demethylation in minor amounts. March and Elliott [601], confirmed by Mellett and Woods [630], administered morphine ^{14}C-labeled on the N-methyl and animals excreted radioactive carbon dioxide.

Axelrod [38] incubated morphine with liver enzymes and isolated normorphine. The free morphine was removed by extraction with ethylene dichloride containing 3% of isoamyl alcohol at pH 9-10. The metabolite was then extracted from the incubation medium saturated with sodium chloride by ethylene dichloride containing 20% of isoamyl alcohol. Normorphine was identified by its R_f value on paper buffered at pH 7.5 (t-amyl alcohol : butyl ether : water 50 : 7 : 43).

Misra et al. [635, 636] administered tritiated morphine to rats and extracted the urine with ethylene dichloride containing 30% of amyl alcohol at pH 10.3. The metabolites thus obtained were chromatographed on paper [(I) butanol : acetic acid : water 4 : 1 : 2] and the single radioactivity peak obtained was eluted and chromatographed again on paper [(II) t-amyl alcohol : butyl ether : water 80 : 7 : 13]. The two peaks that were then observed had the R_f values of morphine and normorphine in several solvent systems. A compound that remained at the origin was hydrolyzed to release morphine (Table 61).

TABLE 61

Product	R_f in system: I	II[a]
Morphine	0.55	0.70
Normorphine	0.55	0.18
Pseudomorphine	0.18	0.04
Morphine glucuronide	0.12	—

[a] Paper buffered at pH 8 by 0.2M phosphate buffer.

The detailed analytical techniques of Misra et al. [636] were applied by Mule et al. [648] to the urine and central nervous system tissues of rats dosed with morphine ^{14}C-labeled on the N-methyl. Only morphine was recovered.

The fact that radioactive carbon dioxide was detectable in the expired gases after the administration of morphine ^{14}C-labeled on the N-methyl to man is inconsistent with the inability to find normorphine in human urine. Rationalization of these inconsistencies led to the suggestion that morphine undergoes trans-N-methylation, but Rapoport negated this hypothesis (unpublished results cited in [647]). Mule and Mannering [647] used Rapoport's technique to check for the absence of trans-N-methylation in the rat. The specific radioactivity of administered labeled morphine was compared to that of morphine isolated from urine and adequately purified by sublimation.

Woods and Chernov [1021] in 1966 isolated and identified a new morphine metabolite, a sulfate. Concentrated cat urine was treated with methanol and the extract chromatographed on paper (butanol : acetic acid : water 4 : 1 : 2). The conjugated metabolites were eluted into water and chromatographed again (butanol : 1.5N ammonium hydroxide 90 : 25). The eluted product was crystallized with a different R_f than the glucuronide. Morphine, characterized by paper chromatography, and the sulfate ion were obtained on hydrolysis.

Dihydromorphine metabolism was studied by Hug and Mellett [440] with the tritiated drug. Free dihydromorphine, and the total dihydromorphine after hydrolysis, were purified by a double extraction at pH 9 with ethylene dichloride containing 30% amyl alcohol. Dihydromorphine and its N-demethylated derivative were separated by chromatography on paper buffered by a

---

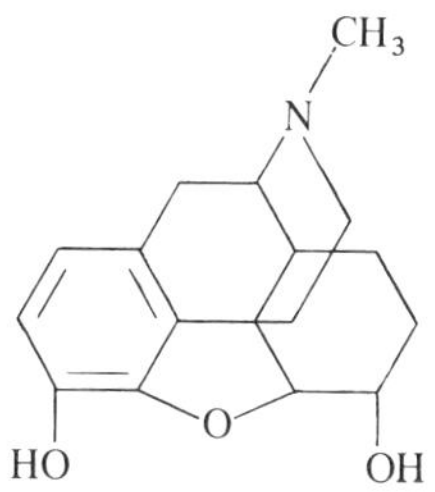

dihydromorphine

0.2M potassium phosphate solution of pH 6 to 10 (t-amyl alcohol : butyl ether : water 80 : 7 : 13) or nonbuffered (butanol : acetic acid : water 4 : 1 : 2) (Table 62).

β-Glucuronidase assays only gave a partial hydrolysis compared to the acid treatment. Thus, two conjugated metabolites were indicated, of which only one was a glucuronide.

Hug et al. [441] studied ^{3}H-labeled dihydromorphine excretion through the kidneys by the stop-flow technique. Dihydromorphine was analyzed in urine by paper chromatography according to the procedures described in [440] to which they added the system propanol : butanol : 0.1N ammonium hydroxide 2 : 1 : 1 on nonbuffered paper (R_f: dihydromorphine 0.83, N-demethylated metabolite 0.65).

Hug and Woods [442] studied the metabolism of <u>nalorphine</u> whose effect on morphine metabolism had already been studied by Mule et al. [648]. Pure tritiated nalorphine was prepared. Nalorphine, either free or total after hydrolysis, was extracted into ethylene dichloride. The absence of normorphine was substantiated by the paper chromatographic techniques of Misra [636]. β-Glucuronidase hydrolysis showed that the essential excretion product was the conjugate glucuronide.

Stitzel et al. [874] investigated the enzymic demethylation of <u>ethylmorphine</u> and recommended a 5 or 20% zinc sulfate solution instead of perchloric or trichloroacetic acid for deproteinization.

The transformations of the synthetic morphine analog, <u>levorphanol</u>, in the dog organism were examined by Fisher and Long [322]. Levorphanol, either free or total after hydrolysis for 30 minutes in the autoclave, was

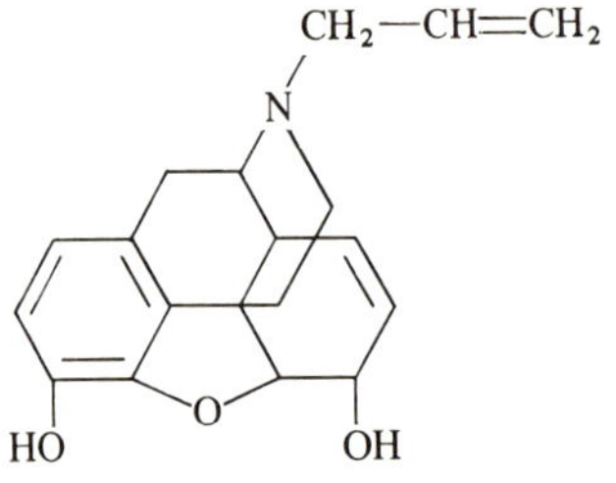

---

nalorphine

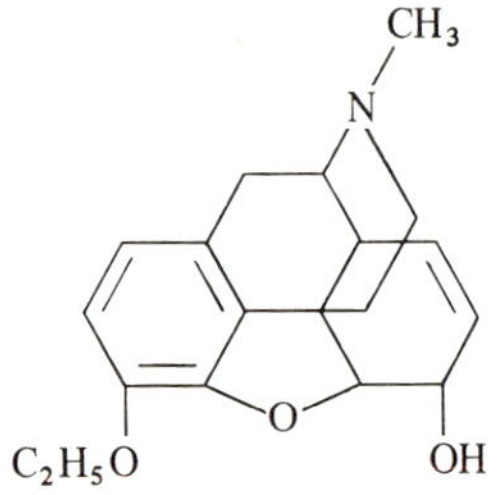

ethylmorphine

TABLE 62

Product	R_f					Nonbuffered paper
	Paper buffered at pH:					
	5.7	7	8	9	10	
Dihydromorphine	0.41	0.20	0.77	0.86	0.88	0.57
Dihydronormorphine	—	0.03	0.15	0.25	0.37	0.58

determined by Brodie's methyl orange method. Interfering compounds had to be eliminated previously by a series of chloroform extractions. The β-glucuronidase hydrolysis of the conjugated metabolite was studied as a function of time. Since the blank assays without enzyme hydrolyzed at the same rate as the assays with enzyme after 72 hours of hydrolysis at 38°C, a conjugate which was not a glucuronide was indicated.

Shore et al. [842] extracted levorphanol into benzene to improve the analytical specificity. The product liberated by acid hydrolysis was analyzed by counter-current distribution (benzene containing 3% of isoamyl alcohol : 0.1M phosphate buffer, pH 7.4 — 49 transfers). The distribution curve showed a single component identical to levorphanol. The product isolated from this fraction had the melting point of pure levorphanol. The demethylated derivative of levorphanol was not found in the fraction where it was expected.

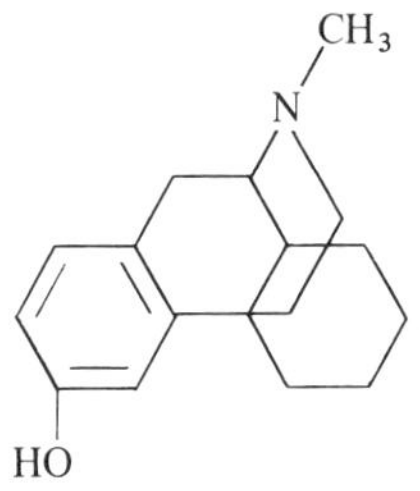

levorphanol

Woods et al. [1023] synthesized levorphanol ^{14}C-labeled on the N-methyl and determined it in biological media by isotopic dilution. A study of the distribution between ethylene dichloride and buffers of various pH values showed that the extraction procedure was specific for levorphanol. The conjugated drug was liberated after 30 minutes of treatment at 110°C with hydrochloric acid.

Levallorphan is the N-allyl analog of levorphanol and it was investigated by Mannering and Schanker [599]. The urine or tissues were hydrolyzed by hydrochloric acid in the autoclave and the solution was extracted with chloroform, before and after pH adjustment to 11-12. The obtained metabolites were chromatographed on paper in accordance with Mannering's technique for morphine and codeine [597] which involved ascending chromatography with the system isoamyl alcohol : acetic acid : water 10 : 1 : 5 equilibrated for three days before use. The separated metabolites were eluted with hot chloroform from the ammonium hydroxide wetted paper and purified by microsublimation under vacuum. Two metabolites were detected from in vitro and in vivo experiments and purified. They were analyzed by UV and IR spectrophotometry, elemental analysis, melting point, and R_f comparison with authentic compounds and various functional reactions. The identity of 3-hydroxymorphinan was confirmed by counter-current distribution. The second metabolite could not be identified. The excretion profiles of both metabolites and the unchanged drug were studied quantitatively by methyl orange titration after paper chromatographic separation. Levallorphan and its two metabolites were mostly present as conjugates.

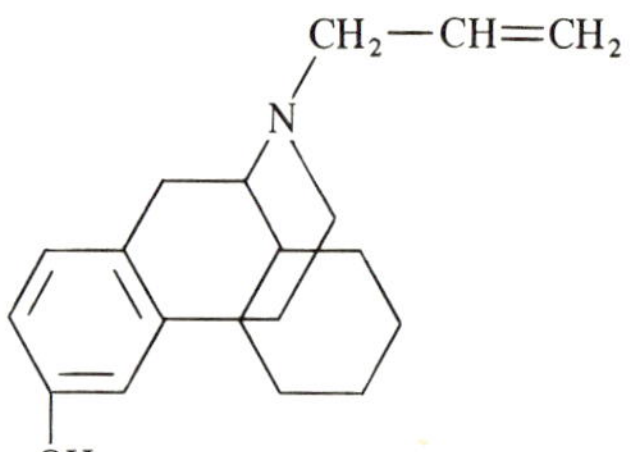

levallorphan

Dextromethorphan has been investigated in the Hoffman La Roche laboratories and Brossi et al. [141] developed a paper chromatographic procedure for the possible morphinan metabolites (Table 63). Two different chromatographies were required to separate the five considered derivatives.

(I) Paper buffered at pH 6.32 by a phosphate buffer (t-amyl alcohol : butyl ether : water 80 : 7 : 13)

(II) Paper buffered at pH 8.09 by a citrate-phosphate buffer (t-amyl alcohol : butyl ether : water 50 : 7 : 43)

The three corresponding demethylated derivatives of dextromethorphan, in addition to the unchanged drug, were thus detected in canine urine after acid hydrolysis.

Willner [1010] extended this study to human metabolism but used the paper chromatography technique of Waldi [964]. Paper was dipped into a 20% formamide solution in acetone and the developing solvent had the composition: hexane : diethylamine : benzol (Eb 180-230° C) 1 : 1 : 8 (Table 64).

TABLE 63

Product	R_f in system:	
	I	II
Morphine	0.13	0.68-0.73
3-Hydroxymorphinan	0.37-0.41	0.55-0.59
3-Methoxymorphinan	0.45-0.49	0.74-0.78
3-Hydroxy-N-methylmorphinan	0.37-0.41	0.84-0.88
Dextromethorphan	0.47-0.52	0.88-0.92

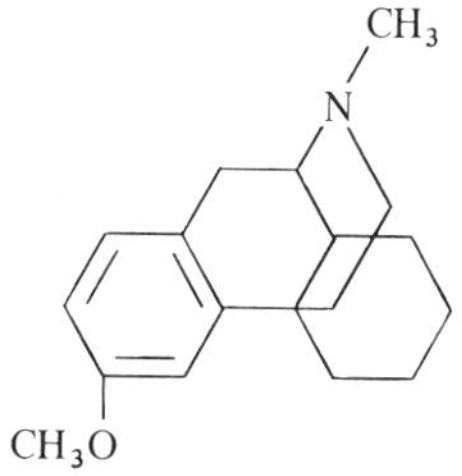

dextromethorphan

TABLE 64

Product	R_f
Dextromethorphan	0.66
3-Methoxymorphinan	0.73
3-Hydroxy-N-methylmorphinan	0.61
3-Hydroxymorphinan	0.63

Only two of the three demethylation products were detected: 3-hydroxy-N-methylmorphinan and 3-hydroxymorphinan.

II. XANTHINE DERIVATIVE ALKALOIDS

The metabolic research of Kruger and Schmidt [527, 528] on the xanthine derivative alkaloids dates from the first years of this century. An increase in the excretion of phosphotungstic acid-reducing compounds was observed after theophylline administration and the presence of mono- or dimethyluric acid was deduced from this fact without experimental verification. Weinfeld [992] finally identified 1-methyl- and 1,3-dimethyluric acids by paper chromatography.

Brodie et al. [131] published the first extensive study in 1952. Unchanged theophylline was extracted into chloroform at pH 7 after the urine had been saturated with sodium chloride. It was identified by counter-current distribution (chloroform:1M phosphate buffer, pH 6.7 — 8 transfers) and determined by UV spectrophotometry. The methylated uric acid metabolites were

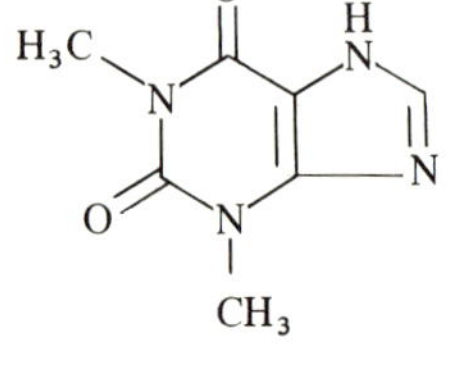

---

theophylline

extracted by isobutyl alcohol in acid medium after the uric acid had been destroyed by uricase. These metabolites were transferred to an aqueous phase on addition of petroleum ether to the alkalinized alcoholic solution. Counter-current distribution (isobutyl alcohol : 0.1N hydrochloric acid) showed a single component which behaved like 1,3-dimethyluric acid. The corresponding fraction was purified by chromatography on Dowex-1 (phosphate) with 0.1N hydrochloric acid as the eluent. The pure crystallized product was identified as 1,3-dimethyluric acid by its solubility and its UV and IR spectra. This metabolite was quantified by counter-current distribution and represented more than 50% of the administered dose.

Cornish and Christman [222] applied similar techniques to theophylline, theobromine, and caffeine. Urine was first percolated on Dowex-2 (H^+). The methylxanthines were removed by water washing. Uric acid and the methyluric acids were further eluted by 0.01N hydrochloric acid. Both eluates were chromatographed on paper (butanol : acetic acid : water 4 : 1 : 1). The separated components were identified by their R_f and determined by UV spectrophotometry after elution. It was also possible to precipitate the silver salts of the various methylxanthines by varying the pH values. This paper [222] was the only one of analytical interest with respect to theobromine.

The only other reported paper on caffeine is that of Axelrod and Reichenthal [43], who determined unchanged caffeine by its UV absorption after benzene extraction at pH 7-8. The specificity of this procedure was checked by counter-current distribution. Although the authors found only 1% of the unchanged caffeine in the urine, they did not isolate its metabolites.

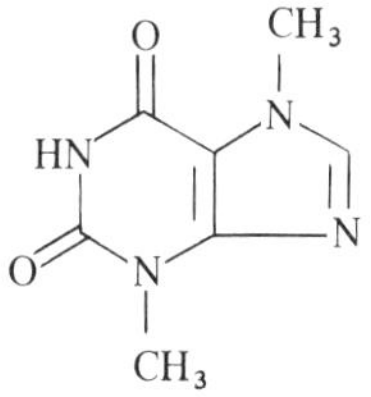

theobromine

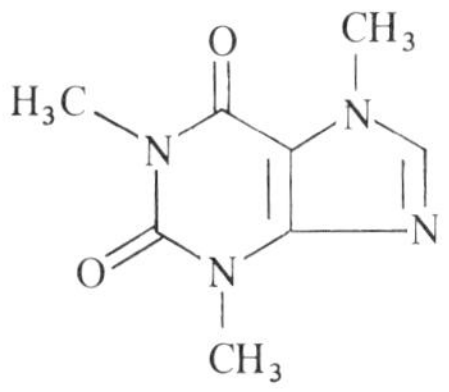

caffeine

The caffeine homolog 1-hexyl-3,7-dimethylxanthine has a lipophilic character which results in a different metabolism. Mohler et al. [638] investigated and identified its urinary metabolites in man and various animals. Concentrated urine was extracted with chloroform and the extract chromatographed on silica gel thin layer (chloroform : ethanol 9 : 1). Four metabolites were detected by comparison with known compounds. The extract was then chromatographed on a silica gel column which was developed with the same solvent. A pure metabolite was obtained and crystallized. Its elemental analysis, UV spectrum, and chemical reactions led to the assignment of 1-(4',5'-dihydroxyhexyl)-3,7-dimethylxanthine. A second metabolite was isolated from the solution in which the first was crystallized. It was purified on silica gel and then alumina columns. It was identified by its melting point and R_f comparison to 1-(5'-hydroxyhexyl)-3,7-dimethylxanthine.

After extraction in alkaline medium, the urine was adjusted to pH 2 and treated with ethyl ether to remove benzoic acid. Extraction with chloroform and addition of methanol resulted in the crystallization of a derivative the elemental analysis, UV and IR spectra, and chemical reactions of which corresponded to 1-(3'-carboxypropyl)-3,7-dimethylxanthine.

In addition, 1-(5'-oxohexyl)-3,7-dimethylxanthine and the unchanged drug were characterized by comparative thin-layer chromatography. The glucuronides were precipitated by the method of Kamil et al. [480] and hydrolyzed by β-glucuronidase and by acid treatment. Two of the preceding metabolites were identified among the hydrolysis products. All of the methods of synthesis of the described metabolites were delineated.

The same authors [639] separated the three main metabolites and the unchanged drug by thin-layer chromatography as above, eluted them from each

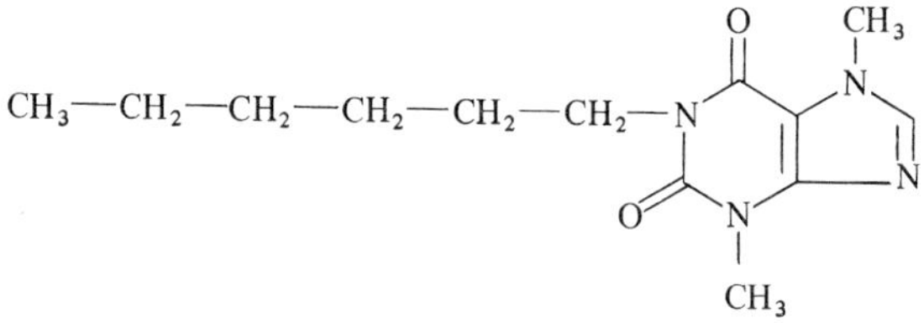

1-hexyl-3,7-dimethylxanthine

fraction into ethanol, and quantitatively determined them by spectrophotometry. The glucuronides were hydrolyzed by β-glucuronidase (8 hours at 37°C, pH 4.6).

III. RESERPINE

Numerof et al. [680] prepared a radioactive <u>reserpine</u> by reacting methyl reserpate with trimethoxybenzoic acid, ^{14}C-labeled on the carbonyl. The urine, feces, and the various organs of treated mice were extracted with hot chloroform, hot acetone, and ethanol. The pooled evaporated extracts were chromatographed on paper (butanol : benzene : 1.5N ammonium hydroxide/ 1.5N ammonium carbonate buffer 80 : 5 : 15) to separate the reserpine from trimethoxybenzoic acid. A large part of the radioactivity was found in the urine as trimethoxybenzoic acid.

Sheppard et al. used a reserpine ^{14}C-labeled on the 4-methoxyl of trimethoxybenzoic acid. They demonstrated that demethylation as well as trimethoxybenzoic acid formation occurred in the rat [836] and in vitro [837]. Detection was by paper chromatography (5N ammonium hydroxide : butanol : isopropanol 3 : 2 : 1) and a complex procedure of fractional extractions from organs was described.

Maggiolo and Haley [590] used these techniques [836] with tritiated reserpine.

Glazko et al. [364, 365] separated reserpine from methyl reserpate by extracting the former with ethylene dichloride at pH 3.7-4.0 and the latter at pH 8. Determination was achieved after conversion of both products into fluorescent derivatives by trichloroacetic acid in the presence of sodium

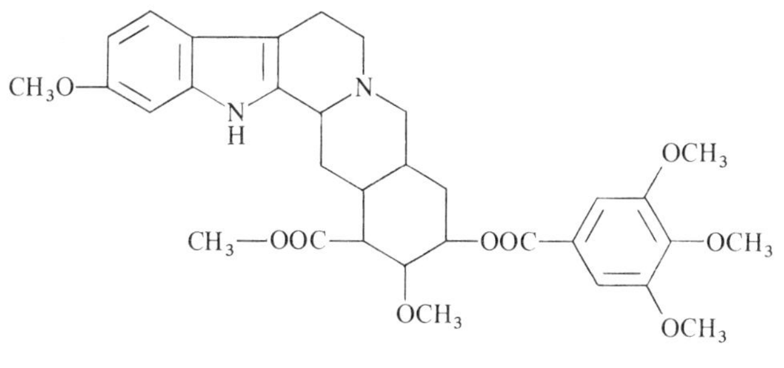

reserpine

nitroprusside. Methyl reserpate was identified as a metabolite by paper chromatography (isopropanol : 1% phenol aqueous solution 8 : 2) (Table 65).

Further proof was provided by counter-current distribution (0.6M citrate-phosphate buffer, pH 6 : ethylene dichloride — 24 transfers). The distribution curve of the metabolite was analogous to that of the authentic sample and the purified fraction showed the same UV and IR spectra. Paper chromatography also revealed methyl reserpate in the in vitro experiments.

IV. OTHER ALKALOIDS

Quinetine, a product of the oxidation of the vinyl side chain of quinine, was presumed by many in the years 1869-1944, to be a urinary metabolite of dosed patients. Other workers postulated that it was merely an artifact of the analytical techniques.

Kelsey et al. [504], after quinine incubation with a rabbit liver preparation, isolated a crystallized metabolite in 1944. It was extracted into ethyl ether from the ammonium hydroxide alkalinized medium. Its molecular weight was determined by cryometry. Mead and Koepfli [628] showed that this metabolite was 2'-hydroxyquinine by a series of chemical reactions as well as by UV spectrophotometry.

Brodie et al. [132] extracted the metabolites in ethyl ether from 48 l of urine at pH 10. The metabolites were then transferred to an aqueous acid phase and fractionated, after ammonium hydroxide addition, by successive

---

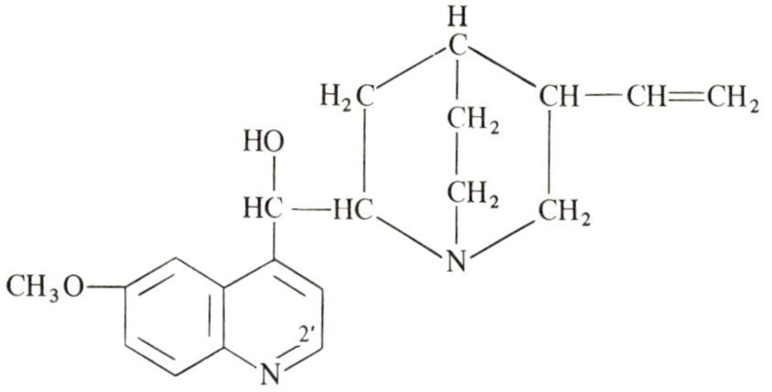

quinine and quinidine

TABLE 65

Product	R_f
Reserpine	0.85
Methyl reserpate	0.75
Reserpic acid	0.30

extractions with heptane, chloroform, and isoamyl alcohol. The extracted components were analyzed by counter-current distribution (simplified eight transfers procedure according to Craig et al. [229]), by UV spectrophotometry, and by elemental analysis, and the pK_a values were measured. Quinine was found in the heptane fraction, 2'-hydroxyquinine and 2-hydroxyquinine in the chloroform fraction and a dihydroxyquinine in the isoamyl alcohol fraction. The same methods were applied to quinidine.

Brodie et al. [132] applied the same analytical procedure to cinchonine and cinchonidine metabolites, subsequent to the studies of Earle et al. [288]. 2'-Hydroxycinchonine was extracted into ethyl ether at pH 10 and colorimetrically determined by the methyl orange method after correcting for the cinchonine present.

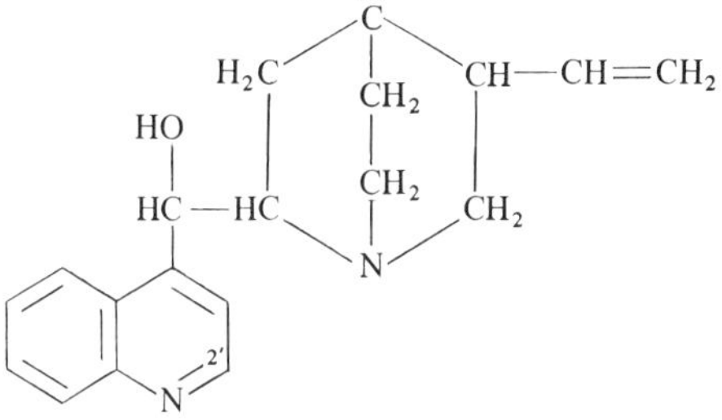

cinchonine
and cinchonidine

The fate of papaverine in man and various animals was considered by Axelrod et al. [44]. Unchanged papaverine was extracted from alkaline medium by heptane and determined by UV spectrophotometry at 251 and 266 mµ to correct for interfering substances. A water : heptane distribution assay at various pH's demonstrated the specificity of this procedure. Phenolic metabolites were isolated, after acid hydrolysis for 1 hour at 100° C, by ethyl ether extraction at pH 7-8. The extracted products were analyzed by counter-current distribution (0.1M citrate buffer, pH 3.9 : ethylene dichloride — 24 transfers). Two phenolic products were thus separated. One of these, on a new distribution at pH 4.5 under the same conditions, was broken into two components. The other also underwent a new distribution at pH 3.3 but appeared as an homogeneous compound that represented 70% of the total phenolic metabolites. The isolated product was identified as 4'-hydroxypapaverine by comparison with an authentic sample on the basis of its melting point, UV and fluorescence spectra, and paper chromatographic behavior (butanol : acetic acid 100 : 4, water saturated, ascending). These phenolic metabolites were excreted as conjugates. Autoclaving at 109° C in 1N hydrochloric acid quantitatively cleaved them, whereas β-glucuronidase only hydrolyzed 75% in 18 hours at 37° C.

Subsequently, Tsukamoto et al. [941] examined brucine, which has two adjacent methoxyls, as does papaverine, to see if a similar demethoxylation occurs in vivo. Acid hydrolyzed, sodium chloride saturated urine was adjusted to pH 9 and extracted with chloroform. The evaporated extract was redissolved into chloroform, reextracted by 1N sulfuric acid, and then extracted again by chloroform after alkalinization. This mixture of crude metabolites was analyzed by thin-layer chromatography on silica gel activated

---

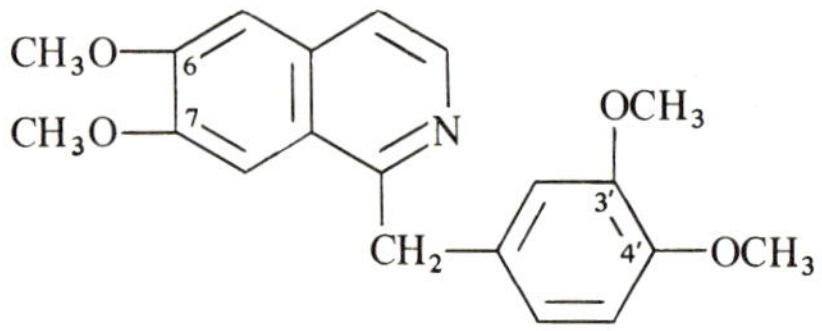

papaverine

for 1 hour at 100°C [(I) chloroform : acetone : diethylamine 5 : 4 : 1, (II) chloroform : diethylamine 9 : 1] and two metabolites were observed. These also appeared on paper chromatography (butanol : acetic acid 50 : 1, water saturated paper buffered at pH 6 by a 0.1M citrate buffer). One of these two metabolites was identified as 2-hydroxy-3-methoxystrychnine on comparison with the authentic sample. The other was extracted from the crude metabolites in methanol and crystallized. Its elemental analysis, UV and IR spectra, and methoxyl determination showed that this second metabolite was an isomer of 2-methoxy-3-hydroxystrychnine. The fraction that contained the nonphenolic bases also revealed another metabolite and traces of unchanged brucine by paper chromatography (Table 66). β-Glucuronidase hydrolysis demonstrated that the phenolic metabolites were excreted as glucuronides.

Watabe et al. [972] pursued in vitro the investigations performed in vivo by the same team [941]. After brucine incubation with a rabbit liver enzymic preparation (homogenate centrifuged at 9,000g for 20 minutes, or microsomes and soluble fraction separated at 108,000g for 1 hour), the alkalinized medium was extracted with chloroform. Nicotinamide was removed by washing the organic phase with a 0.03M phosphate buffer, pH 5.2. Total phenolic metabolites were spectrophotometrically titrated at 318 mμ and then separated on silica gel thin layer (chloroform : diethylamine 9 : 1). After 0.2N hydrochloric acid elution, they were spectrophotometrically determined. They were identified by comparison with the pure known compounds, 2-methoxy-3-hydroxystrychnine and 2-hydroxy-3-methoxystrychnine.

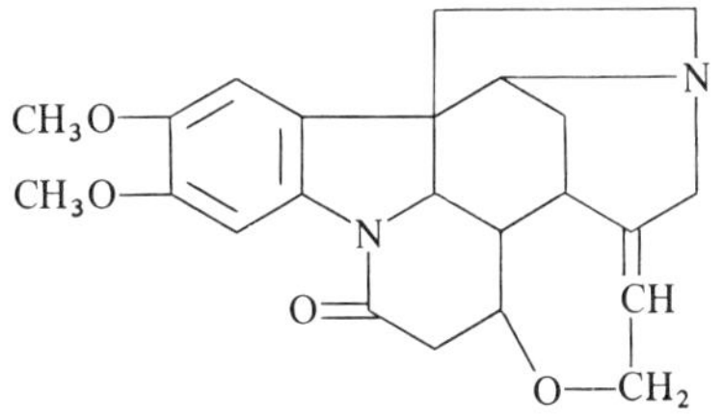

brucine

TABLE 66

Product	R_f Thin layer I	II	Paper
Brucine	—	—	0.49
2-Hydroxy-3-methoxystrychnine	0.23	0.28	0.38
2-Methoxy-3-hydroxystrychnine	0.37	0.58	0.30
Nonphenolic metabolite	—	—	0.44

The metabolic transformations of strychnine were first studied in vitro by Adamson and Fouts [6]. The unchanged drug was extracted into benzene at an alkaline pH and determined by methyl orange colorimetry. The extraction specificity was checked by paper chromatography and comparison of partition coefficients, but the metabolites were not considered.

Tsukamoto et al. [924] also used chromatography on in vitro preparations, where the paper was buffered at pH 4.6 by a 0.1M citrate buffer (butanol: acetic acid 50:1, water saturated, ascending). The incubation medium, after trichloroacetic acid deproteinization, was brought to pH 9, sodium chloride saturated, and chloroform extracted. Paper chromatography of the extract revealed four metabolites. By column chromatography of the extract on alumina (successively benzene, benzene:chloroform 4:1 and 1:1, chloroform, chloroform:methanol 95:5), unchanged strychnine was separated

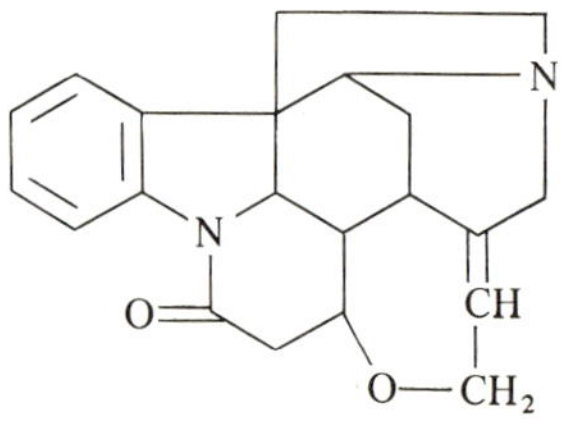

strychnine

from its metabolites. These were then fractionated by counter-current distribution (chloroform : 0.2M citrate buffer, pH 3.6 — 10 transfers), then by paper chromatography as above. The eluted metabolites had an UV spectrum close to strychnine, but were not identifiable. A phenolic metabolite, extracted from the chloroform solution with 1N sodium hydroxide was recognized as 2-hydroxystrychnine by R_f and UV spectra comparison with the synthetic compound.

Noscapine was roughly examined by Nayak et al. [663]. They determined the unchanged drug by fluorometry after ethyl ether extraction and identified it by counter-current distribution (McIlvaine buffer, pH 4.5; cyclohexane : benzene 4:1 — 8 transfers) and by silica gel thin-layer chromatography (ethanol : benzene 1:4).

Atropine metabolism has been studied since as early as 1912, when Fickewirth and Heffter [319] isolated tropine from the urine of atropine-dosed rats by extraction and crystallization.

Evertsbusch and Geiling briefly reported in 1953 [315] the use of biosynthetically obtained ^{14}C-labeled atropine.

Gosselin et al. [381] administered ^{14}C-labeled atropine that had been synthetically prepared by tropanol esterification of α-^{14}C-tropic acid. Chromatography on Amberlite IRC 50 (NH_4^+) separated urine into two fractions. Paper chromatography (water saturated butanol : acetic acid 86:14, ascending) revealed two components in each fraction. One of these was unchanged atropine, but none of the three others was tropic acid.

Kalser et al. [479] investigated the biliary excretion of ^{14}C-labeled atropine by similar procedures. Paper chromatography (butanol : acetic acid :

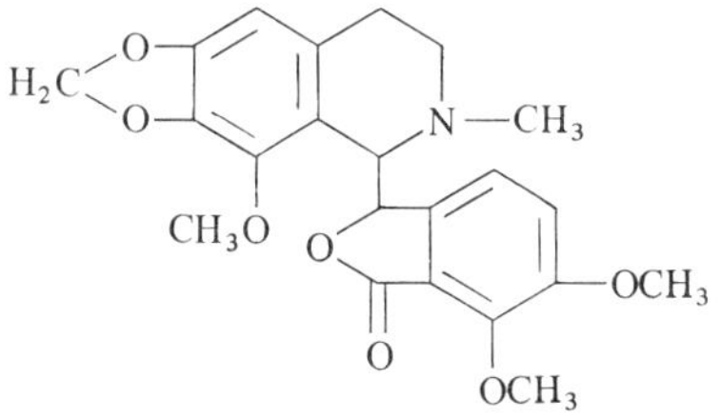

noscapine

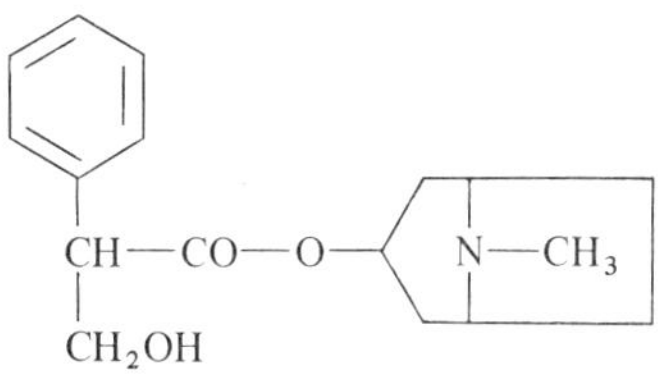

atropine

water 86:14:32.5) showed four metabolites and no unchanged atropine in the bile.

Deptropine undergoes an ether hydrolysis and N-demethylation in vivo. Hespe et al. [422] chromatographed ethyl ether extract of urine or tissues on paper (benzene : butanol : water : acetic acid 9:1:9:1, ascending) or on silica gel thin layer [(I) butanol : 25% ammonium hydroxide 98:2 ,(II) chloroform : acetone : diethylamine : 25% ammonium hydroxide 50:35:10:5]. They used deptropine tritiated in various positions and found, by comparison with known products, the several metabolites that are tabulated in Table 67 and detected the presence of several others, free or conjugated.

The analytical metabolic chemistry of cocaine has not been investigated.

Marsh [606] used ^{14}C-labeled tubocurarine and separated it from chondocurarine and from tubocurarine dimethyl ether by paper chromatography (butanol : acetic acid : water 4:1:5).

Psilocin is derived directly from psilocybin by hydrolysis of the phosphoryl group and it appears to be the biologically active species. Its biotransformation was partly elucidated by Kalberer et al. [477] with the

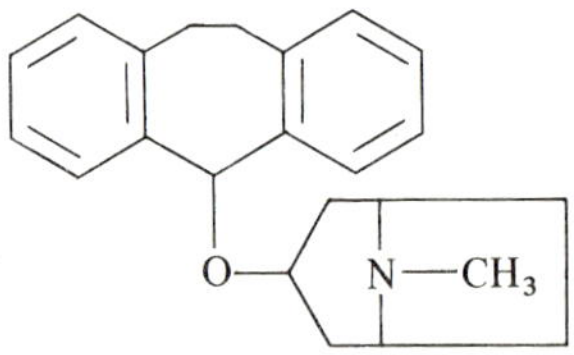

deptropine

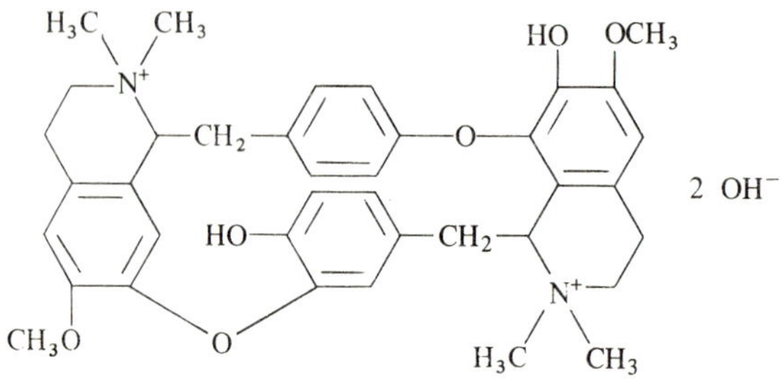

tubocurarine

TABLE 67

	R_f		
Product	On paper	On thin layer	
		I	II
Deptropine	0.85-0.95	0.30-0.36	0.91-0.98
N-Desmethyldeptropine	0.85-0.95	0.21-0.27	0.80-0.86
Tropine	0.85-0.95	0.04-0.14	0.41-0.55
N-Desmethyltropine	0.85-0.95	0.03-0.07	0.10-0.16
Dibenzosuberol	0.30-0.50	0.80-0.90	0.90-0.96

molecule ^{14}C-labeled either in position 2' on the side chain or on the N-methyl. The study determined excreted radioactive carbon dioxide, the unchanged drug by isotopic dilution, and identified 4-hydroxyindolacetic acid by paper chromatography (propanol : water 5 : 1; butanol : acetic acid : water 4 : 1 : 5), which also revealed other unidentified metabolites.

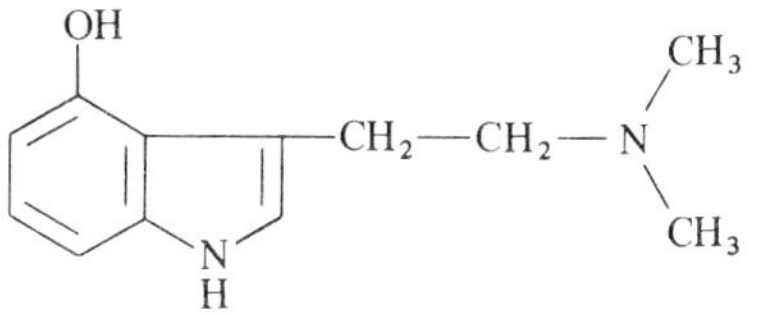

psilocin

Gessner et al. [356] compared psilocybin and bufotenine. The first showed, by paper chromatography of urine (propanol:ammonium hydroxide 4:1; butanol:acetic acid:water 4:1:5), a partial conversion into a glucuronic conjugation product (positive reaction with naphtoresorcinol) and an acid close to 5-hydroxyindolacetic acid. Bufotenine, after urine bidimensional chromatography [(I) propanol:ammonium hydroxide 9:1; (II) 20% potassium chloride aqueous solution — ascending], revealed 5-hydroxyindolacetic acid and an acid metabolite in addition to the unchanged drug, but no glucuronide.

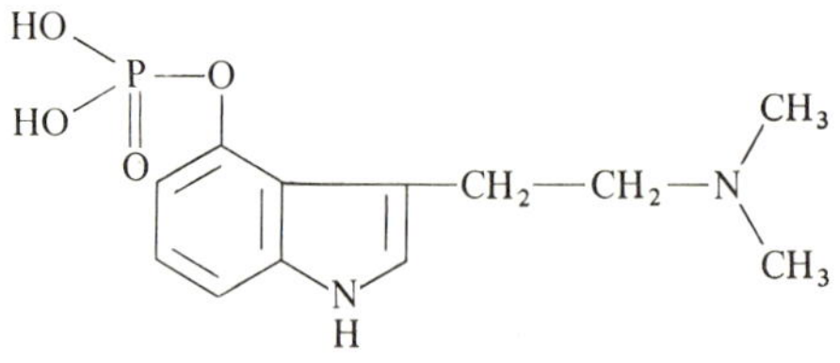

psilocybin

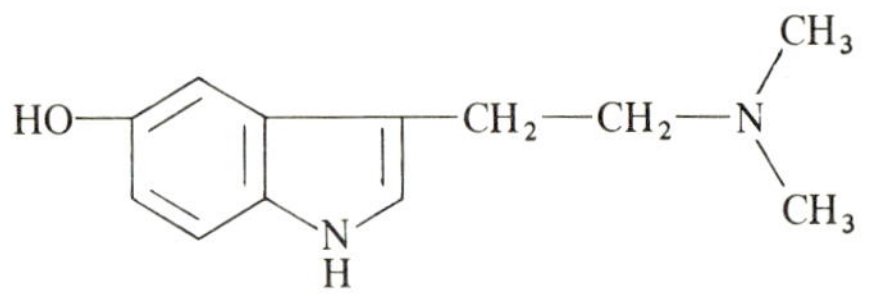

bufotenine

Chapter XVIII

ANTIBIOTICS

The distribution and excretion of antibiotics in the organism have been studied extensively. Actually, the sensitivity and specificity of microbiological assays facilitate such investigations. However, most antibiotics are of complex molecular structure so that characterizations of their in vivo transformations are much more difficult and not yet completed.

I. CHLORAMPHENICOL

Glazko and his colleagues commenced their studies on the metabolism of chloramphenicol [367], an antibiotic with a relatively simple structure, in 1949. Initial determination was by colorimetry after nitro group reduction and Bratton-Marshall reaction. The inactive metabolites were separated from chloramphenicol by ethyl acetate extraction at pH 6.

As much as 90% of the chloramphenicol excreted in human urine [368] was in the form of inactive derivatives that retained the nitro group. A chloramphenicol hydrolysis product [362] was present partly free and partly as a glucuronide. The conjugated metabolite was detected by paper chromatography, aglycone isolation, and periodic acid degradation.

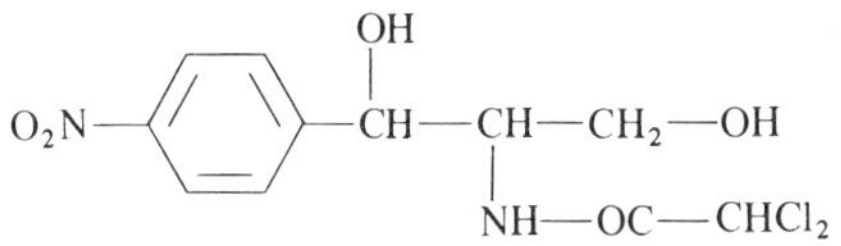

O_2N—⟨benzene⟩—CH—CH—CH_2—OH
          | |
          OH NH—OC—$CHCl_2$

chloramphenicol

Chloramphenicol was excreted in the rat as a glucuronide of chloramphenicol [363]. Paper chromatography (water saturated butanol + 2.5% of phenol and 2% of pyridine) revealed seven metabolites, some with the nitro group but others with an amino group.

Kakemi et al. [473] determined chloramphenicol in blood and urine by colorimetric assay after reaction with isoniazid in alkaline medium. The metabolites did not interfere.

Maruyama and Suzuki [613] also determined chloramphenicol in biological media without interference from its metabolites. The unchanged antibiotic was colorimetrically determined by diazo coupling with 2-diethylaminoethylnaphtylamine after ethyl acetate extraction at pH 6.0. The total nitro compounds (drug + metabolites) were determined by the same colorimetric procedure after phosphoric acid deproteinization without extraction.

Chloramphenicol monosuccinate was separated from chloramphenicol by electrophoresis after ethyl acetate extraction at pH 6. The nitro containing metabolites remained in the aqueous layer [687].

II. TETRACYCLINES

Kelly and Buyske [502] of Lederle administered ^{14}C- or ^{3}H-labeled tetracycline to rats. In addition to radioactivity measurements and microbiological assays, the urine and feces were analyzed by counter-current distribution (butanol : 3% formic acid — 220 transfers), and by paper chromatography with a specially devised technique [501]. Paper was wetted with a 0.1M EDTA disodium salt solution and dried. Two solvent systems were used: butanol : acetic acid : water 4 : 1 : 5 and butanol : ammonium hydroxide : water

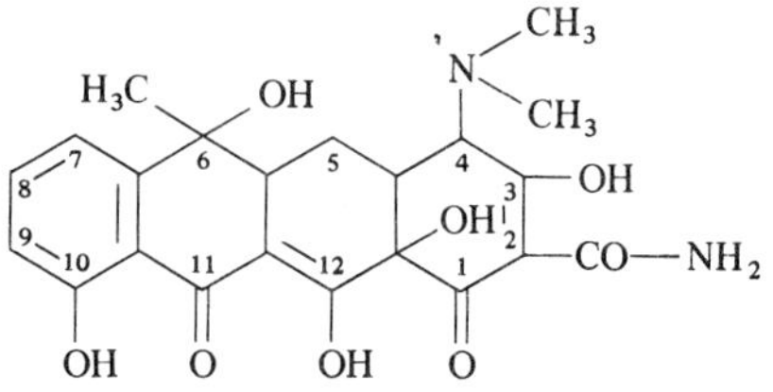

tetracycline

4:1:5. The latter was for separation of the epimers. Counter-current distribution showed a main peak that represented 95% of the excreted radioactivity and corresponded to unchanged tetracycline. Paper chromatography revealed an amount of epitetracycline in this peak, but the same result was obtained with pure tetracycline. Thus, epimerization was concluded to be an artifact due to the formic acid in the developing solvent.

The same authors applied the same techniques to demethylchlortetracycline [503]. Counter-current distribution required the system butanol : acetic acid : water 4:1:5. The results obtained were analogous to those with tetracycline.

Eisner and Wulf [298] studied chlortetracycline metabolism. Urine was paper chromatographed by the technique of Kelly and Buyske [501], where two additional solvent systems were used. Column chromatography on diatomaceous earth was used for feces extracted with 5% formic acid. The stationary phase was the aqueous layer of the system butanol : 0.1M EDTA disodium salt solution, pH 8.3, and the eluent was the butanol phase. Thus, 4-epichlortetracycline was detected in addition to unchanged drug. A small amount of isochlortetracycline was visualized from its blue fluorescence.

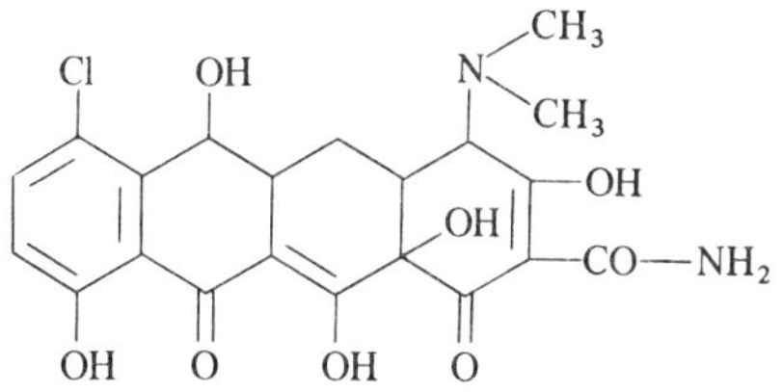

demethylchlortetracycline

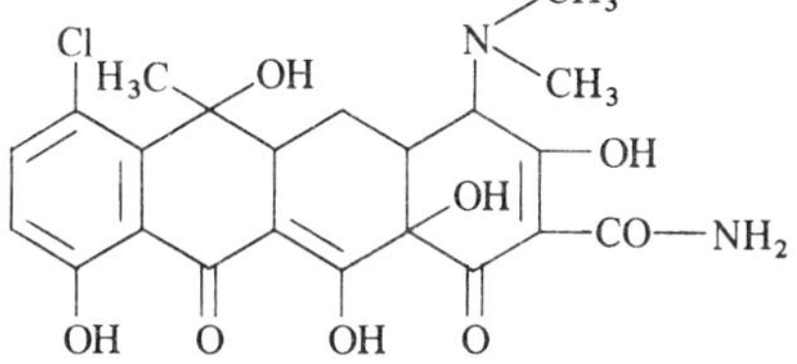

chlortetracycline

Kakemi et al. [474] also found 4-epitetracycline in rat feces but not in urine.

Addison and Clark [7] described a chromatographic technique on Whatman cellulose phosphate cation-exchange paper with water as solvent. The anhydrotetracycline formation that could occur in the Kelly and Buyske procedure [501] was avoided. Tetracycline elution by 10% ammonium chloride solution resulted in 93.1 ± 1.5% recovery. This method confirmed the fact that tetracycline was excreted mostly unchanged with traces of epitetracycline.

The data on paper chromatography of tetracycline-related antibiotics and their possible metabolites are summarized in Table 68.

Caffau and Jacobacci [188] briefly examined the metabolism of several tetracycline antibiotics by means of paper chromatography (butyl acetate; butanol:ammonium hydroxide). Tetracycline gave unchanged drug, 4-epitetracycline, and an unidentified compound. Oxytetracycline gave unchanged drug and a blue-green fluorescent spot. Chlortetracycline yielded a fluorescent spot with sodium arsenite that could be attributed to didesmethylaminoaureominic acid. After alkaline hydrolysis of the urine, isochlortetracycline and the α- and β-aureominic acids were found in addition to this metabolite, although the antibiotic was mostly excreted unchanged. Demethylchlortetracycline was excreted with its corresponding 4-epianhydro derivative.

III. PENICILLINS

Paper chromatography followed by microbiological detection indicates an active metabolite of benzylpenicillin in addition to unchanged antibiotic. Rolinson

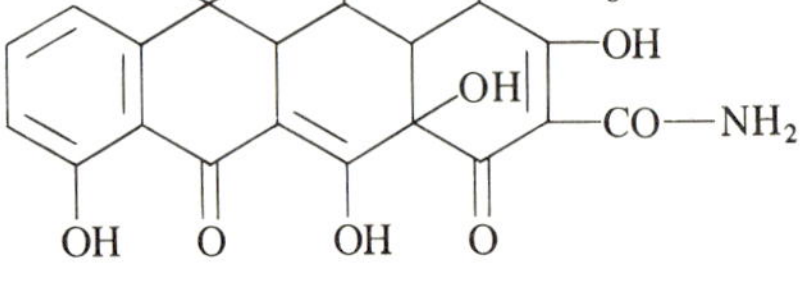

oxytetracycline

and Batchelor [782] compared the in vivo fate of a series of penicillins. Urine or blood was chromatographed on paper buffered at pH 6.2 (water saturated ethyl ether) according to the procedure of Karnovsky and Johnson [487]. Exceptions were meticillin, which was analyzed at pH 5, and ampicillin, where the solvent systems were: butanol:ethanol:water 4:1:5 and butanol:acetic acid:water 4:1:5. One or two active metabolites appeared on microbiological detection, but none was identified.

Suzuki and Tanaka [883] observed that penicillin G was hydrolyzed in an isolated perfused rabbit liver into penicolloic acid. This acid was identified by paper electrophoresis and chromatography (butanol, isobutanol, or isoamyl alcohol saturated with a 0.2M phosphate buffer, pH 6).

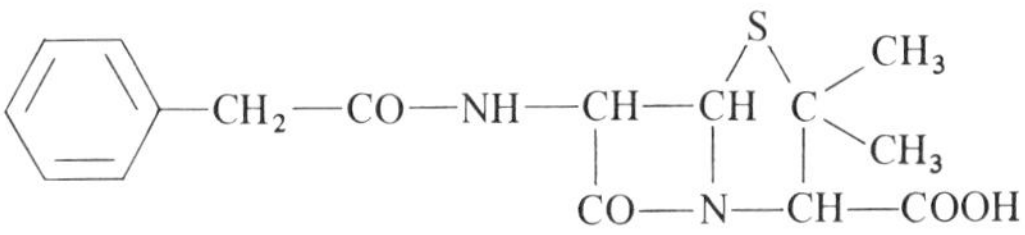

benzylpenicillin

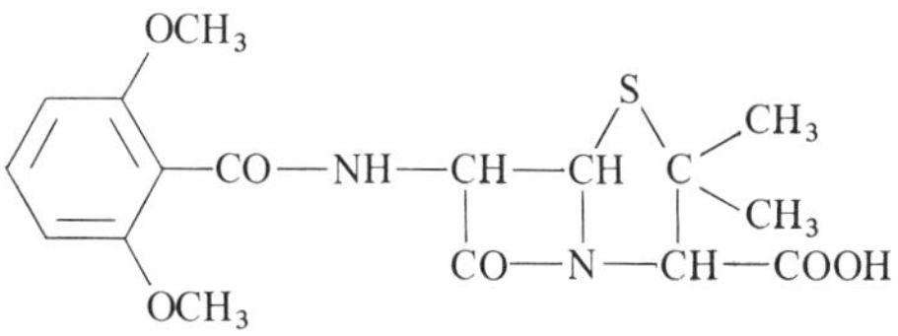

meticillin

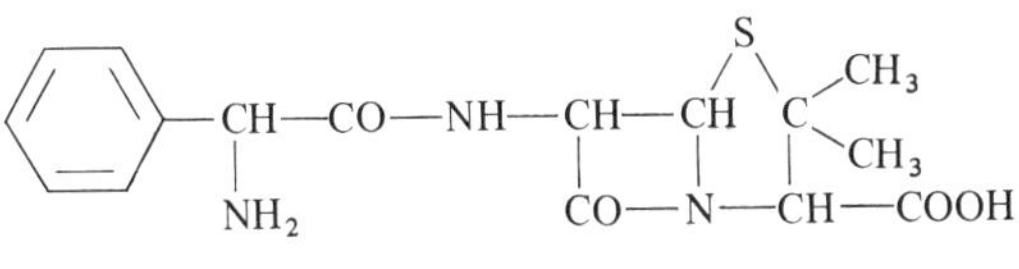

ampicillin

TABLE 68

Product	R_f in system:[a]				
	I	II	III	IV	V
Tetracycline	0.12	0.32	0.59-0.65	0.40	0.59
4-Epitetracycline	0.09	0.26	0.59-0.65	0.15	0.36
Anhydrotetracycline			0.86	0.54-0.62	0.14
4-Epianhydrotetracycline			0.86	0.34-0.40	
Isotetracycline			0.46	0.21	
Chlortetracycline	0.52	0.55	0.77	0.47-0.54	0.61
4-Epichlortetracycline	0.26	0.32	0.77	0.43	
Anhydrochlortetracycline	0.22		0.91	0.83	
4-Epianhydrochlortetracycline	0.74		0.91	0.72	
Isochlortetracycline	0.23	0.37	0.68	0.47-0.54	
4-Epiisochlortetracycline	0.20	0.25	0.68	0.43	
5a-Epichlortetracycline	0.63	0.70	0.80	0.70	
5a-(11a)-Dehydrochlortetracycline	0.30	0.16	0.66	0.57	
Demethylchlortetracycline			0.71	0.36	0.53
4-Epidemethylchlortetracycline			0.71	0.17	0.32
Anhydrodemethylchlortetracycline			0.89	0.70-0.76	
4-Epianhydrodemethylchlortetracycline			0.92	0.44-0.50	
Oxytetracycline			0.59	0.27	0.61
6-Desmethyl-6-deoxytetracycline					0.42
Epi-6-desmethyl-6-deoxytetracycline					0.21
6-Methyleneoxytetracycline					0.46

[a]Systems: I, butyl acetate:5% trichloroacetic acid:0.3M phosphate buffer (pH 2) 5:1:4 [298]; II, butanol:0.1M EDTA solution (pH adjusted at 8.3 with ammonium hydroxide) 1:1 [298]; III, butanol:acetic acid:water 4:1:5 (EDTA-treated paper) [298, 501, 502, 503]; IV, butanol:ammonium hydroxide:water 4:1:5 (EDTA-treated paper) [298, 501, 502, 503]; V, water (Whatman cellulose phosphate cation-exchange paper) [7].

Koga [517] isolated penicilloic acid and penicillamine by electrophoresis from the urine of ^{35}S-labeled penicillin-dosed animals.

Ampicillin paper chromatography (ethyl acetate : isopropanol : water 4 : 2 : 1) did not reveal any active metabolite to Acred et al. [4].

Penicillamine has as a metabolite penicillamine disulfide and other disulfides, which are difficult to separate from methionine and cystine. Borner [102] oxidized penicillamine and its disulfide with performic acid into penicillamic acid and separated it on a Dowex-1 X-8 column by means of a Technicon Autoanalyzer. A ninhydrin determination was performed. Penicillamine was separately determined by colorimetry with the ferric chloride-potassium cyanide reagent.

IV. ERYTHROMYCIN

Lee et al. [544], of Eli Lilly, chromatographed the bile of erythromycin-dosed rats and dogs on paper by an undescribed procedure.* Bioautography revealed one active metabolite.

---

* The chromatographic technique used in Eli Lilly research seems to have never been published, although it had been announced [996] that it would appear under the signatures of Bird and Pugh. However, these authors only reported a separation procedure for chlortetracycline, oxytetracycline, and tetracycline [93] and did not indicate whether it applied to erythromycin.

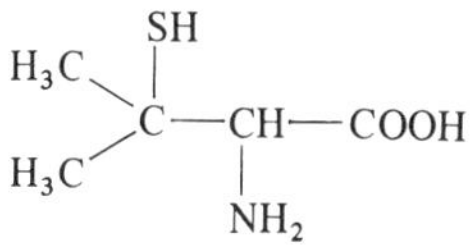

penicillamine

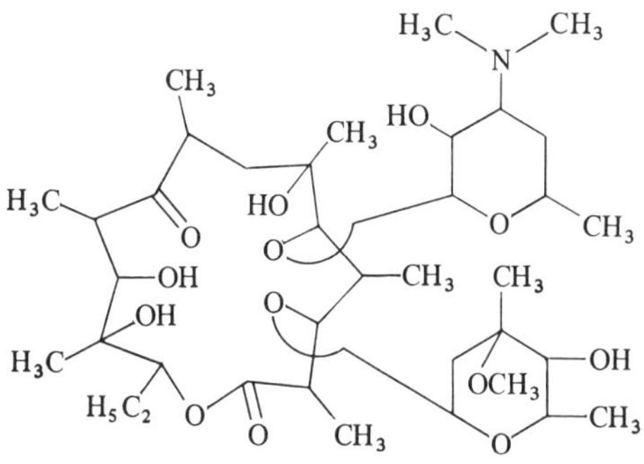

erythromycin

This metabolite was extracted by Welles et al. [996]. The bile was treated with chloroform and the extract separated by counter-current distribution (0.1M phosphate buffer, pH 6.5 : methylisobutylketone : acetone 10 : 10 : 0.5). The metabolite was identified with previously isolated N-desmethylerythromycin by its microbiological activity and its R_f value. Confirmation was obtained by IR spectrophotometry and X-ray diffraction.

Lee et al. [545] using N-^{14}CH$_3$-labeled erythromycin, demonstrated that N-desmethylerythromycin was the only metabolite. Urine, bile, and feces were analyzed by paper chromatography.

V. STREPTOMYCIN, KANAMYCIN

Suzuki et al. [884] showed, with ^{3}H-labeled streptomycin and kanamycin, that these antibiotics were excreted without any important metabolic modification. Microbiological assays agreed with radioactivity measurements.

Caffau and Jacobacci [189] studied the metabolism of a series of oligosaccharide antibiotics: streptomycin, dihydrostreptomycin, kanamycin, and paromomycin. Urine was examined by paper chromatography in various solvent systems, by thin-layer chromatography and paper electrophoresis at pH 5.6 and 8.6. Chemical and microbiological detections were used. Only the unchanged antibiotics were found.

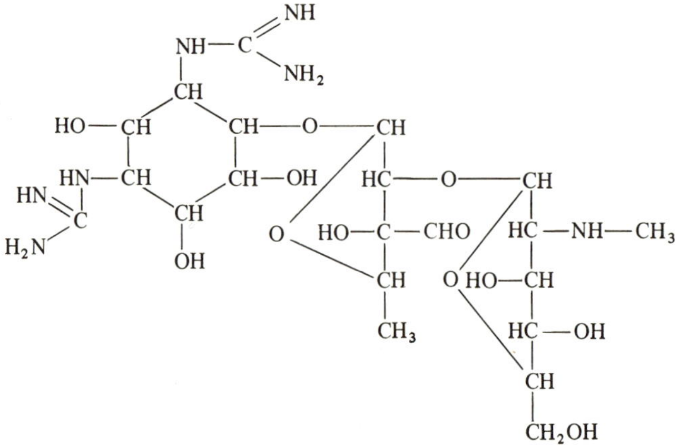

streptomycin

VI. FUSIDIC ACID

Fusidic acid is mainly eliminated by the biliary route. Its metabolism was investigated in cholecystectomized patients by Godtfredsen and Vangedal [369]. Bile was extracted with ethyl acetate and the solution chromatographed on paper (benzene : methanol : water 2 : 1 : 1) and on silica gel thin layer (cyclohexane : chloroform : methanol : acetic acid 10 : 80 : 2.5 : 10; chloroform : acetic acid : methanol 8 : 1 : 1). Seven metabolites were detected, extracted into ethyl acetate, and partitioned (pentane : methanol : water 10 : 8 : 2).

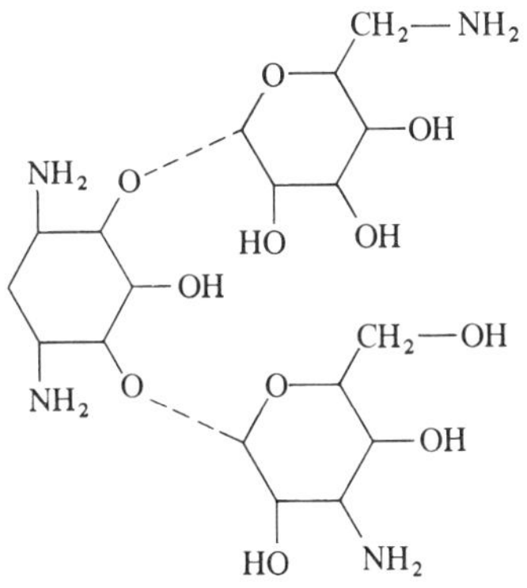

kanamycin

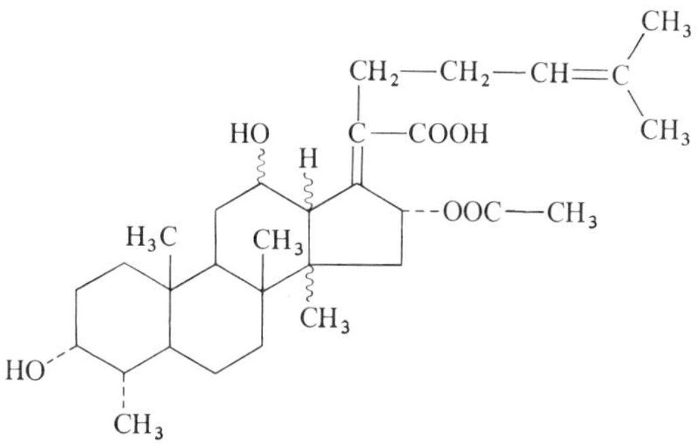

fusidic acid

Pentane removed cholesterol. The methanolic phase contained fusidic acid and five of its metabolites, whereas the aqueous phase contained the two others. The mixture in the methanolic phase was fractionated on a "dry column" of silica gel (chloroform : acetic acid : cyclohexane : methanol 80 : 10 : 10 : 2.5). The column was cut in sections after development and each section eluted with methanol. Two purified metabolites were obtained and examined by elemental analysis and UV and IR spectrophotometry. The NMR and mass spectra were recorded after methylation. The methyl esters were also treated with ozone and the ozonolysis products were identified by thin-layer chromatography. A third metabolite was extracted from the aqueous phase and purified on silica gel "dry column" (chloroform : acetic acid : methanol 8 : 1 : 1), then crystallized and examined by elemental analysis, UV and IR spectrophotometry, and potentiometric titration. On β - glucuronidase hydrolysis (4 hours at 37° C, pH 6), fusidic acid resulted and was recognized by IR spectrophotometry, chromatography, and microbiological assay. The possible structures of the two first metabolites were deduced generally from their spectral characteristics. The four other metabolites were present in small amounts and were not isolated.

VII. PUROMYCIN

The in vitro demethylation of puromycin was followed by Mazel et al. [627] who determined generated formaldehyde by the Cochin and Axelrod technique [209].

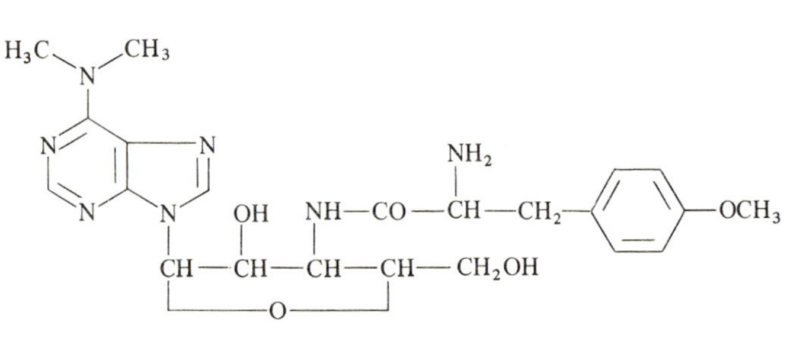

puromycin

VIII. GRISEOFULVIN

Barnes and Boothroyd [49] demonstrated that, unlike other antibiotics, griseofulvin is almost completely metabolized. They used a ^{36}Cl-labeled antibiotic. Animal or human urine was extracted with butanol. After evaporation and redissolution into chloroform, the solution was chromatographed on a silica gel column (chloroform). The radioactive fraction was recrystallized from ethanol. Elemental analysis indicated one carbon and two hydrogens less than griseofulvin. Paper chromatography (benzene : cyclohexane : methanol : water 5 : 5 : 6 : 4, organic phase + 0.5% of acetic acid) agreed with its characterization as desmethylgriseofulvin (R_f 0.15 as compared with 0.9 for griseofulvin). Griseofulvin was regenerated on reaction with diazomethane. Ultraviolet spectra in acid and alkaline solution led to the conclusion that demethylation was in the 6 position.

Kaplan et al. [485] studied the influence of various inhibitors on in vitro griseofulvin metabolism. Unchanged griseofulvin was extracted with ethyl ether. After acidification of the medium, desmethylgriseofulvin was extracted with ether. This was performed before and after β-glucuronidase hydrolysis. Analyses were made by UV spectrophotometry.

Riegelman et al. [764] determined 6-desmethylgriseofulvin in urine by differential spectrophotometry at pH 1.0 and 6.5.

Symchowicz and Wong published a very extensive study of griseofulvin metabolism in vivo in the rat [887] and in vitro in rat tissues [888] with a biosynthetically prepared ^{14}C-labeled antibiotic. Urinary metabolites were extracted into ethyl ether after Glusulase (β-glucuronidase + sulfatase)

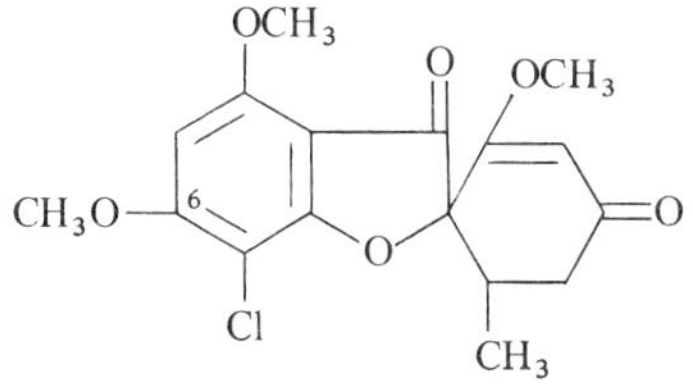

griseofulvin

hydrolysis for 60 minutes at 37° C, pH 5. They were chromatographed on paper (benzene : cyclohexane : methanol : water 5 : 5 : 6 : 4; butanol : ammonium hydroxide 20 : 1) and compared to reference compounds. The 4- and 6-demethylated products, both free and conjugated, were identified in vivo as well as in vitro.

IX. PRISTINAMYCIN

This antibiotic actually is a mixture. Jolles et al. [465, 466] were interested [466] in the metabolism of one of its major components, a cyclic polypeptide. The unchanged product was determined by microbiological assay and chromatography on paper (butanol : acetic acid : water 4 : 1 : 5), on DEAE-cellulose and ECTEOLA-cellulose (propanol : 0.1M phosphate buffer, pH 7, 5 : 2), and on silica gel thin layer (chloroform : methanol : water 6 : 4.3 : 5 and 6 : 5 : 3; chloroform : methanol : 2N ammonium hydroxide 2 : 2 : 1). Elemental amino-acids analysis with the automatic Technicon apparatus characterized the polypetide component of pristinamycin, which actually contains several amino acids that are not normally present in urine. No anomalies attributable to a metabolite could be observed. The peptides characterized by paper chromatography (butanol : 0.1N hydrochloric acid : pyridine 5 : 2 : 3) or on silica gel thin layer (chloroform : methanol : 0.1N ammonium hydroxide 2 : 2 : 1) were compared with specially prepared synthetic polypeptides. The separated products were hydrolyzed and the constituent amino acids analyzed on the Technicon apparatus. A metabolite derived from 3-hydroxypicolic acid was detected. Direct paper and thin-layer chromatographies revealed a fluorescent metabolite that was isolated by three successive paper chromatographies of lyophilized urine. Glycine and 3-hydroxypicolic acid were characterized on hydrolysis and the R_f value of the precursor was close to that of synthetic 3-hydroxypicolylglycine. Definite identification of this metabolite was not obtained.

Chapter XIX

GLYCOSIDES

The study of the metabolism of these structurally complex drugs has evidently presented great analytical difficulties. Nevertheless, Grimmer et al. [386] were able to propose a general procedure to identify and determine small amounts of cardiotonic glycosides and their metabolites in biological media. Urine or organ preparations were first treated with zinc acetate and then extracted with chloroform. The solution was evaporated and the dry residue redissolved into 50% methanol. This methanolic solution was extracted with petroleum ether, then with carbon tetrachloride, and finally with chloroform. This last solvent contained all the glycosides. The chloroform concentrate was chromatographed on a neutral alumina column (methylene chloride with methanol concentration increasing from zero to 50%). The products separated on the column were characterized by chromatography on formamide-impregnated paper with various solvent systems:

- Xylene : methylethylketone 1 : 1, formamide saturated;
- Octanol : water : formamide 4 : 1 : 4;
- 2-Ethyl-1-hexanol : ethyleneglycol monohexyl ether : water : formamide 5 : 5 : 6 : 6.

The separated compounds were eluted by the mixture chloroform : methanol 1 : 1 and determined colorimetrically with xanthydrol.

I. DIGITALIS HETEROSIDES

The metabolism of digitalis cardiotonic heterosides has been studied in the stated intervals by Wright (1954-1962) and Repke (1958-1961).

Wright mostly used paper chromatography. In a study on digoxin by Shepheard, Thorp, and Wright [833], urine was extracted with chloroform after lead acetate precipitation at pH 4.5. The extract was chromatographed, either on formamide-impregnated paper (chloroform : benzene 88 : 12, formamide saturated, horizontal chromatography), or on untreated paper (chloroform : ethyl acetate : benzene 6 : 2 : 2, water saturated). In addition to the administered drug, one metabolite was detected.

Brown, Ranger, and Wright [143] examined digitoxin and lanatoside C metabolism similarly. The solvent system for horizontal chromatography was modified (chloroform : benzene : butanol 78 : 12 : 5, formamide saturated). Brown, Shepheard, and Wright [142] added a third solvent system (ethylhexanol : pentanol : formamide : water 6 : 2 : 4 : 1 — paper impregnated with the organic phase, development with the aqueous phase). Hydrolysis assays were added to chromatographies by Brown and Wright [144].

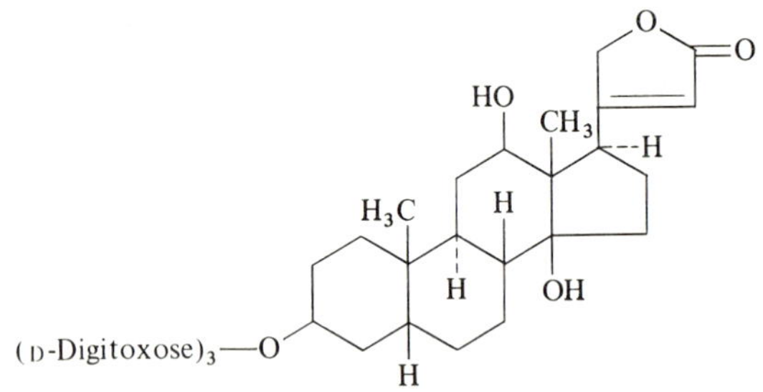

digoxin

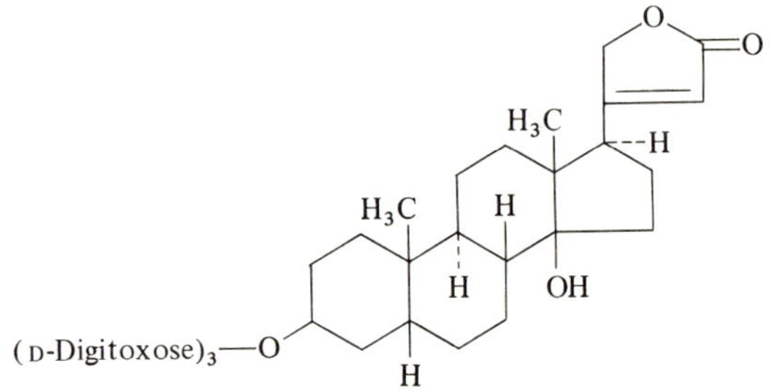

digitoxin

Ashley, Brown, Okita, and Wright [34] extended to man the work done in the rat. They added three new paper chromatographic solvent systems:

- Tetrahydrofuran : chloroform 1 : 1;
- Methylethylketone : isopropyl ether 4 : 1;
- Ethyl hexanol : amyl alcohol : formamide : water 6 : 2 : 2 : 8.

Cox and Wright [224] applied the same techniques to hepatic excretion, but colorimetric determinations could be used with the larger amounts.

Similar studies were conducted on acetyldigitoxin by Martin and Wright [612].

Wright [1025] administered ^{14}C-labeled digoxin in 1962 and identified one of its metabolites as digoxigenin bisdigitoxoside by paper chromatographic separation and isotopic dilution.

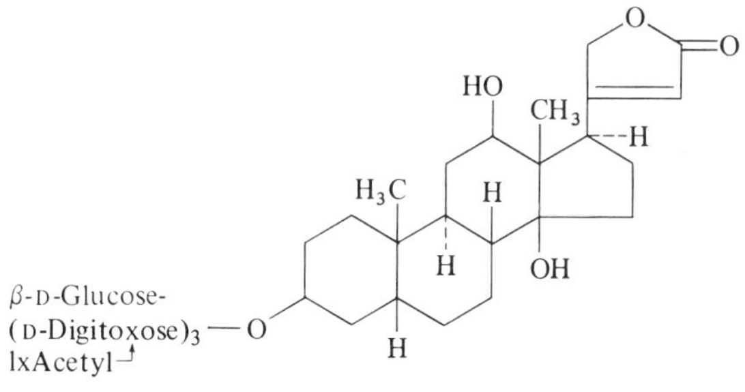

lanatoside C

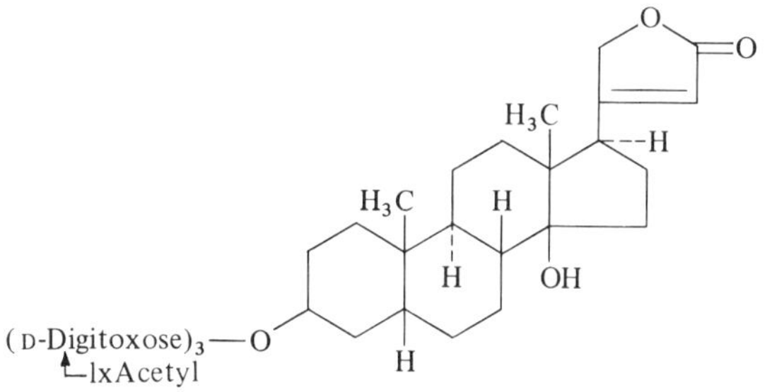

acetyldigitoxin

Repke [752] extracted digitoxin with chloroform from tissues and excreta and, after treatment on alumina, colorimetrically determined the glycoside. Chromatography on formamide-impregnated paper (xylene : methylethylketone 3 : 1 or benzene : methylethylketone 1 : 1) did not detect a metabolite in the rat [753].

The analytical techniques were described in further detail [754], where the chloroform extract was chromatographed on an alumina column and eluted by 2% and 10% ethanol solutions in chloroform. This method showed a very significant retention of digitoxin in the rat organism [755]. Digoxin was the major metabolite of digitoxin in the rat and was determined in various organs [756].

The fractions obtained by alumina column chromatography were further chromatographed on paper for qualitative and quantitative analyses [757] (xylene : methylethylketone 1 : 1; chloroform : isopropyl ether 9 : 1; isopropyl ether : tetrahydrofuran 3 : 2 — all three systems formamide saturated with a formamide-impregnated paper). Repke et al. [757] were thus able to demonstrate that digitoxin was metabolized by reduction of the digitoxose chain and hydroxylation on the carbon in position 12.

Repke studied the products resulting from digitoxin hydroxylation and chain breaking. He isolated nine metabolites and identified five of them [758], using the same procedures for separation and determination. The metabolites were eluted and examined by UV spectrophotometry and compared to authentic compounds. In vitro experiments with liver slices were analyzed similarly [539].

The availability of ^{14}C- or ^{3}H-labeled glycosides offered new research possibilities. Most of the work with these tracers used the separation procedures of Okita et al. [688], which had been published 10 years earlier. A series of partitions between solvents were used to separate products soluble in aqueous ethanol, carbon tetrachloride, and chloroform. The chloroform fraction was analyzed by alumina column chromatography which separated digoxigenin from a mixture of digoxin and digoxigenin mono- and bisdigitoxosides.

This procedure was first applied to monitor digitoxin blood levels in man [689]. It was improved by introducing paper chromatography to complete the separation achieved on the column. Marcus et al. [602] used either chromatography on formamide-impregnated paper [(I) chloroform : benzene :

butanol 78:12:5] according to Brown et al. [142], or a separation on paper impregnated with a 10% amyl alcohol acetonic solution [(II) amyl alcohol: water 1:1] (Table 69).

Similarly, Wong and Spratt [1017] and Lage and Spratt [535] used a final chromatography on formamide-impregnated paper (formamide saturated chloroform). Their technique involved the radioactive monitoring of each fraction. The results were statistically treated on an IBM 7040 computer.

Katzung and Meyers [489] administered highly purified ^{3}H-labeled digitoxin to dogs. Some were equipped with complete biliary fistulas. Urine, feces, and bile were extracted with chloroform and the chloroform soluble and insoluble products were studied by chromatography on formamide-impregnated paper (2,2-dimethyl-4-pentanone:isopropyl ether 4:1; formamide saturated dichloromethane) or on untreated paper (water saturated butanol). Digitoxin and digoxin were the only excretion products found in the chloroform. Hydrochloric acid hydrolysis of water soluble metabolites followed by paper chromatography showed the presence of radioactive digitoxigenin and digitoxose. Enterohepatic recirculation was confirmed.

A sensitive and accurate technique for digitoxin assay in urine, plasma, or feces was proposed by Lukas and Peterson [567] as based on a double isotopic dilution. An amount of tritiated digitoxin was added to the sample, which was first analyzed by paper chromatography (cyclohexane:dioxane:methanol:water 4:4:2:1) to separate digitoxin metabolites. The glycoside was then acetylated with ^{14}C-labeled acetic anhydride. The resulting acetate was purified by a new paper chromatography. Measurement of the radioactivity due to the ^{14}C determined the digitoxin, whereas the ratio of the final tritium (^{3}H) activity over the activity of tritium introduced permitted correction for losses during the analysis.

Thomas and Wright [901] demonstrated metabolic digoxigenin epimerization in the rat on administration of the tritiated drug. They found three components in the chloroform extract of bile, which were separable by thin-layer chromatography (conditions not reported). One of these components was identified as 3-epidigoxigenin by isotopic dilution.

Thin-layer chromatography was introduced in 1965 and simplified the analytical procedures. Abel et al. [2] performed a simple ethanol blood extraction and chromatographed the ethanolic solution on silica gel (ethyl acetate:butanol 9:1). Digoxin was separated from its metabolites:

TABLE 69

Product	R_f in system:	
	I	II
Digoxin	0.56	0.18
Digoxin bisdigitoxoside	0.35	0.25
Digoxigenin	0.29	0.39
Digoxin monodigitoxoside	< 0.29	

digoxigenin mono- and bisdigitoxosides, digoxigenin, and dihydrodigoxigenin. They were eluted and radiometrically determined.

In the same year, Jelliffe [457] had developed a chemical analysis for digitoxin and digoxin in urine. Subsequently, he also used [458] thin-layer chromatography; a chloroform extract of urine was washed with sodium hydroxide and spotted on a silica gel plate and developed with the system: dichloromethane:methanol:formamide 90:9:1. Digitoxin and digoxin were visualized after separation by water spraying. They were eluted and determined by the xanthydrol method.

II. OTHER GLYCOSIDES

Ouabain transformations were examined by Cox, Roxburgh, and Wright [223] with paper chromatography (water saturated butanol — paper

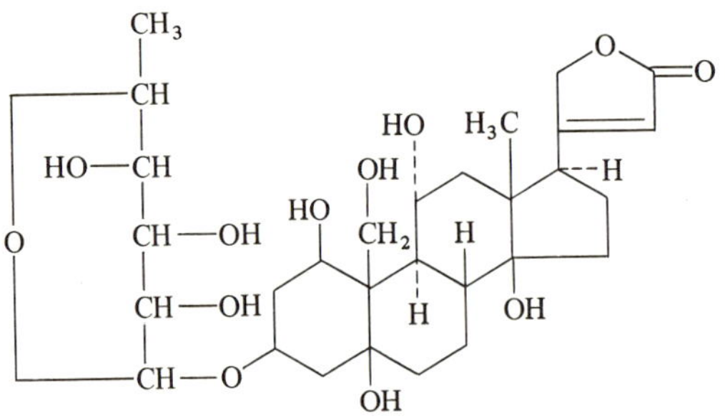

ouabain

OTHER GLYCOSIDES

impregnated with the aqueous phase and development with the butanol phase). After the injection of large doses, two metabolites were observed in addition to the unchanged drug.

Cymarin and its metabolites were chromatographed by Moerman [637] on silica gel thin layer (dichloromethane : methanol : formamide 80 : 19 : 1) and on alumina thin layer (tetrahydrofuran). This separation was preceded by a chloroform extraction and a purification on alumina column (2% and 10% ethanol chloroformic solution). Strophanthidin and cymarol were detected in addition to the unchanged drug.

Lauterbach et al. [538] examined the metabolism of various strophanthidin glycosides: cymarin, convallatoxin and

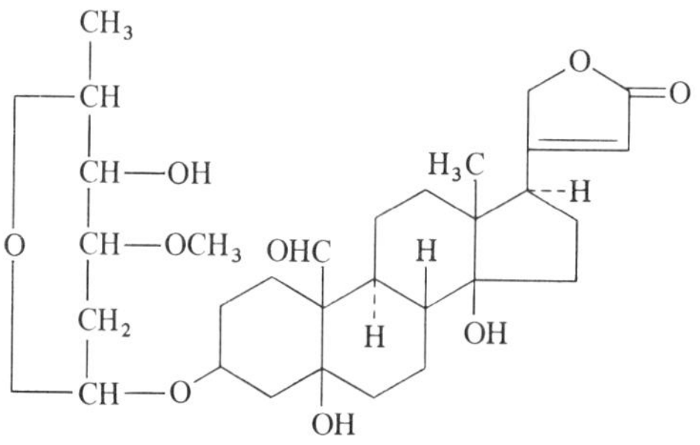

cymarin

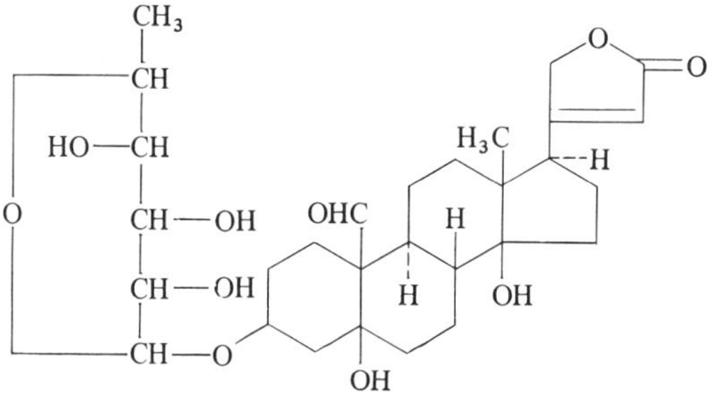

convallatoxin

helveticoside. Incubation with liver slices or homogenates resulted in the reduction of the aldehyde group in position 19 and the formation of the corresponding hydroxylated derivatives. Metabolites were separated by paper chromatography (xylene:methylethylketone 1:1, formamide saturated; chloroform:tetrahydrofuran 1:1, 80% saturated with formamide; water saturated ethyl acetate). They were identified by their R_f, the reactions of the glycoside moiety, and their resistance toward Girard's T reagent.

Zathurecky et al. [1040] mentioned that three helveticoside and convallatoxin metabolites were identified by paper and thin-layer chromatography in various solvents and compared to reference compounds. The separated products were spectrophotometrically determined. These metabolites resulted from the reduction of the aldehyde group to a carbinol and from the hydrolysis of the glycoside bond.

Carlat et al. [193] administered tritiated enoxolone to man. Urine was extracted with ethyl acetate and then at pH 11 with butanol. The butanol fraction that contained the conjugated compounds was treated with β - glucuronidase. The residual aqueous phase was hydrolyzed by sulfuric acid. Chromatography on a Florisil column with ethanol benzenic solutions of increasing concentrations was also applied. Only the unchanged drug could be characterized by its IR spectrum.

Parke et al. [699] took greater precautions in their preparation of tritiated enoxolone. Rat urine, feces, and bile were analyzed by paper

---

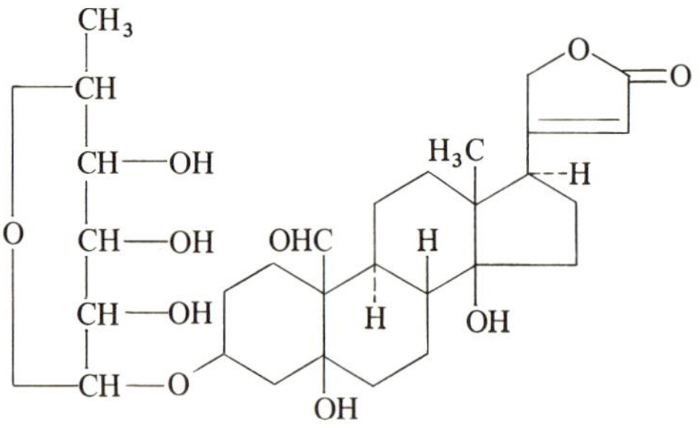

helveticoside

chromatography (chloroform : acetic acid : water 2 : 1 : 1; 70% acetic acid : ethylene dichloride : butanol 10 : 9 : 1; petroleum ether : methanol : water 5 : 4 : 1). Three unidentified metabolites were detected.

Iveson et al. [447] reported $^{14}CO_2$ excretion after administration of carbenoxolone ^{14}C-labeled on the carbons in the 1 and 4 positions of the succinate moiety. Three metabolites were found in the bile and characterized by chromatography and various tests (not defined). They were the 3O-O-glucuronide of carbenoxolone, the 3-O-sulfate ester, and a glucuronide of enoxolone.

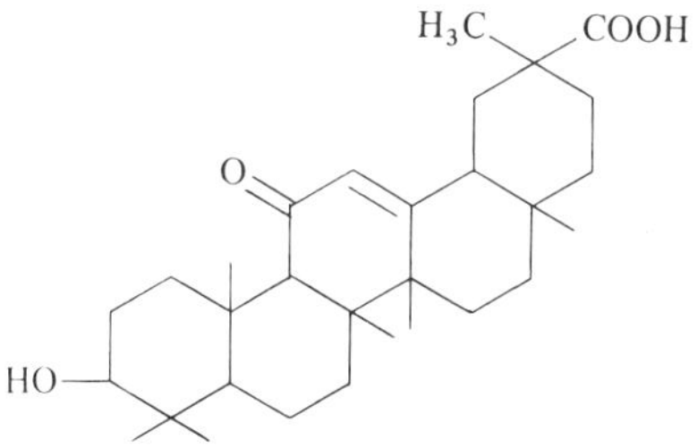

enoxolone

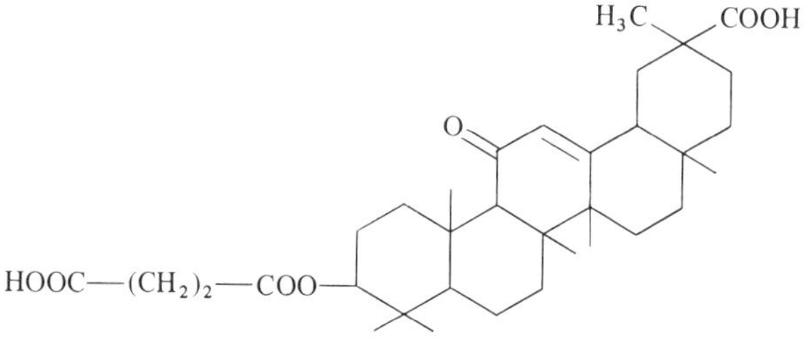

carbenoxolone

The excretion of rutoside, its aglycone quercetin, and related derivatives have been studied by several authors, but its metabolic investigation was initiated by Murray et al. [655]. A pure metabolite was obtained by ethyl ether extraction of urine and barium chloride precipitation. The acid was regenerated from its barium salt and recrystallized from the mixture ethyl ether:petroleum ether. The metabolite was analyzed and compared to 3,4-dihydroxyphenylacetic acid. It had the same melting point and X-ray spectrum.

Kapetanidis and Mirimanoff [484] studied trihydroxyethylrutoside metabolism. Paper and thin-layer chromatography, as well as analysis for excreted phenolic compounds in urine, gave negative results.

Kapetanidis [483] showed that urinary glucuronic acid increased after rutoside administration. The glucuronic acid determination of Bray and Thorpe [117, 111] was improved.

Heparin metabolism is difficult since the structure of this drug is incompletely known. The incorporation of ^{35}S in the molecule did not abet the

---

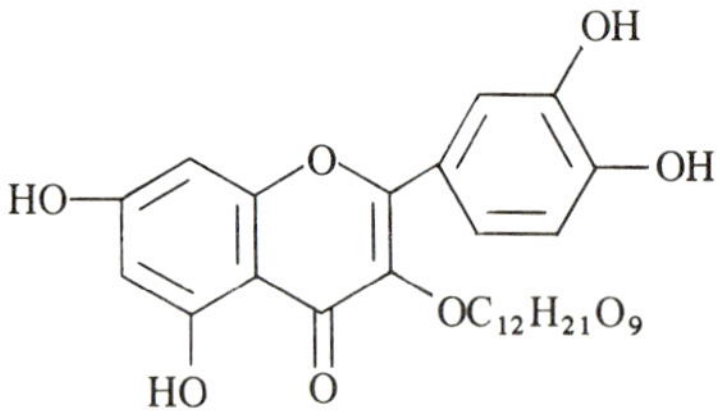

rutoside

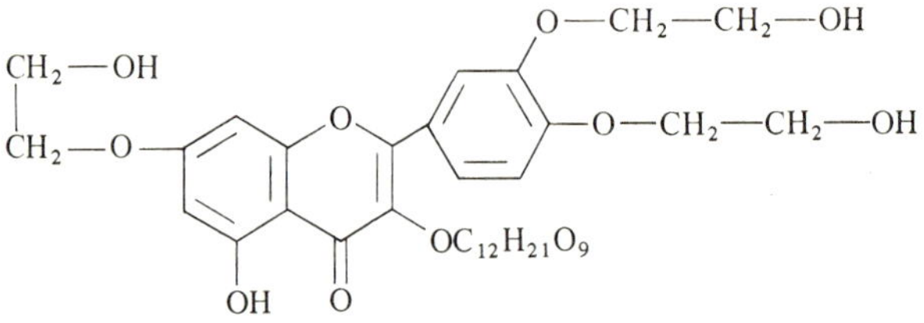

trihydroxyethylrutoside

study of its transformations. The labeled heparin was rapidly degraded to the sulfate and the ^{35}S was reincorporated into the normal mucopolysaccharides of the organism. These were not readily distinguished from administered heparin.

Danishefsky and Eiber [242] administered labeled heparin to the dog. The urine was dialyzed and the heparin precipitated by cetyltrimethylammonium bromide. The precipitate was then extracted by a 5% sodium iodide-alcohol solution which converted heparin and related compounds into their sodium salts and removed the quaternary ammonium ion. The heparin and related compounds were finally analyzed by paper electrophoresis (0.1M acetate buffer, pH 3.6 — 150 V — 18 hours) as well as by paper chromatography (propanol:0.066M phosphate buffer, pH 6.4, 1:3, ascending at 4° C) after Kerby [505]. Several radioactive components appeared but the major one was unchanged heparin.

Day et al. [244] also injected ^{35}S-labeled heparin and isolated sulfomucopolysaccharides by the method of Green and Day [384]. Blood and extracts from organs were hydrolyzed with trypsin in a dialysis tubing. Nondialyzable products were precipitated with acetone and then redissolved into water. On shaking of this solution with trichlorotrifluoroethane the remaining proteins, as well as the cerebroside sulfates, were removed. Heparin was further purified on an ECTEOLA-cellulose column (0.5M to 2.0M sodium chloride solution) after Green [383].

In 1966, McAllister and Demis [572] administered purified heparin to man. Mucopolysaccharides were precipitated by cetyltrimethylammonium bromide and redissolved into 0.05N sodium hydroxide. Paper electrophoresis (0.1M lithium sulfate — 50 V — 17 hours) revealed heparin and a component which moved more slowly. Precipitated mucopolysaccharides were then incubated for one to two days with hyaluronidase at pH 6 to eliminate chondroitin sulfates. Proteins were hydrolyzed by trypsin. The products in solution were precipitated as calcium salts by ethanol. Purified mucopolysaccharides were adsorbed on a AG-1 X 2 column and eluted by a sodium chloride solution of concentration increasing from 0 to 2.0M. Two fractions were obtained: one was identified as heparin, and the other as uroheparin, by paper electrophoresis and chromatography and by various chemical and biological tests. Thus, the major metabolite of heparin was obtained by the elimination of one sulfate group from each heparin tetrasaccharide.

Chapter XX

MISCELLANEOUS DRUGS

I. SULFUR-CONTAINING DRUGS

Snow showed [867] that <u>diethyl disulfide</u> was extensively metabolized, since the largest fraction of the ^{35}S from the administered labeled drug was found as sulfate. Two organic metabolites were detected, and one was methylethylsulfone.

Lowe [564] used a ^{14}C-labeled drug, and a significant fraction of the radioactivity appeared in the expired carbon dioxide with a small amount in the expired ethylmercaptan. The urine was chromatographed on paper in various systems after desalting on ion exchange. One of the two metabolites that were separated was Snow's sulfone. The other was extracted from urine with chloroform and analyzed by counter-current distribution (chloroform : water — 50 transfers). The distribution of the radioactivity indicated three compounds, the partition coefficients of which could be calculated. Thus, it was demonstrated that these metabolites were methylethylsulfone, methylethylsulfoxide, and an unidentified compound.

Peng [706] examined <u>busulfan</u> metabolism with the ^{35}S-labeled drug. In addition to measurement of radioactivity of sulfates, paper chromatography of the concentrated hydrolyzed urine (95% ethanol, ascending) revealed that methanesulfonic acid and unchanged drug were excreted.

$H_5C_2-S-S-C_2H_5$

diethyl disulfide

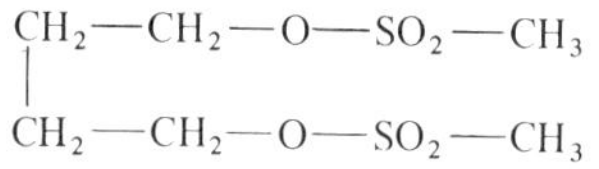

$$\begin{array}{l} CH_2-CH_2-O-SO_2-CH_3 \\ | \\ CH_2-CH_2-O-SO_2-CH_3 \end{array}$$

busulfan

Trams et al. [919] prepared the drug ^{14}C-labeled in its butyl chain in addition to the ^{35}S-derivative. The urinary metabolites were fractionated on a Dowex-50 column (citrate buffer of increasing pH, then sodium hydroxide of increasing strength). At least 17 radioactive compounds were thus separated. The three main ones were chromatographed again on Dowex-50. One of these was identified as the unchanged drug by isotopic dilution and paper chromatography. 1,4-Butanediol was recognized as a minor metabolite after recrystallization with the authentic compound as carrier. ^{35}S-Labeled methanesulfonic acid was isolated as the phenylhydrazonium salt.

Fox et al. [334] also administered ^{35}S-labeled busulfan and found methanesulfonic acid in addition to the unchanged drug by paper chromatography (butanol : dioxane : 2N ammonium hydroxide 4 : 1 : 5; butanol : 2N acetic acid 1 : 1).

Methanesulfonic acid ethyl ester, a moiety of the busulfan molecule, was studied in the rat by Roberts and Warwick [775]. After administration of the ^{14}C-labeled drug, the urine was fractionated by anion-exchange resin (Amberlite IRA-400) or cation-exchange resin (Amberlite IRC-120) before and after acid, alkaline, or enzymic hydrolysis. The fractions separated from unhydrolyzed urine were hydrolyzed in various manners. All fractions were analyzed by mono- or bidimensional paper chromatography in various solvent systems. The metabolites were identified by their behavior in the successive steps of the analytical procedure and their structural assignment was confirmed by the R_f values of synthesized authentic compounds. Also, labeled metabolites were administered to rats and the urines were analyzed by the same procedures used with the drug. Of the many metabolites found in urine, the principal ones were N-acetyl-S-ethylcysteine and conjugated S-ethylcysteine derivatives.

Ellard et al. [303] detected the unchanged ditophal, labeled with ^{35}S, by reverse isotopic dilution and separated the metabolites by countercurrent distribution (chloroform : water — 56 transfers) of a chloroform extract of urine. Methylethylsulfone and methylethylsulfoxide were thus

---

$H_3C-CH_2-O-SO_2-CH_3$
methanesulfonic acid ethyl ester

detected. A third metabolite and inorganic sulfate were also excretion products.

After administration of the ^{35}S-labeled <u>dimethylsulfoxide,</u> Gerhards et al. [353] found only dimethylsulfone and unchanged drug in human and rat urine. They applied silica gel thin-layer chromatography (chloroform : acetone : ethanol 5 : 4 : 1) and confirmed the identification by isotopic dilution.

Dimethylsulfone was isolated in the crystalline state by Williams et al. [1003] who studied by counter-current distribution the metabolites of a steroid administered as a solution in dimethylsulfoxide. The dimethylsulfoxide was identified by its IR spectrum and its melting point.

Distefano and Borgstedt [262] characterized dimethylsulfide in expired air by gas-liquid chromatography and mass spectrometry (the conditions were not reported).

Kolb et al. [518] also detected dimethylsulfide in the expired air by gas-liquid chromatography (the conditions were not reported).

A complete study was undertaken by Hucker et al. [435] with the ^{35}S-labeled drug. Dimethylsulfoxide and dimethylsulfone were separated by paper chromatography (ethyl acetate, ascending), thin-layer chromatography on silica gel (ethyl acetate), or gas-liquid chromatography (3% of Carbowax 20 M on Chromosorb G at 137° C). The dimethylsulfone extracted from urine by chloroform was identified by its IR spectrum, melting point, and elemental analysis [436]. Williams et al. [1002] also separated dimethylsulfoxide and dimethylsulfone by gas-liquid chromatography (30% of butanediol succinate on Chromosorb W at 115° C). The two compounds were identified by their IR spectra.

Hucker and Hoffman [437] separated dimethylsulfoxide and dimethylsulfone by selective extraction of dimethylsulfoxide as a stannic chloride complex.

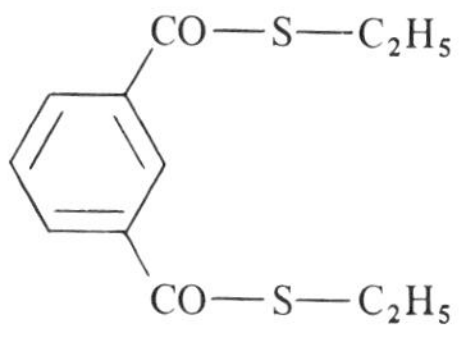

ditophal

$H_3C-SO-CH_3$

dimethylsulfoxide

The initial investigations on <u>diaphenylsulfone</u> metabolism were generally limited to the determination of excretion products by the Bratton-Marshall method.

Titus and Bernstein in 1949 [906] improved the technique by extracting diaphenylsulfone into methylethylketone. They checked the specificity of this extraction with paper chromatography. Two metabolites were characterized by butanol extraction of urine and counter-current distribution (butanol : water — 24 transfers).

Boyer et al. [106] chromatographed urine on paper (water saturated butanol) and found the unchanged drug in addition to one unidentified component.

Bushby and Woiwod [168] isolated the major metabolite of diaphenylsulfone and identified it with a synthesized compound [169]. A conjugated excretion product was separated from the unchanged drug by paper chromatography of urine (water saturated butanol; water saturated phenol; propanol : butanol : water 2:3:5). Acid hydrolysis yielded diaphenylsulfone and glucuronic acid which were detected by paper chromatography. This metabolite was isolated by a double preparative paper chromatography and compared with the synthetic compound, both prior to and after acetylation. It was a N-monoglucuronide.

<u>Disulfiram</u> metabolism has been studied more extensively to answer the question of whether the drug or one of its metabolites reacted with ethanol or one of ethanol's metabolites to induce alcohol intolerance. Domar et al. [265] separated ^{35}S-labeled disulfiram from diethyldithiocarbamic acid by extracting urine with carbon tetrachloride. The cupric salts of both compounds were determined by colorimetry.

Kaslander [488] has demonstrated the presence of a diethyldithiocarbamic acid S-glucuronide in human urine after disulfiram treatment. Twenty-seven

---

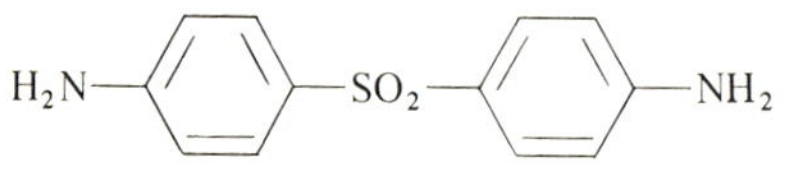

diaphenylsulfone

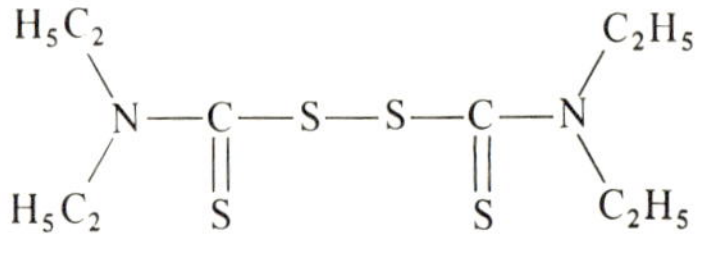

disulfiram

liters were concentrated to 2 l, acidified, and extracted with butanol. The concentrated butanol solution was fractionated by counter-current distribution (butanol:acetic acid:water 4:1:5 — 100 transfers). A fraction that gave a positive reaction with the thione group reagent was concentrated and underwent another counter-current distribution. The product thus purified gave a single spot on paper chromatography (propanol: water 85:15; butanol: acetic acid:water 4:1:1). Paper electrophoresis showed it to be an acid, and the naphthoresorcinol reaction was positive. The indicated diethyldithiocarbamic acid glucuronide was synthetized. The triacetylmethyl ester of this glucuronide was isolated by the method of Kamil et al. [480]. Elemental analysis, counter-current distribution, melting points and IR spectrum confirmed the identity of this metabolite.

Strömme [879] administered the ^{35}S-labeled drug in 1965. Urine, plasma, and liver extracts underwent gel filtration on Sephadex G-25 column using the technique developed by the same author to study disulfiram binding to serum proteins [878]. EDTA (0.01M) buffered at pH 8.5 was used as the elution solvent. All traces of metal were removed first from this buffer by extractions with a 0.1% diphenylthiocarbazone solution in chloroform. Disulfiram, diethyldithiocarbamate, and proteins were thus separated, and the other metabolites were eluted with diethyldithiocarbamate. Glucuronides were hydrolyzed with boiling 7.3M phosphoric acid for 3 1/2 hours. Metabolites were identified by paper chromatography (propanol:water 85:15; butanol:acetic acid:water 4:1:1) and by high voltage electrophoresis (acetic acid:pyridine:water 10:1:89 — 45 V/cm) and compared with the synthetic S-glucuronide (see [488]). The second radioactive metabolite was an inorganic sulfate. Complete stoichiometric balance was obtained with radioactivity measurements.

II. METAL-CONTAINING ORGANIC DRUGS

A. Organomercury Derivatives

Organomercury drugs are amenable to polarographic determination. Weiner and Müller [991] demonstrated that a relationship existed between the half-wave potential of the first mersalyl reduction wave and the pH of the medium. This relationship is not the same for the pure drug in solution and for the compound in dog urine. The relationship in urine was the same as that which had been previously observed with mersalyl in the presence of cysteine. It was thus possible to assign the behavior of mersalyl in urine to its interaction with cysteine.

This polarographic technique was applied by Weiner et al. [990] to follow mersalyl renal excretion in the chicken after injection into the portal vein with similar results.

Müller and Weiner [649] showed the existence of two chlormerodrin and mersalyl excretion products in the dog. The half-wave potential/pH curves exhibited two inflection points for both drugs. One was assigned to the cysteine complex and the other to a complex with N-acetylcysteine. This was confirmed by experiments with reference compounds.

Clarkson et al. [208] demonstrated that chlormerodrin liberates mercuric ions in the kidney. They developed a procedure that distinguished between mercuric ions and unchanged chlormerodrin. It was based upon the ability of ^{203}Hg to diffuse in a mercury vapor atmosphere when it is present as the complexed ion. No diffusion occurs when ^{203}Hg is bound to a carbon atom in an organic compound. The diffusion rate of ^{203}Hg was measured in a Conway cell.

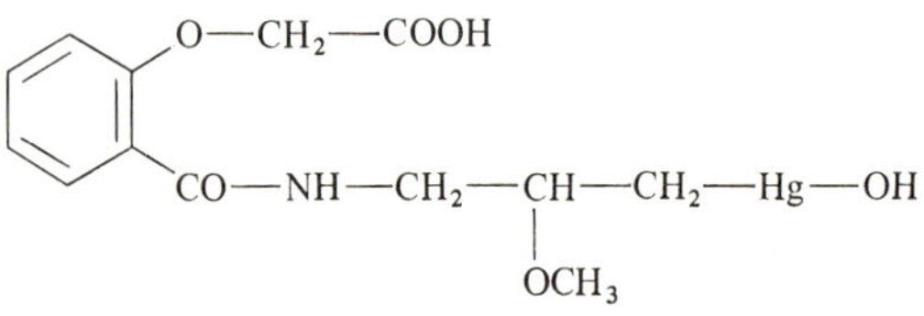

---

mersalyl

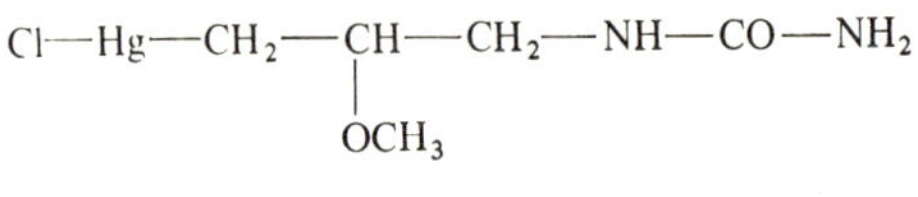

chlormerodrin

Handley and Seibert [397] studied the fate of meralluride. Alumina column chromatography retained the compounds with a free carboxylic group, and they were subsequently eluted with a sodium bicarbonate solution. Two mercurial fractions were separated from dog urine. They were not studied further.

B. Phosphorous-Containing Drugs

Different antineoplastic phosphoramide derivatives have been considered.

Triethylenephosphoramide was first investigated by Craig and Jackson [227], who prepared the ^{32}P-labeled derivative. They were only able to identify an inorganic phosphate in addition to unchanged drug in rat urine by paper chromatography (water saturated butanol). Nadkarni et al. [656, 657] obtained different results in mice with similar techniques. There was no unchanged drug, one intermediary metabolite, and 80% of the dose appeared as inorganic phosphate.

Craig et al. [225, 226] extended their researches to triethylenethiophosphoramide. Paper chromatography (butanol : dioxane : 2N ammonium hydroxide 4:1:5) of the radioactive urinary metabolites showed conversion into triethylenephosphoramide and the occurrence of other metabolites.

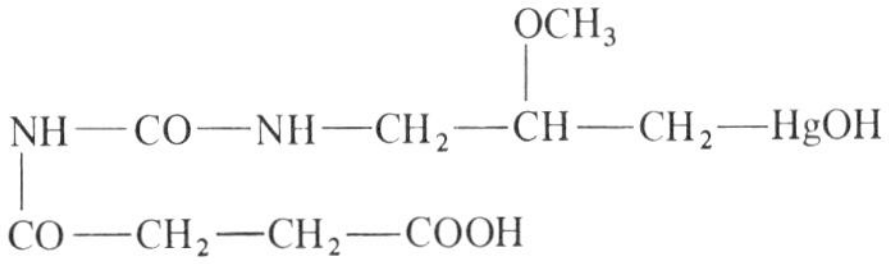

$$\begin{array}{l} \text{NH—CO—NH—CH}_2\text{—CH(OCH}_3\text{)—CH}_2\text{—HgOH} \\ | \\ \text{CO —CH}_2\text{—CH}_2\text{—COOH} \end{array}$$

meralluride

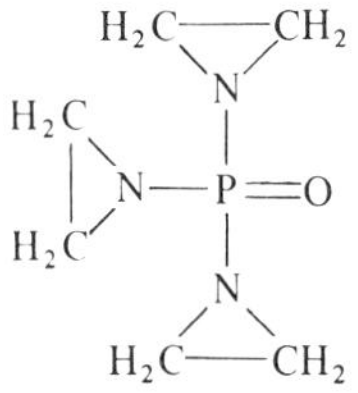

triethylenephosphoramide

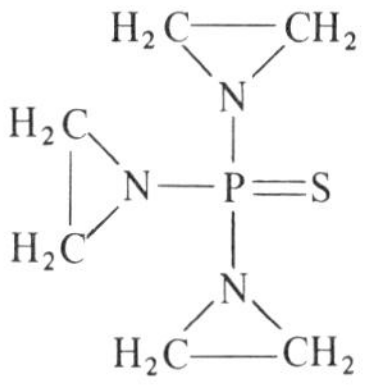

triethylenethiophosphoramide

Cyclophosphamide does not exhibit any cytostatic activity when incubated with tumor cells. However, when administered to rats, a cytostatic compound appears in the serum. This activation phenomenon was shown to be due to the liver microsomal enzymes.

The already known in vitro hydrolysis products of cyclophosphamide, di-(2-chloroethyl)-amine and N-(2-chloroethyl)-aziridine, were characterized as metabolites by Rauen and Dirschka [746] in rat blood. Deproteinized serum was chromatographed on a silica gel column (tetrahydrofuran:water; ethanol:ammonium hydroxide). The isolated products were identified by paper or thin-layer chromatographic comparison with authentic compounds.

Hohorst et al. [428] detected three components which reacted with p-nitrobenzyl-γ-pyridine by paper chromatography of rat serum (butanol: acetic acid:water 6:2:2). Two of these compounds contained phosphorus, the third one, present in small amounts, was identified as di-(2-chloroethyl)-amine. The serum major metabolite had same R_f as the in vitro activation product and seemed to be the cytostatic compound. This compound was isolated by adsorption on Dowex-2 X 8 at pH 9-10 and elution by 1N formic acid. Its R_f and its elemental analysis allowed to identify it as an open-ring cyclophosphamide analog: N,N-bis-(2-chloroethyl)-O-(3-aminopropyl)-phosphoric acid amido ester. The second phosphorous-containing metabolite was not characterized.

The same authors [429] used tritiated cyclophosphamide. They applied an improved chromatography technique of formamide-impregnated paper (20% solution in acetone) with the system benzene:methylethylketone, formamide saturated. Two new metabolites were detected in addition to the unchanged drug, which represented 75% of the activity.

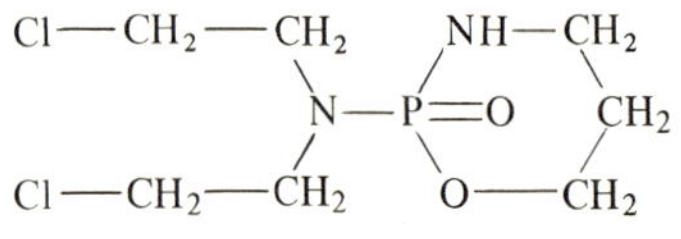

cyclophosphamide

C. Arsenic-Containing Drugs

Overby and Straube [697] demonstrated with arsanilic acid, ^{14}C- and ^{74}As-labeled on the two vicinal atoms, that this drug did not undergo significant in vitro degradation. The ratio of the two isotopes remained constant. They detected [696] several arsenic-containing derivatives in tissues by means of analytical techniques that were described in a separate paper [695]. They involved paper chromatography (acetonitrile : nitric acid : water 78 : 2 : 20; isopropanol : water 7 : 3), paper electrphoresis at various pH, and chromatography on a column of Dowex-50 X-2 (H$^+$) (3% trichloroacetic acid, 10% sodium chloride, and 5% sodium chloride in 0.1% sodium hydroxide).

III. HALOGENATED DRUGS

Chloroform is reduced in vitro to methylene chloride by the hepatic enzymes. Butler [180] was unable to detect this metabolite in the expired air of anesthetized dogs in his described device for collecting volatile compounds in expired air. The accumulated volatile compounds were further analyzed by gas-liquid chromatography.

Paul and Rubinstein [701] administered ^{14}C-labeled chloroform, and gas-liquid chromatography revealed a small amount of ^{14}CO$_2$.

When ^{14}C-labeled halothane was administered, Van Dyke et al. [956] only found minute amounts of radioactive products in urine and expired carbon dioxide. They systematically studied the metabolism of the volatile anesthetics : ethyl ether, chloroform, halothane, and methoxyflurane. ^{14}C- or

---

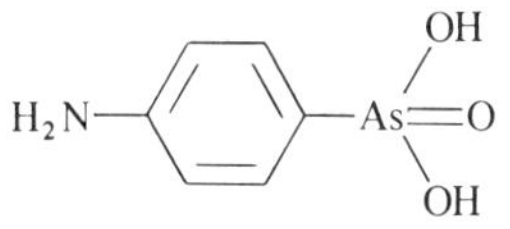

arsanilic acid

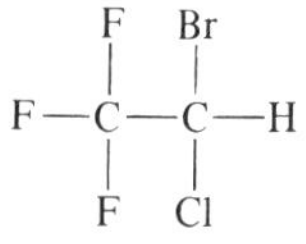

halothane

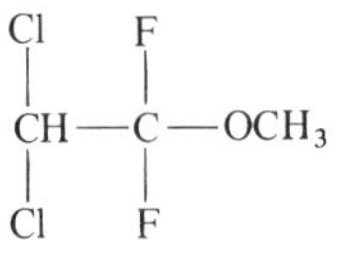

methoxyflurane

^{36}Cl-labeled drugs were administered and the radioactivity of urine, respiratory carbon dioxide, and volatile anesthetics was measured. Expired air was bubbled successively through sodium hydroxide solution and xylene [957]. This study was completed in vitro by chloride determination [955] but urinary metabolites were not identified.

Stier [872] demonstrated that trifluoracetic acid was a metabolite of halothane by chromatography on paper and cellulose thin layer (isopropanol : ammonium hydroxide 4 : 1) and by IR spectrophotometry.

Trifluoroacetic acid was extracted from human urine by adsorption on Dowex-1 X 2 and elution by 1% potassium bromide solution, and it was identified by IR spectrophotometry by Stier and Alter [873].

IV. MISCELLANEOUS COMPOUNDS

The largest fraction of ^{14}C-labeled formaldehyde intraperitonealy administered was found as ^{14}CO$_2$ by Neely [665]. Urine paper chromatography (80% phenol; ammonium hydroxide : ethanol 100 : 1; butanol : acetic acid : water 4 : 1 : 5) and chromatography on a Dowex-50 (H$^+$) column (1N hydrochloric acid) detected some of the ^{14}C in methionine, serine, and an adduct, cysteine-formaldehyde.

Malorny et al. [594] demonstrated formic acid formation after reduction to formaldehyde by chromotropic acid.

Acetylcysteine deacetylation was studied in vitro and in vivo by Sheffner et al. [831]. After administration of the ^{35}S-labeled drug they applied paper chromatography (t-butanol : acetic acid : water 4 : 1 : 1 containing 0.1 g of potassium cyanide and 1.0 g of EDTA disodium salt per liter to avoid sulfhydryl groups oxidation).

Pentaerythrityl tetranitrate, ^{14}C-labeled in positions 1 and 2, was used by DiCarlo et al. [253]. Urine chromatography on silica gel thin layer

---

HS—CH$_2$—CH—COOH
            |
            NH—CO—CH$_3$

acetylcysteine

O$_2$N—O—H$_2$C   CH$_2$—O—NO$_2$
          \ /
          C
          / \
O$_2$N—O—H$_2$C   CH$_2$—O—NO$_2$

pentaerythrityl tetranitrate

(toluene : ethyl acetate 1 : 1; water saturated ethyl acetate) showed pentaerythritol excretion. Some of the metabolites remained bound to the tissue components, so that the balance was incomplete as a function of time. The same technique was applied to study enzymic degradation in human blood [255].

DiCarlo et al. also investigated pentaerythrityl tetranitrate binding to plasma proteins [251]. They applied the method they developed to separate and determine its hydrolysis products [254]. They used a solvent system (toluene : ethyl acetate : butanol : water 10 : 5 : 2 : 2) that separated the unchanged drug and its four metabolites in a single operation on silica gel thin layer (Table 70).

This procedure was applied by the same authors [252] to elucidate pentaerythrityl tetranitrate metabolism in man. Blood metabolites were extracted into dioxane, and, after evaporation of the solution, redissolved into methanol. Since excessive lipids were still present, various chromatographic purification steps were required. These involved a development with the previously mentioned system followed by three successive developments on the same plate with the system butanol : ammonium hydroxide : water 4 : 1 : 3, which carried the metabolites to the top of the plate and left the impurities at the bottom. After elution, new chromatography was performed under normal conditions.

Inosityl nicotinate was largely excreted unchanged. Brattgard et al. [107] verified this by chloroform extraction of blood after administration of the

---

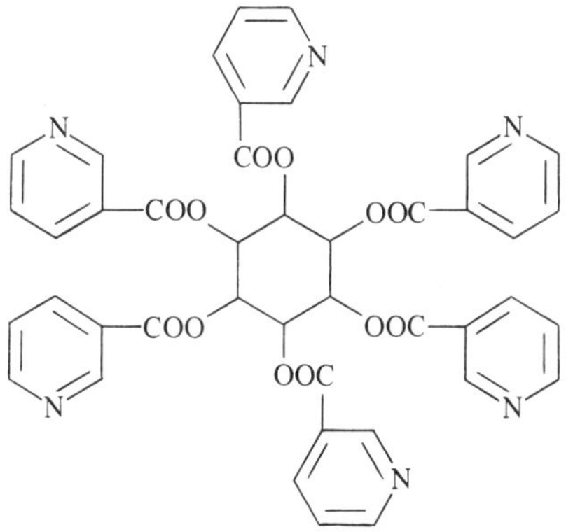

inosityl nicotinate

TABLE 70

Product	R_f
Pentaerythritol	0.00
Pentaerythrityl mononitrate	0.16
Pentaerythrityl dinitrate	0.45
Pentaerythrityl trinitrate	0.60
Pentaerythrityl tetranitrate	0.77

^{14}C-labeled drug. Thin-layer chromatography on binder-free silica gel (chloroform : ethanol 85 : 15) was used.

Glyceryl trinitrate was partly metabolized in vitro into inorganic nitrate which Oberst and Snyder [685] determined by colorimetry. Lorenzetti et al. [562] also titrated nitrates and nitrites in blood after oral and intravenous administration. Needlemann and Krantz [664] showed the formation of glyceryl di- and mononitrate in vitro by thin-layer chromatography on silica gel, activated for 30 minutes at 110°C (benzene : ethyl acetate 4 : 1). The metabolites were identified by R_f and IR spectrum comparison with authentic compounds.

Zicha et al. [1043] separated 3-methyl-3,4-dihydroxy-4-phenyl-1-butyne from its two metabolites by thin-layer chromatography (propionic acid : benzene : xylene 8 : 5 : 3) and they determined it by direct photometry after zinc chloride staining.

Smith and Williams [861] included methylpentynol in a general study on glucuronic conjugation of unsaturated alcohols. Glucuronides were separated and crystallized by the method of Kamil et al. [480], then examined by elemental analysis.

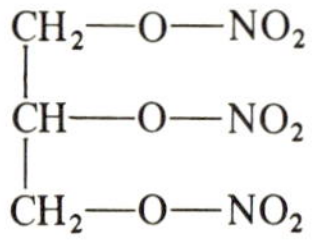

$$\begin{array}{l} CH_2-O-NO_2 \\ | \\ CH-O-NO_2 \\ | \\ CH_2-O-NO_2 \end{array}$$

glyceryl trinitrate

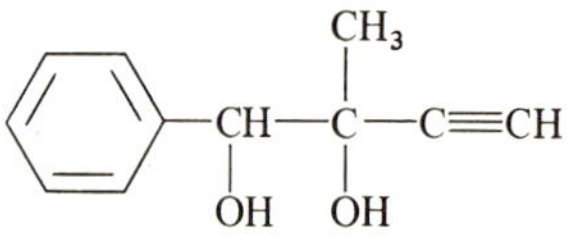

3-methyl-3,4-dihydroxy-4-phenyl-1-butyne

p-Butoxyphenylacethydroxamic acid is an antiinflammatory agent. Its metabolism is complex, as Roncucci et al. [783] separated, after administration of the ^{14}C-labeled drug, 12 metabolites. They applied chromatography on silica gel thin layer (chloroform : methanol 8 : 2 to which four drops of ammonium hydroxide per 150 ml were added) to urine. Three of the excretion products were identified by R_f comparison as the unchanged drug, p-butoxyphenylacetamide, and p-butoxyphenylacetic acid.

Deferoxamine has a high affinity for ferric ions and is used therapeutically as a specific iron chelating agent. The study of its metabolism was undertaken by Keberle [494] who did not give precise information on the analytical methods he used. Three metabolites were isolated in human urine by counter-current distribution. The most important was crystallized and identified and had a carboxyl group in place of the amino group.

Subsequently, Peters et al. [707] described a procedure to determine deferoxamine and its ferric complex in blood and indicated how to avoid the interference of the major metabolite.

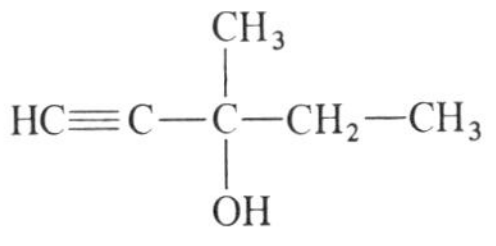

$$HC\equiv C-\underset{\underset{OH}{|}}{\overset{\overset{CH_3}{|}}{C}}-CH_2-CH_3$$

methylpentynol

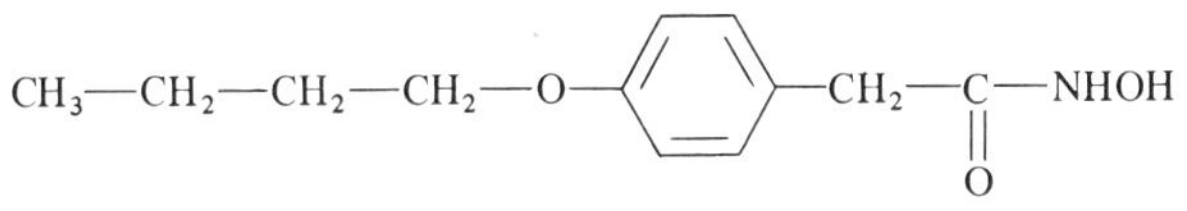

$$CH_3-CH_2-CH_2-CH_2-O-\underset{}{\bigcirc}-CH_2-\underset{\underset{O}{\|}}{C}-NHOH$$

p-butoxyphenylacethydroxamic acid

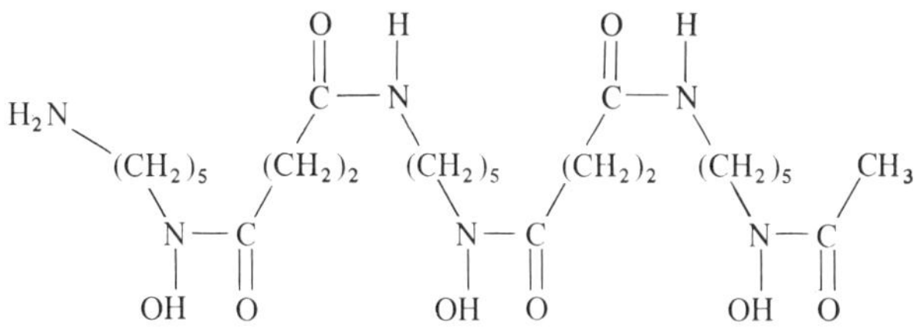

deferoxamine

Alloferin is one of the most complex drug molecules whose metabolism has been studied. Waser and Lüthi [971] prepared its tritiated derivative. The possible metabolites were separated on a thin layer of the mixture silica gel : anthracene 1 : 1 by means of the solvent system acetic acid : hexane : diethylamine 77.5 : 17.5 : 5 and visualized by the fluorographic technique of Lüthi and Waser [570]. Direct autoradiography of tritiated compounds was effected by using their radiation to induce the fluorescence of a convenient substance. The authors used high voltage electrophoresis on a thin layer of silica gel : anthracene for bile and various organ extracts. They found only traces of radioactive metabolites.

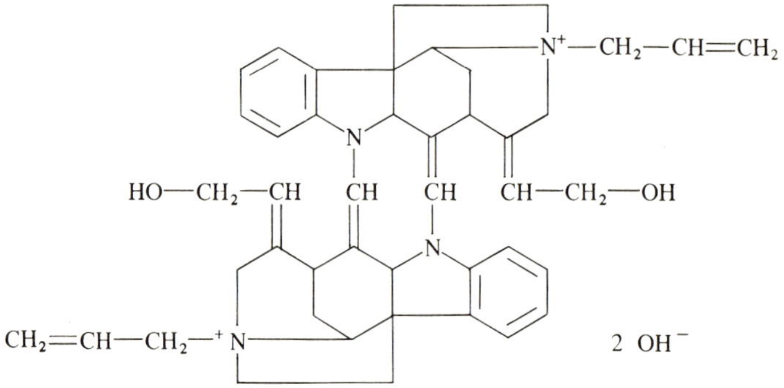

---

alloferin

BIBLIOGRAPHY

(UNTIL JANUARY 1, 1967)

[1] M. Abe, Sci. Rept. Res. Inst. Tohoku. Univ., Ser. C, 8, 1, 1958.

[2] R. M. Abel, R. J. Luchi, G. W. Peskin, H. L. Conn, and L. D. Miller, J. Pharm. Exp. Ther., 150, 463, 1965.

[3] Y. Abiko, K. Onoue, Y. Yamamura, I. Nakazona, and T. Yoshida, J. Biochem. (Tokyo), 48, 838, 1960.

[4] P. Acred, D. M. Brown, D. H. Turner, and M. J. Wilson, Brit. J. Pharmacol., 18, 356, 1962.

[5] R. H. Adamson, S. L. Ague, S. M. Hess, and J. D. Davidson, J. Pharm. Exp. Ther., 150, 322, 1965.

[6] R. H. Adamson and J. R. Fouts, J. Pharm. Exp. Ther., 127, 87, 1959.

[7] E. Addison and R. G. Clark, J. Pharm. Pharmacol., 15, 268, 1963.

[8] T. K. Adler, J. Pharm. Exp. Ther., 106, 371, 1952.

[9] T. K. Adler, J. Pharm. Exp. Ther., 110, 1, 1954.

[10] T. K. Adler, J. M. Fujimoto, E. L. Way, and E. M. Baker, J. Pharm. Exp. Ther., 114, 251, 1955.

[11] T. K. Adler and F. H. Shaw, J. Pharm. Exp. Ther., 104, 1, 1952.

[12] B. W. Agranoff, R. M. Bradley, and J. Axelrod, Proc. Soc. Exp. Biol.,

[13] M. Akagi, Y. Oketani, M. Takada, and T. Suga, Chem. Pharm. Bull., 11, 321, 1963.

[14] M. Akagi, Y. Oketani, and S. Yamane, Chem. Pharm. Bull., 11, 1216, 1963., Med., 96, 261, 1957.

[15] B. Akerman, A. Aström, S. Ross, and A. Telc, Acta Pharmacol. Toxicol., 24, 389, 1966.

[16] A. Albert and C. W. Rees, Biochem, J., 61, 128, 1955.

[17] E. J. Algeri and A. J. McBay, Am. J. Clin. Pathol., 23, 654, 1953.

[18] E. J. Algeri and A. J. McBay, Science, 123, 183, 1956.

[19] J. J. Alleva, J. Medicin. Chem., 6, 621, 1963.

[20] F. T. Allewijn and P. J. Demoen, J. Pharm. Sci., 55, 1028, 1966.

[21] L. G. Allgen, B. Jönsson, B. Nauckhoff, M. L. Andersen, I. Huus, and I. Møller-Nielsen, Experientia, 16, 325, 1960.

[22] E. L. Alpen, H. G. Mandel, V. W. Rodwell, and P. K. Smith, J. Pharm. Exp. Ther., 102, 150, 1951.

[23] M. E. Amundson, M. L. Johnson, and J. A. Manthey, J. Pharm. Sci., 54, 684, 1965.

[24] M. E. Amundson and J. A. Manthey, J. Pharm. Sci., 55, 277, 1966.

[25] M. W. Anders and G. J. Mannering, Anal. Chem., 34, 730, 1962.

[26] R. Angelucci, D. Artini, A. Cresseri, P. N. Giraldi, W. Logemann, G. Nannini, and G. Valzelli, Brit. J. Pharmacol., 24, 274, 1965.

[27] J. Anghileri, N. Recchi, and J. Baruel, Experientia, 22, 78, 1966.

[28] M. D. Armstrong, A. McMillan, and K. N. F. Shaw, Biochim. Biophys Acta, 25, 422, 1957.

[29] M. D. Armstrong, K. N. F. Shaw, and P. E. Wall, J. Biol. Chem., 218, 293, 1956.

[30] A. M. Asatoor, J. Chromatog., 4, 144, 1960.

[31] A. M. Asatoor, J. Chromatog., 7, 415, 1962.

[32] A. M. Asatoor, B. R. Galman, J. R. Johnson, and M. D. Milne, Brit. J. Pharmacol., 24, 293, 1965.

[33] A. M. Asatoor and D. N. S. Kerr, Clin. Chim. Acta, 6, 149, 1961.

[34] J. J. Ashley, B. T. Brown, G. T. Okita, and S. E. Wright, J. Biol. Chem., 232, 315, 1958.

[35] J. Axelrod, J. Pharm. Exp. Ther., 109, 62, 1953.

[36] J. Axelrod, J. Pharm. Exp. Ther., 110, 315, 1954.

[37] J. Axelrod, J. Biol. Chem., 214, 753, 1955.

[38] J. Axelrod, J. Pharm. Exp. Ther., 117, 322, 1956.

[39] J. Axelrod, Science, 126, 400, 1957.

[40] J. Axelrod, L. Aronow, and B. B. Brodie, J. Pharm. Exp. Ther., 106, 166, 1952.

[41] J. Axelrod, J. K. Inscoe, S. Senoh, and B. Witkop, Biochim. Biophys. Acta, 27, 210, 1958.

[42] J. Axelrod, I. J. Kopin, and J. D. Mann, Biochim. Biophys. Acta, 36, 576, 1959.

[43] J. Axelrod and J. Reichenthal, J. Pharm. Exp. Ther., 107, 519, 1953.

[44] J. Axelrod, R. Shofer, J. K. Inscoe, W. M. King, and A. Sjoerdsma, J. Pharm. Exp. Ther., 124, 9, 1958.

[45] J. Axelrod, H. Weil-Malherbe, and R. Tomchick, J. Pharm. Exp. Ther., 127, 251, 1959.

[46] M. Baggiolini and M. H. Bickel, Life Sci., 5, 795, 1966.

[47] M. Baggiolini, M. H. Bickel, and F. S. Messiha, Experientia, 21, 334, 1965.

[48] B. J. Baltes, T. Ellison, L. Levy, and R. Okun, Pharmacologist, 8, 220, 1966.

[49] M. J. Barnes and B. Boothroyd, Biochem. J., 78, 41, 1961.
[50] R. C. R. Barreto, J. Chromatog., 7, 82, 1962.
[51] R. C. R. Barreto, J. Chromatog., 9, 180, 1962.
[52] R. C. R. Barreto, J. Chromatog., 11, 344, 1963.
[53] R. C. R. Barreto and S. O. Sabino, J. Chromatog., 13, 435, 1964.
[54] A. Becher, J. Miksch, P. Rambacher, and A. Schäfer, Klin. Wochschr., 30, 913, 1952.
[55] A. H. Beckett, M. A. Beaven, and A. E. Robinson, J. Pharm. Pharmacol., 12, 203 T, 1960.
[56] A. H. Beckett, M. A. Beaven, and A. E. Robinson, Biochem. Pharmacol., 12, 779, 1963.
[57] A. H. Beckett, R. N. Boyes, and P. J. Appleton, J. Pharm. Pharmacol., 18, 76, 1966.
[58] A. H. Beckett, R. N. Boyes, and J. R. B. Parker, Anaesthesia, 20, 294, 1965.
[59] A. H. Beckett and N. H. Choulis, J. Pharm. Pharmacol., 15, 236 T, 1963.
[60] A. H. Beckett and S. H. Curry, J. Pharm. Pharmacol., 15, 246 T, 1963.
[61] A. H. Beckett, S. H. Curry, and A. G. Bolt, J. Pharm. Pharmacol., 16, 500, 1964.
[62] A. H. Beckett and M. Rowland, J. Pharm. Pharmacol., 16, 27 T, 1964.
[63] A. H. Beckett and M. Rowland, J. Pharm. Pharmacol., 17, 59, 1965.
[64] A. H. Beckett, M. Rowland, and E. J. Triggs, Nature, 207, 200, 1965.
[65] A. H. Beckett and G. R. Wilkinson, J. Pharm. Pharmacol., 17, 104 S, 1965.
[66] R. Beckmann, Arzneimitt. Forsch., 12, 1095, 1962.
[67] R. Beckmann, Arzneimitt. Forsch., 13, 185, 1963.
[68] R. Beckmann, Arzneimitt. Forsch., 15, 761, 1965.
[69] R. Beckmann, Arch. Int. Pharmacodyn., 160, 161, 1966.
[70] R. Beckmann, Arzneimitt. Forsch., 16, 910, 1966.
[71] C. Bedford, A. J. Cummings, and B. K. Martin, Brit. J. Pharmacol., 24, 418, 1965.
[72] G. Beisenherz, F. W. Koss, L. Klatt, and B. Binder, Arch. Int. Pharmacodyn., 158, 380, 1965.
[73] G. Beisenherz, F. W. Koss, L. Klatt, and B. Binder, Arch. Int. Pharmacodyn., 161, 76, 1966.
[74] G. Beisenherz, F. W. Koss, A. Schüle, I. Gebauer, R. Bärisch, and R. Fröde, Arzneimitt. Forsch., 10, 307, 1960.
[75] G. H. Benham, Canad. J. Research, 23 E, 71, 1945.

[76] J. D. Benigni and A. J. Verbiscar, J. Medicin. Chem., 6, 607, 1963.
[77] F. M. Berger, J. Pharm. Exp. Ther., 112, 413, 1954.
[78] E. Bernhammer and K. Krisch, Biochem. Pharmacol., 14, 863, 1965.
[79] K. Bernhard and H. Beer, Helv. Physiol. Acta, 20, 114, 1962.
[80] K. Bernhard, G. Brubacher, and A. H. Lutz, Helv. Chim. Acta, 37, 1839, 1954.
[81] K. Bernhard, M. Just, A. H. Lutz, and J. P. Vuilleumier, Helv. Chim. Acta, 40, 436, 1957.
[82] K. Bernhard, M. Just, J. P. Vuilleumier, and G. Brubacher, Helv. Chim. Acta, 39, 596, 1956.
[83] K. H. Beyer, Z. Anal. Chem., 212, 139, 1965.
[84] K. H. Beyer and J. T. Skinner, J. Pharm. Exp. Ther., 68, 419, 1940.
[85] M. H. Bickel and M. Baggiolini, Helv. Physiol. Acta, 23 C, 77, 1965.
[86] M. H. Bickel and M. Baggiolini, Biochem. Pharmacol., 15, 1155, 1966.
[87] M. H. Bickel, H. J. Weder, and M. Baggiolini, Helv. Physiol. Acta, 24 C, 77, 1966.
[88] A. L. Bieber and A. C. Sartorelli, Cancer Res., 24, 1210, 1964.
[89] A. Bieder, P. Brunel, and L. Mazeau, Ann. Pharm. Fr., 21, 375, 1963.
[90] A. Bieder, P. Brunel, and L. Mazeau, Ann. Pharm. Fr., 24, 493, 1966.
[91] A. Bieder and L. Mazeau, Ann. Pharm. Fr., 20, 211, 1962.
[92] A. Bieder and L. Mazeau, Thérapie, 19, 897, 1964.
[93] H. L. Bird and C. T. Pugh, Antibiot. Chemother., 4, 750, 1954.
[94] R. D. N. Birtley, J. B. Roberts, B. H. Thomas, and A. Wilson, Brit. J. Pharmacol., 26, 393, 1966.
[95] M. W. Blake and P. L. Perlman, J. Pharm. Exp. Ther., 117, 287, 1956.
[96] W. E. Bleidner, J. B. Harmon, W. E. Hewes, T. E. Lynes, and E. C. Hermann, J. Pharm. Exp. Ther., 150, 484, 1965.
[97] W. Block, Arzneimitt. Forsch., 11, 266, 1961.
[98] W. Block and I. Ebigt, Arzneimitt. Forsch., 10, 709, 1960.
[99] A. G. Bolt, I. S. Forrest, and M. T. Serra, J. Pharm. Sci., 55, 1205, 1966.
[100] I. U. Boone, M. Magee, and D. F. Turney, J. Biol. Chem., 221, 781, 1956.
[101] I. U. Boone, V. G. Strang, and B. S. Rogers, Am. Rev. Tuberc., 76, 568, 1957.
[102] K. Borner, Z. Physiol. Chem., 341, 264, 1965.

[103] J. Bösche and G. Schmidt, Arzneimitt. Forsch., 16, 548, 1966.

[104] A. L. A. Boura, W. G. Duncombe, R. D. Robson, and A. McCoubrey, J. Pharm. Pharmacol., 14, 722, 1962.

[105] F. Boyer, M. Saviard, and M. Dechavassine, Ann. Inst. Pasteur, 90, 339, 1956.

[106] F. Boyer, J. Troestler, N. Rist, and J. Tabone, Ann. Inst. Pasteur, 78, 140, 1950.

[107] S. O. Brattgard, R. Brattsand, and J. G. L. Harthon, Arzneimitt. Forsch., 16, 145, 1966.

[108] A. C. Bratton and E. K. Marshall, J. Biol. Chem., 128, 537, 1939.

[109] G. Braun, J. Krapcho, and S. M. Hess, Proc. Soc. Exp. Biol. Med., 118, 983, 1965.

[110] H. G. Bray, G. E. Francis, F. C. Neale, and W. V. Thorpe, Biochem. J., 46, 267, 1950.

[111] H. G. Bray, B. G. Humphris, W. V. Thorpe, K. White, and P. B. Woods, Biochem. J., 52, 412, 1952.

[112] H. G. Bray, S. P. James, and W. V. Thorpe, Nature, 163, 407, 1949.

[113] H. G. Bray, H. J. Lake, and W. V. Thorpe, Biochem. J., 48, 400, 1951.

[114] H. G. Bray, F. C. Neale, and W. V. Thorpe, Biochem. J., 40, 406, 1946.

[115] H. G. Bray, F. C. Neale, and W. V. Thorpe, Biochem. J., 46, 506, 1950.

[116] H. G. Bray, B. E. Ryman, and W. V. Thorpe, Biochem. J., 43, 561, 1948.

[117] H. G. Bray and W. V. Thorpe, Methods Biochem. Anal., 1, 48, 1954.

[118] H. G. Bray, W. V. Thorpe, and K. White, Biochem. J., 46, 271, 1950.

[119] J. W. Bridges, M. R. Kibby, and R. T. Williams, Biochem. J., 96, 829, 1965.

[120] J. W. Bridges and R. T. Williams, J. Pharm. Pharmacol., 15, 565, 1963.

[121] K. J. Broadley and D. J. Roberts, J. Pharm. Pharmacol., 18, 182, 1966.

[122] E. Brochmann-Hansen and A. B. Svendsen, J. Pharm. Sci., 51, 393, 938, 1962.

[123] B. B. Brodie, L. Aronow, and J. Axelrod, J. Pharm. Exp. Ther., 106, 200, 1952.

[124] B. B. Brodie, L. Aronow, and J. Axelrod, J. Pharm. Exp. Ther., 111, 21, 1954.

[125] B. B. Brodie and J. Axelrod, J. Pharm. Exp. Ther., 94, 22, 1948.

[126] B. B. Brodie and J. Axelrod, J. Pharm. Exp. Ther., 94, 29, 1948.

[127] B. B. Brodie and J. Axelrod, J. Pharm. Pharmacol., 97, 58, 1949.

[128] B. B. Brodie and J. Axelrod, J. Pharm. Exp. Ther., 98, 97, 1950.

[129] B. B. Brodie and J. Axelrod, J. Pharm. Exp. Ther., 99, 171, 1950.

[130] B. B. Brodie and J. Axelrod, J. Pharm. Exp. Ther., 106, 200, 1952.

[131] B. B. Brodie, J. Axelrod, and J. Reichenthal, J. Biol. Chem., 194, 215, 1952.

[132] B. B. Brodie, J. E. Baer, and L. C. Craig, J. Biol. Chem., 188, 567, 1951.

[133] B. B. Brodie, J. J. Burns, L. C. Mark, P. A. Lief, E. Bernstein, and E. M. Papper, J. Pharm. Exp. Ther., 109, 26, 1953.

[134] B. B. Brodie, P. A. Lief, and R. Poet, J. Pharm. Exp. Ther., 94, 359, 1948.

[135] B. B. Brodie, L. C. Mark, E. M. Papper, P. A. Lief, E. Bernstein, and E. A. Rovenstine, J. Pharm. Exp. Ther., 98, 85, 1950.

[136] B. B. Brodie and S. Udenfriend, J. Biol. Chem., 158, 705, 1945.

[137] B. B. Brodie and S. Udenfriend, Proc. Soc. Exp. Biol. Med., 74, 845, 1950.

[138] B. B. Brodie, S. Udenfriend, and J. E. Baer, J. Biol. Chem., 168, 299, 1947.

[139] B. B. Brodie, S. Udenfriend, and W. A. Dill, J. Biol. Chem., 168, 335, 1947.

[140] C. W. J. Brooks and E. C. Horning, Anal. Chem., 36, 1540, 1964.

[141] A. Brossi, O. Häfliger, and O. Schnider, Arzneimitt. Forsch., 5, 62, 1955.

[142] B. T. Brown, E. E. Shepheard, and S. E. Wright, J. Pharm. Exp. Ther., 118, 39, 1956.

[143] B. T. Brown, D. Ranger, and S. E. Wright, J. Pharm. Exp. Ther., 113, 353, 1955.

[144] B. T. Brown and S. E. Wright, J. Biol. Chem., 220, 431, 1956.

[145] R. B. Bruce, J. E. Pitts, F. Pinchbeck, and J. Newman, J. Med. Chem., 8, 157, 1965.

[146] R. B. Bruce, L. Turnbull, J. Newman, and J. Pitts, J. Med. Chem., 9, 286, 1966.

[147] C. G. v. Bruck, F. M. Delfs, E. Serick, and V. Wolf, Arzneimitt. Forsch., 10, 621, 1960.

[148] H. Büch, H. Haüser, K. Pfleger, and W. Rüdiger, Arch. Exp. Pathol. Pharmakol., 253, 25, 1966.

[149] D. R. Buhler, J. Pharm. Exp. Ther., 145, 232, 1964.

[150] D. R. Buhler, Biochem. Pharmacol., 14, 371, 1965.

[151] D. R. Buhler, H. Harpootlian, and R. L. Johnston, Biochem. Pharmacol., 15, 1507, 1966.

[152] R. P. Buhs, J. L. Beck, O. C. Speth, J. L. Smith, N. R. Trenner, P. J. Cannon, and J. H. Laragh, J. Pharm. Exp. Ther., 143, 205, 1964.

[153] R. P. Buhs, D. Polin, J. O. Beattie, J. L. Beck, J. L. Smith, O. C. Speth, and N. R. Trenner, J. Pharm. Exp. Ther., 154, 357, 1966.

[154] H. P. Burchfield and R. J. Wheeler, J. Am. Off. Agric. Chem., 49, 651, 1966.

[155] J. J. Burns, B. L. Berger, P. A. Lief, A. Wollack, E. M. Papper, and B. B. Brodie, J. Pharm. Exp. Ther., 114, 289, 1955.

[156] J. J. Burns, R. K. Rose, T. Chenkin, A. Goldman, A. Schulert, and B. B. Brodie, J. Pharm. Exp. Ther., 109, 346, 1953.

[157] J. J. Burns, R. K. Rose, S. Goodwin, J. Reichental, E. C. Horning, and B. B. Brodie, J. Pharm. Exp. Ther., 113, 481, 1955.

[158] J. J. Burns, M. Weiner, G. Simson, and B. B. Brodie, J. Pharm. Exp. Ther., 108, 33, 1953.

[159] J. J. Burns, S. Wexler, and B. B. Brodie, J. Am. Chem. Soc., 75, 2345, 1953.

[160] J. J. Burns, T. F. Yü, A. Ritterband, J. M. Pérel, A. B. Gutman, and B. B. Brodie, J. Pharm. Exp. Ther., 119, 418, 1957.

[161] M. T. Bush, Feder. Proc., 16, 287, 1957.

[162] M. T. Bush, Microchem. J., 1, 269, 1957.

[163] M. T. Bush, Microchem. J., 5, 73, 1961.

[164] M. T. Bush and T. C. Butler, J. Pharm. Exp. Ther., 68, 278, 1940.

[165] M. T. Bush, T. C. Butler, and H. L. Dickison, J. Pharm. Exp. Ther., 108, 104, 1953.

[166] M. T. Bush and P. M. Densen, Anal. Chem., 20, 121, 1948.

[167] M. T. Bush, P. Mazel, and J. Chambers, J. Pharm. Exp. Ther., 134, 110, 1961.

[168] S. R. M. Bushby and A. J. Woiwod, Am. Rev. Tuberc., 72, 123, 1955.

[169] S. R. M. Bushby and A. J. Woiwod, Biochem. J., 63, 406, 1956.

[170] E. Bütikofer, P. Cottier, P. Imhof, H. Keberle, W. Riess, and K. Schmid, Arch. Exp. Pathol. Pharmakol., 244, 97, 1962.

[171] T. C. Butler, J. Pharm. Exp. Ther., 104, 299, 1952.

[172] T. C. Butler, J. Pharm. Exp. Ther., 106, 235, 1952.

[173] T. C. Butler, J. Pharm. Exp. Ther., 108, 11, 1953.

[174] T. C. Butler, J. Pharm. Exp. Ther., 108, 474, 1953.

[175] T. C. Butler, Science, 120, 494, 1954.

[176] T. C. Butler, J. Pharm. Exp. Ther., 113, 178, 1955.

[177] T. C. Butler, J. Pharm. Exp. Ther., 116, 326, 1956.

[178] T. C. Butler, J. Pharm. Exp. Ther., 117, 160, 1956.
[179] T. C. Butler, J. Pharm. Exp. Ther., 119, 1, 1957.
[180] T. C. Butler, J. Pharm. Exp. Ther., 134, 311, 1961.
[181] T. C. Butler, J. Pharm. Exp. Ther., 143, 23, 1964.
[182] T. C. Butler and M. T. Bush, J. Pharm. Exp. Ther., 65, 205, 1939.
[183] T. C. Butler, D. Mahaffee, and C. Mahaffee, Proc. Soc. Exp. Biol. Med., 81, 450, 1952.
[184] T. C. Butler, C. Mahaffee, and W. J. Waddell, J. Pharm. Exp. Ther., 111, 425, 1954.
[185] T. C. Butler and W. J. Waddell, J. Pharm. Exp. Ther., 127, 171, 1959.
[186] H. Büttner, F. Portwich, and J. Seydel, Chemotherapia, 10, 1, 1965.
[187] J. A. Buzard, J. D. Conklin, E. O'Keefe, and M. F. Paul, J. Pharm. Exp. Ther., 131, 38, 1961.
[188] S. Caffau and S. Jacobacci, Minerva Med., 53, 3747, 1962.
[189] S. Caffau and S. Jacobacci, Boll. Ist. Sieroterap. Milano, 43, 57, 1964.
[190] A. D. Campbell, F. K. Coles, L. L. Eubank, and E. G. Huf, J. Pharm. Exp. Ther., 131, 18, 1961.
[191] A. Canas-Rodriguez, Experientia, 22, 472, 1966.
[192] P. Capella and E. C. Horning, Anal. Chem., 38, 316, 1966.
[193] L. E. Carlat, A. W. Margraf, H. H. Weathers, and T. E. Weichselbaum, Proc. Soc. Exp. Biol. Med., 102, 245, 1959.
[194] C. E. Carter, J. Am. Chem. Soc., 72, 1466, 1950.
[195] G. P. Cartoni and F. De Stefano, Giorn. Biochim., 12, 298, 1963.
[196] G. B. Cassano, S. E. Sjöstrand, and E. Hansson, Psychopharmacologia, 8, 1, 1965.
[197] B. Catanese, A. Grasso, and B. Silvestrini, Arzneimitt. Forsch., 16, 1354, 1966.
[198] G. Ceriotti, A. Defranceschi, I. De Carneri, and V. Zamboni, Brit. J. Pharmacol., 8, 356, 1953.
[199] K. D. Charalampous, A. Orengo, K. E. Walker, and J. Kinross-Wright, J. Pharm. Exp. Ther., 145, 242, 1964.
[200] K. D. Charalampous, K. E. Walker, and J. Kinross-Wright, Psychopharmacologia, 9, 48, 1966.
[201] C. B. Christensen, Acta Pharmacol. Toxicol., 24, 139, 1966.
[202] F. Christensen, Acta Pharmacol. Toxicol., 21, 299, 1964.
[203] F. Christensen, Acta Pharmacol. Toxicol., 21, 307, 1964.
[204] F. Christensen, Acta Pharmacol. Toxicol., 22, 141, 1965.
[205] F. Christensen, Acta Pharmacol. Toxicol., 24, 232, 1966.

[206] J. W. Clapp, J. Biol. Chem., 223, 207, 1956.

[207] N. T. Clare, Australian Vet. J., 23, 340, 1947.

[208] T. W. Clarkson, A. Rothstein, and R. Sutherland, Brit. J. Pharmacol., 24, 1, 1965.

[209] J. Cochin and J. Axelrod, J. Pharm. Exp. Ther., 125, 105, 1959.

[210] J. Cochin and J. W. Daly, J. Pharm. Exp. Ther., 139, 154, 1963.

[211] J. Cochin and J. W. Daly, J. Pharm. Exp. Ther., 139, 160, 1963.

[212] Y. Cohen and O. Costerousse, Thérapie, 16, 109, 1961.

[213] Y. Cohen, J. Wepierre, Y. Font du Picard, and J. R. Boissier, Thérapie, 20, 101, 1965.

[214] H. B. Collier, Canad. J. Res., 18 D, 272, 1940.

[215] H. B. Collier, D. E. Allen, and W. E. Swales, Canad. J. Res., 21 D, 151, 1943.

[216] D. F. Colucci and D. A. Buyske, Biochem. Pharmacol., 14, 457, 1965.

[217] A. H. Conney and J. J. Burns, J. Pharm. Exp. Ther., 128, 340, 1960.

[218] A. H. Conney, N. Trousof, and J. J. Burns, J. Pharm. Exp. Ther., 128, 333, 1960.

[219] J. R. Cooper and B. B. Brodie, J. Pharm. Exp. Ther., 114, 409, 1955.

[220] G. Coppi, G. Sekules, and G. Pala, Arzneimitt. Forsch., 16, 601, 1966.

[221] K. Corbett, S. A. Edwards, G. E. Lee, and T. L. Threlfall, Nature, 208, 286, 1965.

[222] H. H. Cornish and A. A. Christman, J. Biol. Chem., 228, 315, 1957.

[223] E. Cox, G. Roxburgh, and S. E. Wright, J. Pharm. Pharmacol., 11, 535, 1959.

[224] E. Cox and S. E. Wright, J. Pharm. Exp. Ther., 126, 117, 1959.

[225] A. W. Craig, B. W. Fox, and H. Jackson, Biochem. J., 69, 16 P, 1958.

[226] A. W. Craig, B. W. Fox, and H. Jackson, Biochem. Pharmacol., 3, 42, 1959.

[227] A. W. Craig and H. Jackson, Brit. J. Pharmacol., 10, 321, 1955.

[228] J. C. Craig, N. Y. Mary, and S. K. Roy, Anal. Chem., 36, 1142, 1964.

[229] L. C. Craig, C. Golumbic, H. Mighton, and E. Titus, J. Biol. Chem., 161, 321, 1945.

[230] J. L. Cramer and B. Scott, Psychopharmacologia, 8, 461, 1966.

[231] A. Cresseri, P. N. Giraldi, W. Logemann, G. Tosolini, and G. Valzelli, Brit. J. Pharmacol., 27, 486, 1966.

[232] A. J. Cummings, J. Pharm. Pharmacol., 15, 212, 1963.

[233] A. J. Cummings and M. L. King, Nature, 209, 620, 1966.

[234] A. J. Cummings and B. K. Martin, Brit. J. Pharmacol., 25, 470, 1965.

[235] A. J. Cummings, B. K. Martin, and R. Renton, Brit. J. Pharmacol., 26, 461, 1966.

[236] A. S. Curry, J. Pharm. Pharmacol., 7, 604, 1955.

[237] A. S. Curry, J. Pharm. Pharmacol., 7, 1072, 1955.

[238] A. S. Curry, Nature, 188, 58, 1960.

[239] W. F. J. Cuthbertson, D. M. Ireland, and W. Wolff, Biochem. J., 55, 669, 1953.

[240] C. E. Dalgliesh, E. C. Horning, M. G. Horning, K. L. Knox, and K. Yarger, Biochem. J., 101, 792, 1966.

[241] J. Daly, J. Axelrod, and B. Witkop, Ann. N.Y. Acad. Sci., 96, 37, 1962.

[242] I. Danishefsky and H. B. Eiber, Arch. Biochem. Biophys., 85, 53, 1959.

[243] C. Davison, J. Wangler, and P. K. Smith, J. Pharm. Exp. Ther., 136, 226, 1962.

[244] M. Day, J. P. Green, and J. D. Robinson, Brit. J. Pharmacol., 18, 625, 1962.

[245] P. G. Dayton, L. E. Sicam, M. Landrau, and J. J. Burns, J. Pharm. Exp. Ther., 132, 287, 1961.

[246] M. Debackere and A. M. Massart-Lëen, Arch. Int. Pharmacodyn., 155, 459, 1965.

[247] F. De Eds, C. W. Eddy, and J. O. Thomas, J. Pharm. Exp. Ther., 64, 250, 1938.

[248] F. De Eds and J. O. Thomas, J. Parasit., 28, 363, 1942.

[249] A. Defranceschi and V. Zamboni, Biochim. Biophys. Acta, 13, 304, 1954.

[250] E. Degkwitz and H. Staudinger, Z. Physiol. Chem., 341, 111, 1965.

[251] F. J. Dicarlo, C. B. Coutinho, N. J. Sklow, L. J. Haynes, and M. C. Crew, Proc. Soc. Exp. Biol. Med., 120, 705, 1965.

[252] F. J. Dicarlo, M. C. Crew, N. J. Sklow, C. B. Coutinho, P. Nonkin, F. Simon, and A. Bernstein, J. Pharm. Exp. Ther., 153, 254, 1966.

[253] F. J. Dicarlo, J. M. Hartigan, C. B. Coutinho, and G. E. Phillips, Proc. Soc. Exp. Biol. Med., 118, 311, 1965.

[254] F. J. Dicarlo, J. M. Hartigan, and G. E. Phillips, Anal. Chem., 36, 2301, 1964.

[255] F. J. Dicarlo, J. M. Hartigan, and G. E. Phillips, Proc. Soc. Exp. Biol. Med., 118, 514, 1965.

[256] F. J. Dicarlo, S. G. Malament, and G. E. Phillips, Toxicol. Appl. Pharmacol., 5, 392, 1963.

[257] F. J. Dicarlo, N. J. Sliver, C. B. Coutinho, L. J. Haynes, and G. E. Phillips, Chemotherapia, 9, 129, 1964.

[258] W. Dietz and K. Soehring, Arch. Pharm., 290/62, 80, 1957.

[259] W. A. Dill, L. Peterson, T. Chang, and A. J. Glazko, Abstr. Papers. Am. Chem. Soc., 149, 30 N, 1965.

[260] W. Diller, E. Krüger-Thiemer, and E. Wempe, Arzneimitt. Forsch., 9, 432, 1959.

[261] J. V. Dingell, F. Sulser, and J. R. Gillette, J. Pharm. Exp. Ther., 143, 14, 1964.

[262] V. Distefano and H. H. Borgstedt, Science, 144, 1137, 1964.

[263] F. Dobson and R. T. Williams, Biochem. J., 40, 215, 1946.

[264] M. Dohrn, Biochem. Z., 43, 240, 1912.

[265] G. Domar, A. Fredga, and H. Linderholm, Acta Chem. Scand., 3, 1441, 1949.

[266] T. Dorfmüller, Dtsch. Med. Wochenschr., 81, 888, 1956.

[267] T. Dorfmüller, Ärztliche Labor., 2, 1, 1956.

[268] T. Dorfmüller, Ärztliche Labor., 3, 8, 1957.

[269] J. F. Douglas, J. Pharm. Exp. Ther., 150, 105, 1965.

[270] J. F. Douglas, B. J. Ludwig, T. Ginsberg, and F. M. Berger, J. Pharm. Exp. Ther., 136, 5, 1962.

[271] J. F. Douglas, B. J. Ludwig, and A. Schlosser, J. Pharm. Exp. Ther., 138, 21, 1962.

[272] J. F. Douglas, B. J. Ludwig, A. Schlosser, and J. Edelson, Biochem. Pharmacol., 15, 2087, 1966.

[273] J. F. Douglas, B. J. Ludwig, and N. Smith, Proc. Soc. Exp. Biol. Med., 112, 436, 1963.

[274] J. S. Douglas and P. J. Nicholls, J. Pharm. Pharmacol., 17, 115 S, 1965.

[275] C. D. Douglass and R. Hogan, Proc. Soc. Exp. Biol. Med., 100, 446, 1959.

[276] J. Drabner, H. Bauer, and W. Schwerd, Arch. Toxikol., 21, 367, 1966.

[277] P. E. Dresel and I. H. Slater, Proc. Soc. Exp. Biol. Med., 79, 286, 1952.

[278] L. G. Dring, R. L. Smith, and R. T. Williams, J. Pharm. Pharmacol., 18, 402, 1966.

[279] J. L. Driscoll, B. J. Gudzinowicz, and H. F. Martin, J. Gas. Chromatog., 2, 109, 1964.

[280] B. Dubnick, G. A. Leeson, R. Leverett, D. F. Morgan, and G. E. Phillips, J. Pharm. Exp. Ther., 140, 85, 1963.

[281] B. Dubnick, D. F. Morgan, C. A. Towne, and G. E. Phillips, J. Pharm. Exp. Ther., 153, 301, 1966.

[282] D. E. Duggan and E. Titus, J. Pharm. Exp. Ther., 130, 375, 1960.

[283] J. Duhault and S. Fenard, Arch. Int. Pharmacodyn., 158, 251, 1965.

[284] B. Duhm, W. Maul, H. Medenwald, K. Patzschke, and L. A. Wegner, Z. Naturforsch., 20 b, 434, 1965.

[285] C. Dumazert and El Ouachi, Ann. Pharm. Fr., 12, 723, 1954.

[286] G. J. Dutton and I. D. E. Storey, Biochem. J., 57, 275, 1954.

[287] D. P. Earle and B. B. Brodie, J. Pharm. Exp. Ther., 91, 250, 1947.

[288] D. P. Earle, W. J. Welch, and J. A. Shannon, J. Clin. Investig., 27, 87, 1948.

[289] H. Eberhardt and M. Debackere, Arzneimitt. Forsch., 15, 929, 1965.

[290] H. Eberhardt, K. J. Freundt, M. V. Clarmann, and M. Bofinger, Arch. Toxikol., 21, 175, 1965.

[291] H. Eberhardt, K. J. Freundt, and J. W. Langbein, Arzneimitt. Forsch., 12, 1087, 1962.

[292] H. Eberhardt, O. W. Lerbs, and K. J. Freundt, Arzneimitt. Forsch., 13, 804, 1963.

[293] I. Eberholst and I. Huus, Arzneimitt. Forsch., 16, 876, 1966.

[294] A. G. Ebert and S. M. Hess, J. Pharm. Exp. Ther., 148, 412, 1965.

[295] A. G. Ebert, G. K. W. Yim, and T. S. Miya, Biochem. Pharmacol., 13, 1267, 1964.

[296] J. Edelson, A. Schlosser, and J. F. Douglas, Biochem. Pharmacol., 14, 901, 1965.

[297] I. B. Eisdorfer and W. C. Ellenbogen, J. Chromatog., 4, 329, 1960.

[298] H. J. Eisner and R. J. Wulf, J. Pharm. Exp. Ther., 142, 122, 1963.

[299] G. B. Elion, S. Bieber, and G. H. Hitchings, Ann. N.Y. Acad. Sci., 60, 297, 1954.

[300] G. B. Elion, S. Callahan, H. Nathan, S. Bieber, S. W. Rundles, and G. H. Hitchings, Biochem. Pharmacol., 12, 85, 1963.

[301] G. B. Elion, A. Kovensky, G. H. Hitchings, E. Metz, and R. W. Rundles, Biochem. Pharmacol., 15, 863, 1966.

[302] G. B. Elion, S. Mueller, and G. H. Hitchings, J. Am. Chem. Soc., 81, 3042, 1959.

[303] G. A. Ellard, J. M. B. Garrod, B. Scales, and G. A. Snow, Biochem. Pharmacol., 14, 129, 1965.

[304] R. I. Ellin and D. E. Easterday, J. Pharm. Pharmacol., 13, 370, 1961.

[305] T. Ellison, L. Gutzait, and E. J. Van Loon, Feder. Proc., 24, 688, 1965.

[306] T. Ellison, L. Gutzait, and E. J. Van Loon, J. Pharm. Exp. Ther., 152, 383, 1966.

[307] A. El Masry, J. N. Smith, and R. T. Williams, Biochem. J., 64, 50, 1956.

[308] J. L. Emmerson and T. S. Miya, J. Pharm. Exp. Ther., 137, 148, 1962.

[309] J. L. Emmerson, T. S. Miya, and G. K. W. Yim, J. Pharm. Exp. Ther., 129, 89, 1960.

[310] I. Enander, A. Sundwall, and B. Sörbo, Biochem. Pharmacol., 7, 226, 1961.

[311] I. Enander, A. Sundwall, and B. Sörbo, Biochem. Pharmacol., 7, 232, 1961.

[312] I. Enander, A. Sundwall, and B. Sörbo, Biochem. Pharmacol., 11, 377, 1962.

[313] D. W. Esplin and D. M. Woodbury, J. Pharm. Exp. Ther., 118, 129, 1956.

[314] U. S. von Euler and I. Orwen, Acta Physiol. Scand., 33, Suppl. 118, 1, 1955.

[315] V. Evertsbusch and E. M. K. Geiling, Feder. Proc., 12, 319, 1953.

[316] S. Fabro, H. Schuhmacher, R. L. Smith, and R. T. Williams, Nature, 201, 1125, 1964.

[317] J. W. Faigle, H. Keberle, W. Riess, and K. Schmid, Experientia, 18, 389, 1962.

[318] G. Ferlemann and W. Vogt, Arch. Exp. Pathol. Pharmakol., 250, 479, 1965.

[319] G. Fickewirth and A. Heffter, Biochem. Ztschr., 40, 36 and 48, 1912.

[320] K. Finger and L. Zicha, Med. Pharm. Exp., 13, 161, 1965.

[321] K. Fischer and W. Specht, Arch. Toxikol., 17, 48, 1958.

[322] A. L. Fisher and J. P. Long, J. Pharm. Exp. Ther., 107, 241, 1953.

[323] V. Fishman and H. Goldenberg, Proc. Soc. Exp. Biol. Med., 104, 99, 1960.

[324] V. Fishman and H. Goldenberg, Proc. Soc. Exp. Biol. Med., 110, 187, 1962.

[325] V. Fishman and H. Goldenberg, Proc. Soc. Exp. Biol. Med., 112, 501, 1963.

[326] V. Fishman and H. Goldenberg, J. Pharm. Exp. Ther., 150, 122, 1965.

[327] V. Fishman, A. Heaton, and H. Goldenberg, Proc. Soc. Exp. Biol. Med., 109, 548, 1962.

[328] W. H. Fishman, Chemistry of Drug Metabolism. Springfield, Ill.: Charles C. Thomas, Inc., 1961.

[329] W. H. Fishman and S. Green, J. Biol. Chem., 215, 527, 1955.
[330] T. L. Flanagan, T. H. Lin, W. J. Novick, I. M. Rondish, C. A. Bocher, and E. J. Van Loon, J. Med. Pharm. Chem., 1, 263, 1959.
[331] O. Folin and V. Ciocalteu, J. Biol. Chem., 73, 627, 1927.
[332] I. S. Forrest, A. G. Bolt, and M. T. Serra, Life Sci., 5, 473, 1966.
[333] I. S. Forrest, F. M. Forrest, and M. Berger, Biochim. Biophys. Acta, 29, 441, 1958.
[334] B. W. Fox, A. W. Craig, and H. Jackson, Biochem. Pharmacol., 5, 27, 1960.
[335] B. W. Fox and W. H. Prusoff, Biochem. Pharmacol., 15, 1317, 1966.
[336] W. O. Foye, R. N. Duvall, W. E. Lange, M. H. Talbot, and E. L. Prien, J. Pharm. Exp. Ther., 125, 198, 1959.
[337] V. Francova, K. Raz, Z. Franc, A. Cerny, and V. Jelinek, Coll. Czech. Chem. Commun., 30, 2631, 1965.
[338] H. H. Frey, Arch. Int. Pharmacodyn., 118, 12, 1959.
[339] H. H. Frey, A. Doenicke, and G. Jäger, Med. Exp., 4, 243, 1961.
[340] H. H. Frey, G. Eberhardt, and J. Rustemeyer, Arch. Toxikol., 18, 189, 1960.
[341] H. H. Frey and M. P. Magnussen, Arzneimitt. Forsch., 16, 612, 1966.
[342] H. H. Frey, F. Sudendey, and D. Krause, Arzneimitt. Forsch., 9, 294, 1959.
[343] A. J. Friedhoff and M. Goldstein, Ann. N.Y. Acad. Sci., 96, 5, 1962.
[344] A. J. Friedhoff and L. E. Hollister, Biochem. Pharmacol., 15, 269, 1966.
[345] J. M. Fujimoto, Feder. Proc., 15, 425, 1956.
[346] J. M. Fujimoto and E. L. Way, Feder. Proc., 13, 356, 1954.
[347] J. M. Fujimoto and E. L. Way, J. Pharm. Exp. Ther., 121, 340, 1957.
[348] J. M. Fujimoto and E. L. Way, J. Am. Pharm. Assoc. Sc. Ed., 47, 273, 1958.
[349] A. T. Fuller, Lancet, 1, 194, 1937.
[350] I. C. Geddes and D. E. Douglas, Feder. Proc., 15, 260, 1956.
[351] E. Gerhards, H. Gibian, and K. H. Kolb, Arzneimitt. Forsch., 14, 394, 1964.
[352] E. Gerhards, H. Gibian, and K. H. Kolb, Z. Physiol. Chem., 343, 150, 1965.
[353] E. Gerhards, H. Gibian, and G. Raspe, Arzneimitt. Forsch., 15, 1295, 1965.
[354] E. Gerhards and K. H. Kolb, Arzneimitt. Forsch., 15, 1375, 1965.

[355] E. Gerhards, K. H. Kolb, and P. E. Schulze, Arch. Pharmakol. Exp. Pathol., 255, 200, 1966.

[356] P. K. Gessner, P. A. Khairallah, W. M. McIsaac, and I. H. Page, J. Pharm. Exp. Ther., 130, 126, 1960.

[357] J. R. Gillette, J. V. Dingell, F. Sulser, R. Kuntzman, and B. B. Brodie, Experientia, 17, 417, 1961.

[358] J. R. Gillette and J. J. Kamm, J. Pharm. Exp. Ther., 130, 262, 1960.

[359] D. R. Gilligan, J. Clin. Invest., 24, 301, 1945.

[360] A. Giotti and E. W. Maynert, J. Pharm. Exp. Ther., 101, 296, 1951.

[361] B. Glasson and A. Benakis, Helv. Physiol. Acta, 19, 323, 1961.

[362] A. J. Glazko, W. A. Dill, and M. C. Rebstock, J. Biol. Chem., 183, 679, 1950.

[363] A. J. Glazko, W. A. Dill, and L. M. Wolf, J. Pharm. Exp. Ther., 104, 452, 1952.

[364] A. J. Glazko, W. A. Dill, L. M. Wolf, and A. Kazenko, Feder. Proc., 14, 58, 1955.

[365] A. J. Glazko, W. A. Dill, L. M. Wolf, and A. Kazenko, J. Pharm. Exp. Ther., 118, 377, 1956.

[366] A. J. Glazko, W. A. Dill, L. M. Wolf, and A. Kazenko, J. Pharm. Exp. Ther., 121, 119, 1957.

[367] A. J. Glazko, L. M. Wolf, and W. A. Dill, Arch. Biochem., 23, 411, 1949.

[368] A. J. Glazko, L. M. Wolf, W. A. Dill, and A. C. Bratton, J. Pharm. Exp. Ther., 96, 445, 1949.

[369] W. O. Godtfredsen and S. Vangedal, Acta Chem. Scand., 20, 1599, 1966.

[370] O. Gold and H. Stormann, Arzneimitt. Forsch., 14, 1108, 1964.

[371] H. Goldenberg and V. Fishman, Proc. Soc. Exp. Biol. Med., 108, 178, 1961.

[372] H. Goldenberg and V. Fishman, Biochem. Pharmacol., 14, 365, 1965.

[373] H. Goldenberg, V. Fishman, Å. Heaton, and R. Burnett, Proc. Soc. Exp. Biol. Med., 115, 1044, 1964.

[374] S. Goldschmidt and R. Wehr, Z. Physiol. Chem., 308, 9, 1957.

[375] M. Goldstein and B. Anagnoste, Biochim. Biophys. Acta, 107, 166, 1965.

[376] R. Gönnert, Bull. Wld. Health Orgn., 25, 702, 1961.

[377] McC. Goodall and H. Alton, Biochem. Pharmacol., 14, 1595, 1965.

[378] McC. Goodall, H. Alton, and L. Rosen, Biochem. Pharmacol., 13, 703, 1964.

[379] E. Gordis, Biochem. Pharmacol., 15, 2124, 1966.

[380] J. H. Gorvin and G. Brownlee, Nature, 179, 1248, 1957.
[381] R. E. Gosselin, J. D. Gabourel, S. C. Kalser, and J. H. Wills, J. Pharm. Exp. Ther., 115, 217, 1955.
[382] E. L. Graves, T. J. Elliott, and W. Bradley, Nature, 162, 257, 1948.
[383] J. P. Green, Nature, 186, 472, 1960.
[384] J. P. Green and M. Day, Biochem. Pharmacol., 3, 190, 1960.
[385] L. A. Greenberg and D. Lester, J. Pharm. Exp. Ther., 88, 87, 1946.
[386] G. Grimmer, W. Küssner, and K. Lingner, Arzneimitt. Forsch., 10, 28, 1960.
[387] E. G. Gross and V. Thompson, J. Pharm. Exp. Ther., 68, 413, 1940.
[388] B. J. Gudzinowicz, J. Gas. Chromatog., 4, 110, 1966.
[389] J. R. Gwilt, A. Robertson, and E. W. McChesney, J. Pharm. Pharmacol., 15, 440, 1963.
[390] E. Hackenthal, Arzneimitt. Forsch., 15, 1075, 1965.
[391] J. Halberkann and F. Fretwurst, Arq. Inst. Biol. São Paulo, 11, 149, 1940.
[392] J. Halberkann and F. Fretwurst, Z. Physiol. Chem., 285, 97, 1950.
[393] C. R. Hall, V. Cordova, and F. Rieders, Pharmacologist, 7, 148, 1965.
[394] L. Hamilton and G. B. Elion, Ann. N.Y. Acad. Sci., 60, 304, 1954.
[395] C. H. Hammar and W. Prellwitz, Klin. Wschr., 44, 1010, 1966.
[396] A. Hampton and A. R. P. Paterson, Biophys. Biochim. Acta, 114, 185, 1966.
[397] C. A. Handley and R. A. Seibert, J. Pharm. Exp. Ther., 117, 253, 1956.
[398] H. J. Hansen, W. G. Giles, and S. B. Nadler, Proc. Soc. Exp. Biol. Med., 113, 163, 1963.
[399] H. J. Hansen, J. P. Vandevoorde, W. G. Giles, and S. B. Nadler, Proc. Soc. Exp. Biol. Med., 115, 713, 1964.
[400] S. W. F. Hanson, G. T. Mills, and R. T. Williams, Biochem. J., 38, 274, 1944.
[401] E. Hansson, P. Hoffmann, and L. Kristerson, Acta Pharmacol. Toxicol., 22, 213, 1965.
[402] J. Harley-Mason and A. H. Laird, J. Chem. Soc., 2629, 1959.
[403] J. Harley-Mason, A. H. Laird, and J. R. Smythies, Confinia Neurologica, 18, 152, 1958.
[404] R. E. Harman, M. A. P. Meisinger, G. E. Davis, and F. A. Kuehl, J. Pharm. Exp. Ther., 143, 215, 1964.

[405] S. C. Harris, M. L. Searle, and A. C. Ivy, J. Pharm. Exp. Ther., 89, 92, 1947.

[406] R. L. Hartles and R. T. Williams, Biochem. J., 41, 206, 1947.

[407] R. L. Hartles and R. T. Williams, Biochem. J., 44, 335, 1949.

[408] W. T. Haskins and G. W. Luttermoser, J. Pharm. Exp. Ther., 109, 201, 1953.

[409] E. P. Hausner, C. L. Shafer, M. Corson, D. Johnson, T. Trujillo, and W. Langham, Circulation, 3, 171, 1951.

[410] A. Haüssler and H. Wicha, Arzneimitt. Forsch., 15, 81, 1965.

[411] A. Heller, J. E. Kasik, L. Clark, and L. J. Roth, Anal. Chem., 33, 1755, 1961.

[412] E. S. Henderson, R. H. Adamson, C. Denham, and V. T. Oliverio, Cancer Res., 25, 1008, 1965.

[413] E. S. Henderson, R. H. Adamson, and V. T. Oliverio, Cancer Res., 25, 1018, 1965.

[414] W. Hennig and H. Weiler, Arzneimitt. Forsch., 5, 60, 1955.

[415] U. Henriksen, I. Huus, and R. Kopf, Arch. Int. Pharmacodyn., 109, 39, 1957.

[416] B. Herrmann, Helv. Physiol. Acta, 21, 402, 1963.

[417] B. Herrmann, Arzneimitt. Forsch., 14, 219, 1964.

[418] B. Herrmann and R. Pulver, Arch. Int. Pharmacodyn., 126, 454, 1960.

[419] B. Herrmann, W. Schindler, and R. Pulver, Med. Exp., 1, 381, 1959.

[420] G. Hertting, Biochem. Pharmacol., 13, 1119, 1964.

[421] W. Hespe, A. M. de Ross, and W. T. Nauta, Arch. Int. Pharmacodyn., 156, 180, 1965.

[422] W. Hespe, A. M. de Ross, and W. T. Nauta, Arch. Int. Pharmacodyn., 164, 397, 1966.

[423] S. Hess, H. Weissbach, B. G. Redfield, and S. Udenfriend, J. Pharm. Exp. Ther., 124, 189, 1958.

[424] J. Hirtz, Mises au point de chimie analytique (Paris: Masson et Cie, édit.), 17, 103, 1968.

[425] H. Hoffmann and K. Roller, Arzneimitt. Forsch., 14, 1001, 1964.

[426] I. Hoffmann, O. Nieschulz, K. Popendicker, and E. Tauchert, Arzneimitt. Forsch., 9, 133, 1959.

[427] K. Hoffmann, H. Keberle, and H. J. Schmid, Helv. Chim. Acta, 40, 387, 1957.

[428] H. J. Hohorst, A. Ziemann, and N. Brock, Arzneimitt. Forsch., 15, 432, 1965.

[429] H. J. Hohorst, A. Ziemann, and N. Brock, Arzneimitt. Forsch., 16, 1529, 1966.

[430] G. Hollunger, Acta Pharmacol. Toxicol., 17, 356, 1960.

[431] G. Hollunger, Acta Pharmacol. Toxicol., 17, 365, 1960.

[432] E. C. Horning, M. G. Horning, W. J. A. Vanden Heuvel, K. L. Knox, B. Holmstedt, and J. W. Brooks, Anal. Chem., 36, 1546, 1964.

[433] G. Hübner and E. Pfeil, Z. Physiol. Chem., 296, 225, 1954.

[434] H. B. Hucker, Pharmacologist, 4, 171, 1962.

[435] H. B. Hucker, P. M. Ahmad, and E. A. Miller, J. Pharm. Exp. Ther., 154, 176, 1966.

[436] H. B. Hucker, P. M. Ahmad, E. A. Miller, and R. Brobyn, Nature, 209, 619, 1966.

[437] H. B. Hucker and E. A. Hoffman, Experientia, 22, 855, 1966.

[438] H. B. Hucker and C. C. Porter, Feder. Proc., 20, 172, 1961.

[439] E. G. Huf, F. K. Coles, and L. L. Eubank, Proc. Soc. Exp. Biol. Med., 102, 276, 1959.

[440] C. C. Hug and L. B. Mellett, J. Pharm. Exp. Ther., 149, 446, 1965.

[441] C. C. Hug, L. B. Mellett, and E. J. Cafruny, J. Pharm. Exp. Ther., 150, 259, 1965.

[442] C. C. Hug and L. A. Woods, J. Pharm. Exp. Ther., 142, 248, 1963.

[443] H. B. Hughes, J. Pharm. Exp. Ther., 109, 444, 1953.

[444] I. Hynie, J. König, and K. Kácl, J. Chromatog., 19, 192, 1965.

[445] R. M. J. Ings, G. L. Law, and E. W. Parnell, Biochem. Pharmacol., 15, 515, 1966.

[446] K. Irrgang, Arzneimitt. Forsch., 15, 688, 1965.

[447] P. Iveson, D. V. Parke, and R. T. Williams, Biochem. J., 100, 28 P, 1966.

[448] H. Iwainsky, Arzneimitt. Forsch., 7, 745, 1957.

[449] H. Iwainsky, I. Sehrt, and M. Grunert, Arzneimitt. Forsch., 15, 193, 1965.

[450] J. V. Jackson and M. S. Moss, Nature, 192, 553, 1961.

[451] E. Jacobsen and I. Gad, Arch. Exp. Pathol. Pharmakol., 196, 280, 1940.

[452] M. Jaffe, Ber. Dtsch. Chem. Ges., 34, 2737, 1901.

[453] M. Jaffe, Ber. Dtsch. Chem. Ges., 35, 2891, 1902.

[454] M. Jaffe and P. Hilbert, Hoppe-Seyl. Ztschr., 12, 295, 1888.

[455] O. R. Jagenburg and K. Toczko, Biochem. J., 92, 639, 1964.

[456] G. V. James, Biochem. J., 34, 640, 1940.

[457] R. W. Jelliffe, Circulation, 32, Suppl. II, 119, 1965.

[458] R. W. Jelliffe, D. Jackson, C. J. Page, and S. Morgan, J. Lab. Clin. Med., 67, 694, 1966.

[459] D. G. Johns, A. T. Lannotti, A. C. Sartorelli, B. A. Booth, and J. R. Bertino, Proc. Am. Ass. Cancer Res., 6, 33, 1965.

[460] D. E. Johnson and H. P. Burchfield, Lectures Gas Chromatography. 1964. Agric. Biol. Appl. Plenum Press, New York, 1965.

[461] D. E. Johnson, J. D. Millar, and H. P. Burchfield, Life Sci., 2, 959, 1963.

[462] D. E. Johnson, C. F. Rodriguez, and H. P. Burchfield, Biochem. Pharmacol., 14, 1453, 1965.

[463] D. E. Johnson, C. F. Rodriguez, and H. P. Burchfield, Abstr. Papers. Am. Chem. Soc., 149, 29 N, 1965.

[464] D. E. Johnson, C. F. Rodriguez, and W. Schlameus, J. Gas Chromatog., 3, 345, 1965.

[465] G. Jolles, B. Terlain, and J. P. Thomas, Nature, 207, 199, 1965.

[466] G. Jolles, B. Terlain, and J. P. Thomas, Thérapie, 20, 1471, 1965.

[467] G. Jommi, P. Manitto, and M. A. Silanos, Arch. Biochem. Biophys., 108, 334, 1964.

[468] G. Jommi, P. Manitto, and M. A. Silanos, Arch. Biochem. Biophys., 108, 562, 1964.

[469] B. M. Jones, R. J. M. Ratcliffe, and S. G. E. Stevens, J. Pharm. Pharmacol., 17, 52 S, 1965.

[470] E. S. Josephson, J. Greenberg, D. J. Taylor, and H. L. Bami, J. Pharm. Exp. Ther., 103, 7, 1951.

[471] E. S. Josephson, D. J. Taylor, J. Greenberg, and A. P. Ray, Proc. Soc. Exp. Biol. Med., 76, 700, 1951.

[472] M. Jurovcik, K. Raska, Z. Sormova, and F. Sorm, Coll. Czech. Chem. Commun., 30, 3370, 1965.

[473] K. Kakemi, T. Arita, and S. Ohashi, Sci. Sect. Am. Pharm. Assoc. Preprints Papers. Las Vegas, 1962, B-IV.

[474] K. Kakemi, T. Arita, H. Sezaki, and T. Nadai, Yakugaku Zasshi., 83, 871, 1963.

[475] K. Kakemi, T. Arita, H. Sezaki, M. Nakano, and T. Kiriyama, Yakugaku Zasshi., 82, 195, 1962.

[476] K. Kakemi, T. Arita, H. Yamashina, and R. Konishi, Yakugaku Zasshi., 82, 540, 1962.

[477] F. Kalberer, W. Kreis, and J. Rutschmann, Biochem. Pharmacol., 11, 261, 1962.

[478] S. C. Kalser, E. J. Kelvington, and M. M. Randolph, Pharmacologist, 8, 220, 1966.

[479] S. C. Kalser, E. J. Kelvington, M. M. Randolph, and D. M. Santomenna, J. Pharm. Exp. Ther., 147, 252, 1965.

[480] I. A. Kamil, J. N. Smith, and R. T. Williams, Biochem. J., 50, 235, 1951.

[481] J. J. Kamm and E. J. Van Loon, Clin. Chem., 12, 789, 1966.

[482] P. O. Kane, Nature, 195, 495, 1962.
[483] I. Kapetanidis, Pharm. Acta Helv., 40, 331, 1965.
[484] I. Kapetanidis and A. Mirimanoff, Pharm. Acta Helv., 39, 269, 1964.
[485] S. A. Kaplan, S. Riegelman, and K. H. Lee, J. Pharm. Sci., 55, 14, 1966.
[486] E. M. Kapp and A. F. Coburn, J. Biol. Chem., 145, 549, 1942.
[487] M. L. Karnovsky and M. J. Johnson, Anal. Chem., 21, 1125, 1949.
[488] J. Kaslander, Biochim. Biophys. Acta, 71, 730, 1963.
[489] B. G. Katzung and F. H. Meyers, J. Pharm. Exp. Ther., 154, 575, 1966.
[490] P. N. Kaul, J. Pharm. Pharmacol., 14, 237, 1962.
[491] S. Kawai, T. Nagatsu, T. Imanari, and Z. Tamura, Chem. Pharm. Bull., 14, 618, 1966.
[492] J. Kawamata and T. Hiratani, Med. J. Osaka Univ., 6, 417, 1955.
[493] J. Kawamata and K. Kashiwagi, Med. J. Osaka Univ., 6, 119, 1955.
[494] H. Keberle, Ann. N.Y. Acad. Sci., 119, 758, 1964.
[495] H. Keberle and K. Hoffmann, Experientia, 12, 21, 1956.
[496] H. Keberle and K. Hoffmann, Helv. Chim. Acta, 39, 767, 1956.
[497] H. Keberle, W. Riess, and K. Hoffmann, Arch. Int. Pharmacodyn., 142, 117, 1963.
[498] H. Keberle, W. Riess, K. Schmid, and K. Hoffmann, Arch. Int. Pharmacodyn., 142, 125, 1963.
[499] H. Keberle, K. Schmid, K. Hoffmann, J. P. Vuilleumier, and K. Bernhard, Helv. Chim. Acta, 42, 417, 1959.
[500] R. E. Keller and W. C. Ellenbogen, J. Pharm. Exp. Ther., 106, 77, 1952.
[501] R. G. Kelly and D. A. Buyske, Antibiot. and Chemother., 10, 604, 1960.
[502] R. G. Kelly and D. A. Buyske, J. Pharm. Exp. Ther., 130, 144, 1960.
[503] R. G. Kelly, L. A. Kanegis, and D. A. Buyske, J. Pharm. Exp. Ther., 134, 320, 1961.
[504] F. E. Kelsey, E. M. K. Geiling, F. K. Oldham, and E. H. Dearborn, J. Pharm. Exp. Ther., 80, 391, 1944.
[505] G. P. Kerby, Proc. Soc. Exp. Biol. Med., 83, 263, 1953.
[506] M. Kiese and G. Renner, Arch. Exp. Pathol. Pharmakol., 252, 480, 1966.
[507] N. Kirshner, McC. Goodall, and L. Rosen, Proc. Soc. Exp. Biol. Med., 98, 627, 1958.
[508] H. Kleinsorge, K. Thalmann, and K. Rösner, Arzneimitt. Forsch., 9, 121, 1959.

[509] J. Klicka, A. D. Notation, and F. Ungar, Abstr. Papers Am. Chem. Soc., 152, 259 C, 1966.

[510] A. Klutch, M. Harfenist, and A. H. Conney, J. Med. Chem., 9, 63, 1966.

[511] B. K. Koe and R. Pinson, J. Med. Chem., 7, 635, 1964.

[512] B. A. Koechlin and L. d'Arconte, Analyt. Biochem., 5, 195, 1963.

[513] B. A. Koechlin and V. Iliev, Ann. N.Y. Acad. Sci., 80, 864, 1959.

[514] B. A. Koechlin, F. Rubio, S. Palmer, T. Gabriel, and R. Duschinsky, Biochem. Pharmacol., 15, 435, 1966.

[515] B. A. Koechlin, M. A. Schwartz, G. Krol, and W. Oberhansli, J. Pharm. Exp. Ther., 148, 399, 1965.

[516] B. A. Koechlin, M. A. Schwartz, and W. E. Oberhaensli, J. Pharm. Exp. Ther., 138, 11, 1962.

[517] T. Koga, Nagasaki Igakkai Zasshi, 34, 2225, 1959.

[518] K. H. Kolb, G. Jänicke, M. Kramer, P. E. Schulze, and G. Raspe, Arzneimitt. Forsch., 15, 1292, 1965.

[519] I. J. Kopin, Science, 131, 1372, 1960.

[520] I. J. Kopin, J. Axelrod, and E. Gordon, J. Biol. Chem., 236, 2109, 1961.

[521] F. L. Kozelka and C. H. Hine, J. Pharm. Exp. Ther., 77, 175, 1943.

[522] B. Kramer, Biochem. Pharmacol., 11, 299, 1962.

[523] M. Kraml, T. Dobson, K. Sestanj, M. A. Davis, and D. Dvornik, Abstr. Papers Am. Chem. Soc., 152, 50 P, 1966.

[524] H. A. Krebs, W. O. Sykes, and W. C. Bartley, Biochem. J., 41, 622, 1947.

[525] W. Kreis, R. Bloch, D. L. Warkentin, and J. H. Burchenal, Biochem. Pharmacol., 12, 1165, 1963.

[526] W. Kreis, S. B. Piepho, and H. V. Bernhard, Experientia, 22, 431, 1966.

[527] M. Kruger and P. Schmidt, Ber. Dtsch. Chem. Ges., 32, 2677, 1899.

[528] M. Kruger and P. Schmidt, Arch. Exp. Pathol. Pharmakol., 45, 259, 1901.

[529] R. Kuhn, Psychopharmacologia, 8, 201, 1965.

[530] L. Kum-Tatt, J. Pharm. Pharmacol., 13, 759, 1961.

[531] R. Kuntzman, A. Klutch, I. Tsai, and J. J. Burns, J. Pharm. Exp. Ther., 149, 29, 1965.

[532] K. Kuroda, J. Pharm. Exp. Ther., 137, 156, 1962.

[533] H. Kutt, W. Winters, R. Scherman, and F. McDowell, Arch. Neurol., 11, 649, 1964.

[534] E. H. Labrosse, J. Axelrod, and S. S. Kety, Science, 128, 593, 1958.

[535] G. L. Lage and J. L. Spratt, J. Pharm. Exp. Ther., 149, 248, 1965.
[536] W. E. Lange and S. A. Bell, J. Pharm. Sci., 55, 386, 1966.
[537] M. E. Latham and E. W. Elliott, J. Pharm. Exp. Ther., 101, 259, 1951.
[538] F. Lauterbach, D. Nitz, and K. Prescher, Arch. Exp. Pathol. Pharmakol., 247, 71, 1964.
[539] F. Lauterbach and K. Repke, Arch. Exp. Pathol. Pharmakol., 239, 196, 1960.
[540] D. Lawrow, Ber. Dtsch. Chem. Ges., 33, 2345, 1900.
[541] D. Lawrow, Z. Physiol. Chem., 32, 114, 1901.
[542] M. Ledvina and D. Baladova, Ceskosl. Farm., 14, 457, 1965.
[543] M. Ledvina and D. Baladova, Pharmazie, 21, 227, 1966.
[544] C. C. Lee, R. C. Anderson, H. L. Bird, and K. K. Chen, Antibiotics Annual, 1953-1954, 493.
[545] C. C. Lee, R. C. Anderson, and K. K. Chen, J. Pharm. Exp. Ther., 117, 274, 1956.
[546] C. C. Lee, L. W. Trevoy, J. W. T. Spinks, and L. B. Jaques, Proc. Soc. Exp. Biol. Med., 74, 151, 1950.
[547] H. M. Lee, E. G. Scott, and A. Pohland, J. Pharm. Exp. Ther., 125, 14, 1959.
[548] J. L. Leeling, B. M. Phillips, and O. E. Fancher, J. Pharm. Sci., 54, 1736, 1965.
[549] K. Lehmann, Arzneimitt. Forsch., 15, 812, 1965.
[550] H. Lehner, H. Lauener, and J. Schmutz, Arzneimitt. Forsch., 14, 89, 1964.
[551] B. Lester and L. A. Greenberg, J. Pharm. Exp. Ther., 90, 68, 1947.
[552] R. M. Levine and B. B. Clark, J. Pharm. Exp. Ther., 114, 63, 1955.
[553] S. C. C. Lin and E. L. Way, J. Pharm. Exp. Ther., 150, 309, 1965.
[554] T. H. Lin, S. W. Reynolds, I. M. Rondish, and E. J. Van Loon, Proc. Soc. Exp. Biol. Med., 102, 602, 1959.
[555] K. P. Link, D. Berg, and W. M. Barker, Science, 150, 378, 1965.
[556] W. H. Linkenheimer, S. J. Stolzenberg, and L. A. Wozniak, J. Pharm. Exp. Ther., 149, 280, 1965.
[557] M. Lipson, Austr. J. Exp. Biol. Med. Sci., 18, 269, 1940.
[558] I. London and R. B. Poet, Proc. Soc. Exp. Biol. Med., 94, 191, 1957.
[559] T. L. Loo and R. H. Adamson, Biochem. Pharmacol., 11, 170, 1962.
[560] T. L. Loo and R. H. Adamson, J. Med. Chem., 8, 513, 1965.
[561] T. L. Loo, M. E. Michael, A. J. Garceau, and J. C. Reid, J. Am. Chem. Soc., 81, 3039, 1959.

[562] O. J. Lorenzetti, A. Tye, and J. W. Nelson, J. Pharm. Sci., 55, 105, 1966.

[563] L. H. Louis, S. S. Fajan, J. W. Conn, W. A. Struck, J. B. Wright, and J. L. Johnson, J. Am. Chem. Soc., 78, 5701, 1956.

[564] J. S. Lowe, Biochem. Pharmacol., 3, 163, 1960.

[565] B. J. Ludwig, J. F. Douglas, L. S. Powell, M. Meyer, and F. M. Berger, J. Med. Pharm. Chem., 3, 53, 1961.

[566] B. J. Ludwig, H. Luts, and W. A. West, J. Am. Chem. Soc., 77, 5751, 1955.

[567] S. D. Lukas and R. E. Peterson, J. Clin. Invest., 45, 782, 1966.

[568] J. S. Lundy and A. E. Osterberg, Proc. Mayo Clin., 4, 386, 1924.

[569] U. Lüthi and P. G. Waser, Arch. Int. Pharmacodyn., 156, 319, 1965.

[570] U. Lüthi and P. G. Waser, Nature, 205, 1190, 1965.

[571] A. R. Maas, P. L. Carey, R. E. Hamilton, and A. E. Heming, Proc. Soc. Exp. Biol. Med., 103, 154, 1960.

[572] B. M. McAllister and D. J. Demis, Nature, 212, 293, 1966.

[573] E. W. McChesney, J. Pharm. Exp. Ther., 89, 368, 1947.

[574] E. W. McChesney, W. D. Conway, W. F. Banks, J. E. Rogers, and J. M. Shekosky, J. Pharm. Exp. Ther., 151, 482, 1966.

[575] E. W. McChesney, J. P. McAuliff, A. R. Surrey, and A. J. Olivet, Feder. Proc., 13, 97, 1954.

[576] W. M. McIsaac and M. Kanda, J. Pharm. Exp. Ther., 143, 7, 1964.

[577] W. M. McIsaac and R. T. Williams, Biochem. J., 66, 369, 1957.

[578] R. D. Mackenzie and W. R. McGrath, Proc. Soc. Exp. Biol. Med., 109, 511, 1962.

[579] F. G. Mac Mahon and L. A. Woods, J. Pharm. Exp. Ther., 103, 354, 1951.

[580] R. E. McMahon, J. Am. Chem. Soc., 80, 411, 1958.

[581] R. E. McMahon, J. Am. Chem. Soc., 81, 5199, 1959.

[582] R. E. McMahon, J. Org. Chem., 24, 1834, 1959.

[583] R. E. McMahon, J. Pharm. Exp. Ther., 130, 383, 1960.

[584] R. E. McMahon, H. W. Culp, and F. J. Marshall, J. Pharm. Exp. Ther., 149, 436, 1965.

[585] R. E. McMahon and N. R. Easton, J. Pharm. Exp. Ther., 135, 128, 1962.

[586] R. E. McMahon, F. J. Marshall, and H. W. Culp, J. Pharm. Exp. Ther., 149, 272, 1965.

[587] R. E. McMahon, F. J. Marshall, H. W. Culp, and W. N. Miller, Biochem. Pharmacol., 12, 1207, 1963.

[588] R. E. McMahon, J. S. Welles, and H. M. Lee, J. Am. Chem. Soc., 82, 2864, 1960.

[589] W. D. McNally, W. L. Bergman, and J. F. Polli, J. Lab. Clin. Med., 32, 913, 1947.
[590] C. Maggiolo and T. J. Haley, Proc. Soc. Exp. Biol. Med., 115, 149, 1964.
[591] L. Maître and M. Staehelin, Nature, 206, 723, 1965.
[592] S. Makisumi, K. Tanaka, and M. Kido, Igaku Kenkyu, 27, 598, 1957.
[593] R. Mallein, J. Rondelet, and M. Boucherat, Thérapie, 21, 1579, 1966.
[594] G. Malorny, N. Rietbrock, and M. Schneider, Arch. Exp. Pathol. Pharmakol., 250, 419, 1965.
[595] H. G. Mandel, N. M. Cambosos, and P. K. Smith, J. Pharm. Exp. Ther., 112, 495, 1954.
[596] H. G. Mandel, V. W. Rodwell, and P. K. Smith, J. Pharm. Exp. Ther., 106, 433, 1952.
[597] G. J. Mannering, A. C. Dixon, E. M. Baker, and T. Asami, J. Pharm. Exp. Ther., 111, 142, 1954.
[598] G. J. Mannering, A. C. Dixon, N. V. Carroll, and O. B. Cope, J. Lab. Clin. Med., 44, 292, 1954.
[599] G. J. Mannering and L. S. Schanker, J. Pharm. Exp. Ther., 124, 296, 1958.
[600] R. W. Manthei, Feder. Proc., 15, 455, 1956.
[601] C. H. March and H. W. Elliott, Proc. Soc. Exp. Biol. Med., 86, 494, 1954.
[602] F. I. Marcus, G. J. Kapadia, and G. G. Kapadia, J. Pharm. Exp. Ther., 145, 203, 1964.
[603] V. Marks, Clin. Chim. Acta, 6, 724, 1961.
[604] W. Marme, Z. Klin. Med., 4, 241, 1883.
[605] W. Marme, Deutsch. Med. Wschr., 9, 33, 1883.
[606] D. F. Marsh, J. Pharm. Exp. Ther., 105, 299, 1952.
[607] E. K. Marshall, Physiol. Rev., 19, 240, 1939.
[608] E. K. Marshall, A. C. Bratton, H. J. White, and J. T. Litchfield, Bull. Johns Hopkins Hosp., 67, 163, 1940.
[609] E. K. Marshall, W. C. Cutting, and K. Emerson, Science, 85, 202, 1937.
[610] E. K. Marshall, K. Emerson, W. C. Cutting, and B. Babbitt, J. Am. Med. Assoc., 108, 953, 1937.
[611] H. F. Martin, J. L. Driscoll, and B. J. Gudzinowicz, Anal. Chem., 35, 1901, 1963.
[612] J. F. Martin and S. E. Wright, J. Pharm. Exp. Ther., 128, 329, 1960.
[613] M. Maruyama and Y. Suzuki, Takamine Kenkyusho Nempo, 10, 158, 1958.

[614] G. B. Maughan, K. A. Evelyn, and J. S. L. Browne, J. Biol. Chem., 126, 567, 1938.
[615] E. W. Maynert, J. Biol. Chem., 195, 397, 1952.
[616] E. W. Maynert, J. Biol. Chem., 195, 403, 1952.
[617] E. W. Maynert, J. Pharm. Exp. Ther., 130, 275, 1960.
[618] E. W. Maynert, J. Pharm. Exp. Ther., 150, 118, 1965.
[619] E. W. Maynert, J. Pharm. Exp. Ther., 150, 476, 1965.
[620] E. W. Maynert and J. M. Dawson, J. Biol. Chem., 195, 389, 1952.
[621] E. W. Maynert and L. Losin, J. Pharm. Exp. Ther., 115, 275, 1955.
[622] E. W. Maynert and H. B. Van Dyke, Pharmacol. Rev., 1, 217, 1949.
[623] E. W. Maynert and H. B. Van Dyke, Science, 110, 661, 1949.
[624] E. W. Maynert and H. B. Van Dyke, J. Pharm. Exp. Ther., 98, 174, 1950.
[625] E. W. Maynert and H. B. Van Dyke, J. Pharm. Exp. Ther., 98, 180, 1950.
[626] E. W. Maynert and H. B. Van Dyke, J. Pharm. Exp. Ther., 98, 184, 1950.
[627] P. Mazel, A. Kerza-Kwiatecki, and J. Simanis, Biochim. Biophys. Acta, 114, 72, 1966.
[628] J. Mead and J. B. Koepfli, J. Biol. Chem., 154, 507, 1944.
[629] B. Melander, G. Gliniecke, B. Granstrand, G. Hanshoff, C. Jakobsson, and M. Palm, Third Int. Congress Chemotherapy (Stuttgart), 1963, p. 893.
[630] L. B. Mellett and L. A. Woods, Proc. Soc. Exp. Biol. Med., 106, 221, 1961.
[631] C. Meythaler, G. Dworak, and E. Schmid, Arzneimitt. Forsch., 16, 800, 1966.
[632] G. Milhaud, J. P. Aubert, and F. Boyer, C. R. Acad. Sci., 240, 2090, 1955.
[633] W. L. Miller, J. J. Krake, M. J. Vanderbrook, and L. M. Reineke, Ann. N.Y. Acad. Sci., 71, 118, 1957.
[634] P. M. Miranda, N. G. Zampaglione, and J. L. Way, Pharmacologist, 8, 180, 1966.
[635] A. L. Misra, S. J. Mule, and L. A. Woods, Nature, 190, 82, 1961.
[636] A. L. Misra, S. J. Mule, and L. A. Woods, J. Pharm. Exp. Ther., 132, 317, 1961.
[637] E. Moerman, Arch. Int. Pharmacodyn., 156, 489, 1965.
[638] W. Mohler, I. Bletz, and M. Reiser, Arch. Pharmaz., 299, 448, 1966.
[639] W. Mohler, K. Popendiker, and M. Reiser, Arzneimitt. Forsch., 16, 1524, 1966.

[640] G. Mohnike, G. Wittenhagen, and W. Langenbeck, Naturwiss., 45, 13, 1958.

[641] W. Mohrschulz, Arch. Pharm. Ber. Dtsch. Pharm. Ges., 289/61, 508, 1956.

[642] A. M. Morgan, E. B. Truitt, and J. M. Little, J. Am. Pharm. Assoc. Sci. Ed., 46, 374, 1957.

[643] K. A. H. Mörner, Hoppe-Seylers Z., 13, 12, 1889.

[644] K. A. H. Mörner, Jber. Fortschr. Tierchem., 19, 80, 1889.

[645] J. A. Morrison, Arch. Int. Pharmacodyn., 157, 385, 1965.

[646] V. Morvillo and S. Garattini, Atti. Soc. Lombarda Sci. Med. e Biol., 7, 102, 1952.

[647] S. J. Mule and G. J. Mannering, Bull. Stupéfiants, 17, 27, 1965.

[648] S. J. Mule, L. A. Woods, and L. B. Mellett, J. Pharm. Exp. Ther., 136, 242, 1962.

[649] O. H. Müller and I. M. Weiner, J. Pharm. Exp. Ther., 118, 461, 1956.

[650] T. Murata, Chem. Pharm. Bull., 8, 629, 1960.

[651] T. Murata, Chem. Pharm. Bull., 9, 146, 1961.

[652] T. Murata, Chem. Pharm. Bull., 9, 167, 1961.

[653] T. Murata, Chem. Pharm. Bull., 9, 334, 1961.

[654] T. Murata, Chem. Pharm. Bull., 9, 335, 1961.

[655] C. W. Murray, A. N. Booth, F. DeEds, and F. T. Jones, J. Am. Pharm. Assoc. Sci. Ed., 43, 361, 1954.

[656] M. V. Nadkarni, E. I. Goldenthal, and P. K. Smith, Proc. Am. Assoc. Cancer Res., 1, 39, 1953.

[657] M. V. Nadkarni, E. I. Goldenthal, and P. K. Smith, Cancer Res., 17, 97, 1957.

[658] V. Nair, Biochem. Pharmacol., 3, 78, 1959.

[659] K. Nakamura, Y. Masuda, K. Nakatsuji, and T. Hiroka, Arch. Pharmakol. Exp. Pathol., 254, 406, 1966.

[660] M. Nakao, J. Biochem. (Tokyo), 44, 327, 1957.

[661] M. Nakao and I. Yanagisawa, J. Biochem. (Tokyo), 44, 433, 1957.

[662] S. A. Narrod, A. L. Wilk, and C. T. G. King, J. Pharm. Exp. Ther., 147, 380, 1965.

[663] K. P. Nayak, E. Brochmann-Hansen, and E. L. Way, J. Pharm. Sci., 54, 191, 1965.

[664] P. Needlemann and J. C. Krantz, Biochem. Pharmacol., 14, 1225, 1965.

[665] W. B. Neely, Biochem. Pharmacol., 13, 1137, 1964.

[666] N. Neff, G. V. Rossi, G. D. Chase, and J. L. Rabinowitz, J. Pharm. Exp. Ther., 144, 1, 1964.

[667] E. Nelson, I. O'Reilly, and T. Chulski, Clin. Chim. Acta, 5, 774, 1960.

[668] O. Nemecek, K. Macek, M. Queisnerova, and Z. J. Vejdelek, Arzneimitt. Forsch., 16, 1339, 1966.

[669] M. Newell, M. Argus, and F. E. Ray, Biochem. Pharmacol., 5, 30, 1960.

[670] P. J. Nicholls, J. Pharm. Pharmacol., 18, 46, 1966.

[671] W. Nielsch and L. Giefer, Arzneimitt. Forsch., 9, 636, 1959.

[672] W. Nielsch and L. Giefer, Arzneimitt. Forsch., 9, 700, 1959.

[673] S. K. Niyogi, Nature, 202, 1225, 1964.

[674] S. K. Niyogi, V. F. Cordova and F. Rieders, Nature, 206, 716, 1965.

[675] S. K. Niyogi and F. Rieders, Pharmacologist, 7, 148, 1965.

[676] E. L. Noach, D. M. Woodbury, and L. S. Goodman, J. Pharm. Exp. Ther., 122, 301, 1958.

[677] J. Novotny, R. Smetana, and H. Raskova, Biochem. Pharmacol., 14, 1537, 1965.

[678] H. Nowak, G. Schorre, and R. Struller, Arzneimitt. Forsch., 16, 407, 1966.

[679] P. T. Nowell, C. A. Scott, and A. Wilson, Brit. J. Pharmacol., 19, 498, 1962.

[680] P. Numerof, M. Gordon, and J. M. Kelly, J. Pharm. Exp. Ther., 115, 427, 1955.

[681] F. W. Oberst, J. Lab. Clin. Med., 24, 318, 1938.

[682] F. W. Oberst, J. Pharm. Exp. Ther., 69, 240, 1940.

[683] F. W. Oberst, U. S. Publ. Health Repts., Suppl. n° 158, 50, 1940.

[684] F. W. Oberst, J. Pharm. Exp. Ther., 73, 401, 1941.

[685] F. W. Oberst and F. H. Snyder, J. Pharm. Exp. Ther., 93, 444, 1948.

[686] A. G. Ogston and J. E. Stanier, Biochem. J., 49, 591, 1951.

[687] A. Ohara, M. Maruyama, and M. Yamasaki, Takamine Kenkyusho Nempo, 10, 178, 1958.

[688] G. T. Okita, F. E. Kelsey, P. J. Talso, L. B. Smith, and E. M. K. Geiling, Circulation, 7, 161, 1953.

[689] G. T. Okita, P. J. Talso, J. H. Curry, F. D. Smith, and E. M. K. Geiling, J. Pharm. Exp. Ther., 113, 376, 1955.

[690] K. Okumura, Yakugaku Kenkyu, 34, 36, 1962.

[691] V. T. Oliverio, Anal. Chem., 33, 263, 1961.

[692] V. T. Oliverio, R. H. Adamson, E. S. Henderson, and J. D. Davidson, J. Pharm. Exp. Ther., 141, 149, 1963.

[693] V. T. Oliverio and J. D. Davidson, J. Pharm. Exp. Ther., 137, 76, 1962.

[694] K. Opitz and M. L. Weischer, Arzneimitt. Forsch., 16, 1311, 1966.
[695] L. R. Overby, S. F. Bocchieri, and R. L. Fredrickson, J. Ass. Off. Agr. Chem., 48, 17, 1965.
[696] L. R. Overby and R. L. Fredrickson, Toxicol. Appl. Pharmacol., 7, 855, 1965.
[697] L. R. Overby and L. Straube, Toxicol. Appl. Pharmacol., 7, 850, 1965.
[698] H. Ozawa and A. Kiyomoto, Igaku To Seibutsugaku, 27, 110, 1953.
[699] D. V. Parke, S. Pollock, and R. T. Williams, J. Pharm. Pharmacol., 15, 500, 1963.
[700] D. V. Parke and R. T. Williams, J. Chem. Soc., 1950, 1757.
[701] B. B. Paul and D. Rubinstein, J. Pharm. Exp. Ther., 141, 141, 1963.
[702] W. D. Paul, M. M. Al Hashimi, and J. I. Routh, Feder. Proc., 10, 232, 1951.
[703] F. Pechtold, Arzneimitt. Forsch., 14, 972, 1964.
[704] E. A. Peets and D. A. Buyske, Biochem. Pharmacol., 13, 1403, 1964.
[705] E. A. Peets, W. M. Sweeney, V. A. Place, and D. A. Buyske, Am. Rev. Respirat. Dis., 91, 51, 1965.
[706] C. T. Peng, J. Pharm. Exp. Ther., 120, 229, 1957.
[707] G. Peters, H. Keberle, K. Schmid, and H. Brunner, Biochem. Pharmacol., 15, 93, 1966.
[708] J. H. Peters, K. S. Miller, and P. Brown, Analyt. Biochem., 12, 379, 1965.
[709] J. H. Peters, K. S. Miller, and P. Brown, J. Pharm. Exp. Ther., 150, 298, 1965.
[710] B. M. Phillips and T. S. Miya, J. Pharm. Sci., 53, 1098, 1964.
[711] J. N. Plampin and C. K. Cain, J. Med. Chem., 6, 247, 1963.
[712] N. P. Plotnikoff, H. W. Elliott, and E. L. Way, J. Pharm. Exp. Ther., 104, 377, 1952.
[713] N. P. Plotnikoff, E. L. Way, and H. W. Elliott, J. Pharm. Exp. Ther., 117, 414, 1956.
[714] A. Pohland, H. R. Sullivan, and H. M. Lee, Abstracts Papers Am. Chem. Soc., 136, 15-O, 1959.
[715] C. C. Porter, J. Pharm. Exp. Ther., 120, 447, 1957.
[716] C. C. Porter and D. C. Titus, J. Pharm. Exp. Ther., 139, 77, 1963.
[717] G. S. Porter and J. Beresford, J. Pharm. Pharmacol., 18, 223, 1966.
[718] H. S. Posner, Abstracts Papers Am. Chem. Soc., 136, 81 C, 1959.

BIBLIOGRAPHY

[719] H. S. Posner, R. Culpan, and J. Levine, J. Pharm. Exp. Ther., 141, 377, 1963.

[720] G. D. Potter and J. L. Guy, Proc. Soc. Exp. Biol. Med., 116, 658, 1964.

[721] L. F. Prescott, R. P. Buhs, J. O. Beattie, O. C. Speth, N. R. Trenner, and L. Lasagna, Circulation, 34, 308, 1966.

[722] F. R. Preuss and C. S. Ebenezer, Naturwiss., 52, 430, 1965.

[723] F. R. Preuss, H. M. Hassler, and R. Köpf, Arzneimitt. Forsch., 16, 395, 1966.

[724] F. R. Preuss, H. M. Hassler, and R. Köpf, Arzneimitt. Forsch., 16, 401, 1966.

[725] F. R. Preuss and R. Kopp, Arzneimitt. Forsch., 9, 256, 1959.

[726] F. R. Preuss and R. Kopp, Arzneimitt. Forsch., 9, 785, 1959.

[727] F. R. Preuss and H. W. Kopsch, Arzneimitt. Forsch., 16, 858, 1966.

[728] F. R. Preuss and E. Mayer, Arch. Toxikol., 18, 243, 1960.

[729] F. R. Preuss and E. Mayer, Arch. Pharm., 298, 781, 1965.

[730] F. R. Preuss and E. Mayer, Arzneimitt. Forsch., 15, 747, 1965.

[731] F. R. Preuss and K. M. Voigt, Naturwiss., 51, 537, 1964.

[732] F. R. Preuss and G. Willig, Arzneimitt. Forsch., 13, 155, 1963.

[733] F. R. Preuss and G. Willig, Arzneimitt. Forsch., 13, 234, 1963.

[734] F. R. Preuss, G. Willig, and H. Friebolin, Arch. Pharm., 296, 157, 1963.

[735] O. Pribilla, Arzneimitt. Forsch., 14, 723, 1964.

[736] O. Pribilla, Arzneimitt. Forsch., 15, 1148, 1965.

[737] S. Price, H. F. Martin, and B. J. Gudzinovicz, Biochem. Pharmacol., 13, 659, 1964.

[738] W. H. Prusoff, J. J. Jaffe, and H. Günther, Biochem. Pharmacol., 3, 110, 1960.

[739] R. Pulver and K. N. Von Kaulla, Schweiz. Med., Wchenschr., 78, 956, 1948.

[740] G. P. Quinn, P. A. Shore, and B. B. Brodie, J. Pharm. Exp. Ther., 127, 103, 1959.

[741] J. Raaflaub and D. E. Schwartz, Experientia, 21, 44, 1965.

[742] J. B. Ragland and V. J. Kinross-Wright, Anal. Chem., 36, 1356, 1964.

[743] K. Raska, M. Jurovcik, Z. Sormova, and F. Sorm, Coll. Czech. Chem. Commun., 30, 3001, 1965.

[744] K. Raska, M. Jurovcik, Z. Sormova, and F. Sorm, Coll. Czech. Chem. Commun., 31, 2803, 1966.

[745] J. Ratcliffe and P. Smith, Chem. and Ind., 925, 1959.
[746] H. M. Rauen and U. Dirschka, Arzneimitt. Forsch., 14, 159, 1964.
[747] J. Raventos, J. Pharm. Pharmacol., 6, 217, 1954.
[748] H. M. Redetzki, J. E. Redetzki, and A. L. Elias, Biochem. Pharmacol., 15, 425, 1966.
[749] D. J. Reed and F. N. Dost, Proc. Am. Assoc. Cancer Res., 7, 57, 1966.
[750] C. N. Remy, J. Biol. Chem., 238, 1078, 1963.
[751] B. R. Rennick, M. Z. Pryor, and B. G. Basch, J. Pharm. Exp. Ther., 148, 270, 1965.
[752] K. Repke, Naturwiss., 44, 619, 1957.
[753] K. Repke, Naturwiss., 45, 20, 1958.
[754] K. Repke, Arch. Exp. Pathol. Pharmakol., 233, 261, 1958.
[755] K. Repke, Arch. Exp. Pathol. Pharmakol., 233, 271, 1958.
[756] K. Repke, Arch. Exp. Pathol. Pharmakol., 236, 242, 1959.
[757] K. Repke, S. Klesczewski, and L. Roth, Arch. Exp. Pathol. Pharmakol., 237, 34, 1959.
[758] K. Repke, L. Roth, and S. Klesczewski, Arch. Exp. Pathol. Pharmakol., 237, 155, 1959.
[759] R. K. Richards and J. D. Taylor, Anesthesiology, 17, 414, 1956.
[760] G. Richarz, W. Schoetensack and M. Vogel, Arzneimitt. Forsch., 10, 676, 1960.
[761] D. Richter, Biochem. J., 32, 1763, 1938.
[762] J. C. Rickards, G. E. Boxer, and C. C. Smith, J. Pharm. Exp. Ther., 98, 380, 1950.
[763] J. Rieder, Arzneimitt. Forsch., 15, 1134, 1965.
[764] S. Riegelman, W. L. Epstein, and R. J. Dayan, Sci. Sect. Am. Pharm. Assoc. Preprints Papers Las Vegas, 1962 B-IX.
[765] W. Riess, K. Schmid, and H. Keberle, Klin. Wschr., 43, 740, 1965.
[766] R. F. Riley, J. Am. Chem. Soc., 72, 5712, 1950.
[767] R. F. Riley, J. Pharm. Exp. Ther., 99, 329, 1950.
[768] R. F. Riley and F. M. Berger, Arch. Biochem., 20, 159, 1949.
[769] P. Ritter and M. Jermann, Arzneimitt. Forsch., 16, 1647, 1966.
[770] D. J. Roberts, J. Pharm. Pharmacol., 17, 769, 1965.
[771] D. J. Roberts, Biochem. Pharmacol., 15, 63, 1966.
[772] J. B. Roberts, B. H. Thomas, and A. Wilson, Brit. J. Pharmacol., 25, 234, 1965.
[773] J. B. Roberts, B. H. Thomas, and A. Wilson, Brit. J. Pharmacol., 25, 763, 1965.

[774] J. B. Roberts, B. H. Thomas, and A. Wilson, Biochem. Pharmacol., 15, 71, 1966.
[775] J. J. Roberts and G. P. Warwick, Biochem. Pharmacol., 1, 60, 1958.
[776] A. E. Robinson, J. Pharm. Pharmacol., 18, 19, 1966.
[777] D. Robinson and R. T. Williams, Biochem. J., 62, 23 P, 1956.
[778] E. J. Robinson and M. L. Crossley, Arch. Biochem. Biophys., 1, 415, 1943.
[779] C. F. Rodriguez and D. E. Johnson, Life Sci., 5, 1283, 1966.
[780] C. F. Rodriguez and D. E. Johnson, Abstr. Papers, Am. Chem. Soc., 152, 43 B, 1966.
[781] W. I. Rogers, D. W. Yesair, and C. J. Kensler, J. Pharm. Exp. Ther., 152, 139, 1966.
[782] G. N. Rolinson and F. R. Batchelor, Antimicrobial Agents and Chemotherapy, 1962, 654.
[783] R. Roncucci, M. J. Simon, G. Lambelin, N. P. Buu-Hoï, and J. Thiriaux, Biochem. Pharmacol., 15, 1563, 1966.
[784] D. Rosi, G. Peruzzotti, E. W. Dennis, D. A. Berberian, H. Freele, and S. Archer, Nature, 208, 1005, 1965.
[785] J. J. Ross, R. L. Young, and A. R. Maass, Science, 128, 1279, 1958.
[786] L. J. Roth, E. Leifer, J. R. Hogness, and W. H. Langham, J. Biol. Chem., 178, 963, 1949.
[787] L. J. Roth and R. W. Manthei, Proc. Soc. Exp. Biol. Med., 81, 566, 1952.
[788] W. Rüdiger and H. Büch, Arch. Exp. Pathol. Pharmakol., 251, 107, 1965.
[789] H. W. Ruelius, M. L. Lee, and H. E. Alburn, Arch. Biochem. Biophys., 111, 376, 1965.
[790] W. D. Rupp and R. E. Handschumacher, Biochem. Pharmacol., 12, 13, 1963.
[791] N. P. Salzman and B. B. Brodie, J. Pharm. Exp. Ther., 118, 46, 1956.
[792] N. P. Salzman, N. C. Moran, and B. B. Brodie, Nature, 176, 1122, 1955.
[793] H. G. Sammons, J. Shelswell, and R. T. Williams, Biochem. J., 35, 557, 1941.
[794] E. J. Sarcione and L. Stutzman, Proc. Soc. Exp. Biol. Med., 101, 766, 1959.
[795] J. P. Scannell and G. H. Hitchings, Proc. Soc. Exp. Biol. Med., 122, 627, 1966.
[796] D. Schachter, Feder. Proc., 15, 479, 1956.

[797] D. Schachter, J. Clin. Invest., 36, 297, 1957.
[798] D. Schachter, D. J. Kass, and T. J. Lannon, J. Biol. Chem., 234, 201, 1959.
[799] D. Schachter and J. G. Manis, J. Clin. Invest., 37, 800, 1958.
[800] F. Schatz and V. Jahn, Arzneimitt. Forsch., 16, 866, 1966.
[801] O. Schaumann, Arch. Exp. Pathol. Pharmakol., 239, 311, 1960.
[802] R. W. Schayer, Arch. Biochem., 28, 371, 1950.
[803] R. R. Scheline, R. L. Smith, and R. T. Williams, J. Med. Pharm. Chem., 4, 109, 1961.
[804] W. Schindler, Helv. Chim. Acta, 43, 35, 1960.
[805] E. Schmerzler, W. Yu, M. I. Hewitt, and I. J. Greenblatt, J. Pharm. Sci., 55, 1955, 1966.
[806] G. Schmidt and B. Arnold, Arch. Toxikol., 16, 50, 1956.
[807] W. C. Schneider, J. Biol. Chem., 161, 293, 1945.
[808] H. Schuhmacher, R. L. Smith, R. B. L. Stagg, and R. T. Williams, Pharm. Acta Helv., 39, 394, 1964.
[809] H. Schuhmacher, R. L. Smith, and R. T. Williams, Brit. J. Pharmacol., 25, 324, 1965.
[810] H. Schuhmacher, R. L. Smith, and R. T. Williams, Brit. J. Pharmacol., 25, 338, 1965.
[811] R. Schüppel, Arch. Pharmakol. Exp. Pathol., 255, 71, 1966.
[812] R. Schüppel and K. Soehring, Pharm. Acta Helv., 40, 105, 1965.
[813] D. E. Schwartz, Experientia, 22, 212, 1966.
[814] D. E. Schwartz, H. Bruderer, J. Rieder, and A. Brossi, Biochem. Pharmacol., 13, 777, 1964.
[815] D. E. Schwartz, H. Bruderer, J. Rieder, and A. Brossi, Biochem. Pharmacol., 15, 645, 1966.
[816] M. A. Schwartz, J. Pharm. Exp. Ther., 130, 157, 1960.
[817] M. A. Schwartz, Proc. Soc. Exp. Biol. Med., 107, 613, 1961.
[818] M. A. Schwartz, B. A. Koechlin, E. Postma, S. Palmer, and G. Krol, J. Pharm. Exp. Ther., 149, 423, 1965.
[819] M. A. Schwartz and E. Postma, J. Pharm. Sci., 55, 1358, 1966.
[820] C. A. Scott, P. T. Nowell, and A. Wilson, J. Pharm. Pharmacol., 14, 31 T, 1962.
[821] J. V. Scudi, Science, 91, 486, 1940.
[822] J. V. Scudi and S. J. Childress, J. Biol. Chem., 218, 587, 1956.
[823] R. A. Seibert, C. E. Williams, and R. A. Huggins, Science, 120, 222, 1954.
[824] N. Seiler, Z. Physiol. Chem., 341, 105, 1965.
[825] N. Seiler, G. Werner, and M. Wiechmann, Naturwiss., 50, 643, 1963.

[826] N. Seiler and M. Wiechmann, Z. Physiol. Chem., 337, 229, 1964.
[827] N. P. Sen and P. L. McGeer, Biochem. Biophys. Res. Commun., 13, 390, 1963.
[828] E. Serick and W. Loop, Chemotherapia, 10, 74, 1965-1966.
[829] J. Seydel, H. Büttner, and F. Portwich, Klin. Wschr., 43, 1060, 1965.
[830] N. T. Shahidi, H. Wyler, W. H. Hitzig, and A. Dreiding, Clin. Res., 14, 327, 1966.
[831] A. L. Sheffner, E. M. Medler, K. R. Bailey, D. G. Gallo, A. J. Mueller, and H. P. Sarett, Biochem. Pharmacol., 15, 1523, 1966.
[832] J. Shelswell and R. T. Williams, Biochem. J., 34, 528, 1940.
[833] E. E. Shepheard, R. H. Thorp, and S. E. Wright, J. Pharm. Exp. Ther., 112, 133, 1954.
[834] D. M. Shepherd and G. B. West, Nature, 169, 797, 1952.
[835] H. Sheppard, B. S. d'Asaro, and A. J. Plummer, J. Am. Pharm. Assoc. Sci. Ed., 45, 681, 1956.
[836] H. Sheppard, R. C. Lucas, and W. H. Tsien, Arch. Int. Pharm. Ther., 103, 256, 1955.
[837] H. Sheppard and W. H. Tsien, Proc. Soc. Exp. Biol. Med., 90, 437, 1955.
[838] J. Shibasaki and E. Nakamura, J. Pharm. Soc. Japan, 86, 1211, 1966.
[839] J. Shibasaki, E. Nakamura, Y. Sakamoto, and Y. Kurokawa, J. Pharm. Soc. Japan, 86, 699, 1966.
[840] J. Shibasaki, Y. Sakamoto, and E. Nakamura, J. Pharm. Soc. Japan, 85, 87, 1965.
[841] A. L. Shlyakhman, V. I. Grishina, and S. E. Shnol, Voprosy Med. Khim., 5, 422, 1959.
[842] P. A. Shore, J. Axelrod, A. M. Hogben, and B. B. Brodie, J. Pharm. Exp. Ther., 113, 192, 1955.
[843] O. Siblikova, J. Lavicka, G. Vachek, B. Tesarek, O. Vitulova, and M. Cerna, Arzneimitt. Forsch., 11, 1106, 1961.
[844] J. A. F. de Silva, B. A. Koechlin, and G. Bader, J. Pharm. Sci., 55, 692, 1966.
[845] J. A. F. de Silva, M. A. Schwartz, J. Stefanovic, J. Kaplan, and L. d'Arconte, Anal. Chem., 36, 2099, 1964.
[846] B. Silvestrini, B. Catanese, G. Corsi, and P. Ridolfi, J. Pharm. Pharmacol., 16, 38, 1964.
[847] Z. Sipal and A. Jindra, Pharmazie, 20, 376, 1965.
[848] A. Sivak, W. I. Rogers, I. Wodinsky, and C. J. Kensler, J. Nat. Cancer Inst., 33, 457, 1964.
[849] A. Sjoerdsma, A. Vendsalu, and K. Engelman, Circulation, 28, 492, 1963.

[850] H. K. Slotta and J. Müller, Z. Physiol. Chem., 238, 14, 1936.

[851] C. C. Smith, J. Pharm. Exp. Ther., 116, 67, 1956.

[852] D. L. Smith, A. A. Forist, and W. E. Dulin, J. Med. Chem., 8, 350, 1965.

[853] D. L. Smith, A. A. Forist, and G. C. Gerritsen, J. Pharm. Exp. Ther., 150, 316, 1965.

[854] D. L. Smith and M. F. Grostic, Abstr. Papers, Am. Chem. Soc., 152, 49 P, 1966.

[855] D. L. Smith, H. H. Keasling, and A. A. Forist, J. Med. Chem., 8, 520, 1965.

[856] D. L. Smith, T. J. Vecchio, and A. A. Forist, Metab. Clin. Exp., 14, 229, 1965.

[857] J. N. Smith and R. T. Williams, Biochem. J., 42, 351, 1948.

[858] J. N. Smith and R. T. Williams, Biochem. J., 42, 538, 1948.

[859] J. N. Smith and R. T. Williams, Biochem. J., 44, 239, 1949.

[860] J. N. Smith and R. T. Williams, Biochem. J., 44, 250, 1949.

[861] J. N. Smith and R. T. Williams, Biochem. J., 56, 618, 1954.

[862] P. K. Smith, H. L. Gleason, C. G. Stoll, and S. Ogorzalec, J. Pharm. Exp. Ther., 87, 237, 1946.

[863] R. L. Smith, R. A. D. Williams, and R. T. Williams, Life Sci., 1, 333, 1962.

[864] R. L. Smith and R. T. Williams, J. Med. Pharm. Chem., 4, 97, 1961.

[865] R. L. Smith and R. T. Williams, J. Med. Pharm. Chem., 4, 147, 1961.

[866] R. L. Smith and R. T. Williams, J. Med. Pharm. Chem., 4, 163, 1961.

[867] G. A. Snow, Biochem. J., 65, 77, 1957.

[868] E. Spector, Nature, 189, 751, 1961.

[869] E. Spector and F. E. Shideman, J. Pharm. Exp. Ther., 116, 54, 1956.

[870] E. Spector and F. E. Shideman, Biochem. Pharmacol., 2, 182, 1959.

[871] L. H. Sternbach, B. A. Koechlin, and E. Reeder, J. Org. Chem., 27, 4671, 1962.

[872] A. Stier, Biochem. Pharmacol., 13, 1544, 1964.

[873] A. Stier and H. Alter, Anaesthesist., 15, 154, 1966.

[874] R. E. Stitzel, F. E. Greene, R. Furner, and H. Conaway, Biochem. Pharmacol., 15, 1001, 1966.

[875] D. Stolnikow, Hoppe-Seyl. Z., 8, 238, 1883-1884.

[876] R. G. Strachan, M. A. P. Meisinger, W. V. Ruyle, R. Hirschmann, and T. Y. Shen, J. Med. Chem., 7, 799, 1964.

[877] W. H. Straub, D. F. Flanagan, R. Aaron, and J. C. Rose, Proc. Soc. Exp. Biol. Med., 116, 1119, 1964.

[878] J. H. Strömme, Biochem. Pharmacol., 14, 381, 1965.

[879] J. H. Strömme, Biochem. Pharmacol., 14, 393, 1965.

[880] R. Strufe, Med. Chem. Abhandl. Med.-Chem. Forschungstätten Farbenfabriken Bayer AG., 7, 337, 1963.

[881] A. L. Stubbs and R. T. Williams, Biochem. J., 41, XLIX, 1947.

[882] C. Y. Sung and A. P. Truant, J. Pharm. Exp. Ther., 112, 432, 1954.

[883] T. Suzuki and F. Tanaka, Yakugaku Zasshi, 79, 410, 1959.

[884] Y. Suzuki, T. Takeuchi, and T. Komai, J. Antibiotics, 15 (series A), 67, 1962.

[885] A. B. Svendsen and E. Brochmann-Hanssen, J. Pharm. Sci., 51, 494, 1962.

[886] S. Symchowicz, W. D. Peckham, C. A. Korduba, and P. L. Perlman, Biochem. Pharmacol., 11, 499, 1962.

[887] S. Symchowicz and K. K. Wong, Biochem. Pharmacol., 15, 1595, 1966.

[888] S. Symchowicz and K. K. Wong, Biochem. Pharmacol., 15, 1601, 1966.

[889] E. Takabatake, J. Pharm. Soc. Japan, 76, 511, 1956.

[890] E. Takabatake, Pharm. Bull. (Japan), 5, 260, 1957.

[891] M. Takeda, H. Yoshimura, and H. Tsukamoto, J. Pharm. Soc. Japan, 86, 1191, 1966.

[892] V. Tamminen and A. R. Alha, Suomen Kemistilehti, B 32, 119, 1959.

[893] I. Tanaka, Japan Safety Forces Med. J., 3, 788, 1956.

[894] I. Tanaka, Japan Safety Forces Med. J., 3, 811, 1956.

[895] A. L. Tatum, Physiol., Revs., 19, 472, 1939.

[896] A. L. Tatum, Ann. Rev. Physiol., 2, 359, 1940.

[897] J. D. Taylor, R. K. Richards, and D. L. Tabern, J. Pharm. Exp. Ther., 104, 93, 1952.

[898] R. K. Thauer, G. Stöffler, and H. Uehleke, Arch. Exp. Pathol. Pharmakol., 252, 32, 1965.

[899] K. Theimer and R. Stadler, Arzneimitt. Forsch., 16, 1068, 1966.

[900] R. C. Thomas and G. J. Ikeda, J. Med. Chem., 9, 507, 1966.

[901] R. E. Thomas and S. E. Wright, J. Pharm. Pharmacol., 17, 459, 1965.

[902] G. F. Thompson and K. Smith, Anal. Chem., 37, 1591, 1965.

[903] V. Thompson and E. G. Gross, J. Pharm. Exp. Ther., 72, 138, 1941.

[904] W. V. Thorpe and R. T. Williams, Nature, 146, 686, 1940.
[905] E. K. Tillson, G. S. Schuchardt, J. K. Fishman, and K. H. Beyer, J. Pharm. Exp. Ther., 111, 385, 1954.
[906] E. Titus and J. Bernstein, Ann. N.Y. Acad. Sci., 52, 719, 1949.
[907] E. Titus, L. C. Craig, C. Golumbic, H. R. Mighton, I. M. Wempen, and R. C. Elderfield, J. Org. Chem., 13, 39, 1948.
[908] E. Titus, S. Ulick, and A. P. Richardson, J. Pharm. Exp. Ther., 93, 129, 1948.
[909] E. Titus and H. Weiss, J. Biol. Chem., 214, 807, 1955.
[910] D. J. Tocco, R. P. Buhs, H. D. Brown, A. R. Matzuk, H. E. Mertel, R. E. Harman, and N. R. Trenner, J. Med. Chem., 7, 399, 1964.
[911] D. J. Tocco, J. R. Egerton, W. Bowers, V. W. Christensen, and C. Rosenblum, J. Pharm. Exp. Ther., 149, 263, 1965.
[912] D. J. Tocco, C. Rosenblum, C. M. Martin, and H. J. Robinson, Toxicol. Appl. Pharmacol., 9, 31, 1966.
[913] Y. Tochino, Wakayama Igaku, 6, 421, 1955.
[914] Y. Tochino, Wakayama Med. Repts., 3, 27, 1956.
[915] S. Toki and T. Takenouchi, Chem. Pharm. Bull., 13, 606, 1965.
[916] S. Toki, K. Toki, and H. Tsukamoto, Chem. Pharm. Bull., 10, 708, 1962.
[917] K. Toki, S. Toki, and H. Tsukamoto, J. Biochem. (Japan), 53, 43, 1963.
[918] S. L. Tompsett, J. Pharm. Pharmacol., 16, 207, 1964.
[919] E. G. Trams, M. V. Nadkarni, V. Dequattro, G. D. Maengwyn-Davis, and P. K. Smith, Biochem. Pharmacol., 2, 7, 1959.
[920] H. Tsukamoto, H. Ide, and E. Takabatake, Chem. Pharm. Bull., 8, 236, 1960.
[921] H. Tsukamoto, K. Kato, and K. Tatsumi, Pharm. Bull. (Japan), 5, 570, 1957.
[922] H. Tsukamoto, K. Kato, and K. Yoshida, Chem. Pharm. Bull., 12, 731, 1964.
[923] H. Tsukamoto and Y. Kuroiwa, Chem. Pharm. Bull., 7, 731, 1959.
[924] H. Tsukamoto, K. Oguri, T. Watabe, and H. Yoshimura, J. Biochem. (Tokyo), 55, 394, 1964.
[925] H. Tsukamoto, E. Takabatake, and T. Ariyoshi, Pharm. Bull. (Japan), 3, 459, 1955.
[926] H. Tsukamoto, E. Takabatake, and H. Yoshimura, Pharm. Bull. (Japan), 2, 201, 1954.
[927] H. Tsukamoto, S. Toki, and K. Kaneda, Chem. Pharm. Bull., 7, 651, 1959.
[928] H. Tsukamoto and A. Yamamoto, Pharm. Bull. (Japan), 3, 427, 1955.

[929] H. Tsukamoto, A. Yamamoto, and O. Kamata, Pharm. Bull. (Japan), 5, 565, 1957.
[930] H. Tsukamoto and H. Yoshimura, Pharm. Bull. (Japan), 3, 397, 1955.
[931] H. Tsukamoto, H. Yoshimura, and H. Ide, Chem. Pharm. Bull., 11, 9, 1963.
[932] H. Tsukamoto, H. Yoshimura, H. Ide, and S. Mitsui, Chem. Pharm. Bull., 11, 427, 1963.
[933] H. Tsukamoto, H. Yoshimura, and K. Tatsumi, Chem. Pharm. Bull., 11, 421, 1963.
[934] H. Tsukamoto, H. Yoshimura, and K. Tatsumi, Chem. Pharm. Bull., 11, 1134, 1963.
[935] H. Tsukamoto, H. Yoshimura, and S. Toki, Pharm. Bull. (Japan), 3, 239, 1955.
[936] H. Tsukamoto, H. Yoshimura, and S. Toki, Pharm. Bull. (Japan), 4, 364, 1956.
[937] H. Tsukamoto, H. Yoshimura, and S. Toki, Pharm. Bull. (Japan), 4, 368, 1956.
[938] H. Tsukamoto, H. Yoshimura, and S. Toki, Pharm. Bull. (Japan), 4, 371, 1956.
[939] H. Tsukamoto, H. Yoshimura, and S. Toki, Chem. Pharm. Bull., 6, 15, 1958.
[940] H. Tsukamoto, H. Yoshimura, and S. Toki, Chem. Pharm. Bull., 6, 88, 1958.
[941] H. Tsukamoto, H. Yoshimura, T. Watabe, and K. Oguri, Biochem. Pharmacol., 13, 1577, 1964.
[942] H. Tsukamoto and M. Yoshimura, Chem. Pharm. Bull., 9, 584, 1961.
[943] M. Ueda and K. Kuribayashi, J. Pharm. Soc. Japan, 84, 1104, 1964.
[944] T. Uno and M. Kono, J. Pharm. Soc. Japan, 80, 201, 1960.
[945] T. Uno, M. Kono, and M. Furutani, J. Pharm. Soc. Japan, 81, 192, 1961.
[946] T. Uno, M. Kono, and H. Kondo, J. Pharm. Soc. Japan, 81, 1509, 1961.
[947] T. Uno, T. Kushima, and M. Fujimoto, Chem. Pharm. Bull., 13, 261, 1965.
[948] T. Uno and Y. Okasaki, J. Pharm. Soc. Japan, 80, 1682, 1960.
[949] T. Uno and M. Ueda, J. Pharm. Soc. Japan, 80, 1785, 1960.
[950] T. Uno and M. Ueda, J. Pharm. Soc. Japan, 82, 759, 1962.
[951] T. Uno, H. Yasuda, and Y. Sekine, Chem. Pharm. Bull., 11, 872, 1963.
[952] T. Uno and S. Yutaka, Chem. Pharm. Bull., 14, 687, 1966.
[953] W. J. A. Vandenheuvel, E. O. A. Haahti, and E. C. Horning, Clin. Chem., 8, 351, 1962.

[954] H. B. Van Dyke, J. V. Scudi, and D. L. Tabern, J. Pharm. Exp. Ther., 90, 364, 1947.

[955] R. A. Van Dyke and M. B. Chenoweth, Biochem. Pharmacol., 14, 603, 1965.

[956] R. A. Van Dyke, M. B. Chenoweth, and E. R. Larsen, Nature, 204, 471, 1964.

[957] R. A. Van Dyke, M. B. Chenoweth, and A. Van Poznak, Biochem. Pharmacol., 13, 1239, 1964.

[958] A. Venkataraman, P. R. Venkataraman, and H. B. Lewis, J. Biol. Chem., 173, 641, 1948.

[959] P. Venkataraman, L. Eidus, K. Ramachandran, and S. P. Tripathy, Tubercle (G.-B.), 46, 262, 1965.

[960] E. Vidic, Arzneimitt. Forsch., 7, 314, 1957.

[961] W. Vogt, G. Schmidt, and T. Dakhil, Arch. Exp. Pathol. Pharmakol., 250, 488, 1965.

[962] W. J. Waddell, J. Pharm. Exp. Ther., 149, 23, 1965.

[963] W. J. Waddell and T. C. Butler, J. Pharm. Exp. Ther., 132, 291, 1961.

[964] D. Waldi, Arch. Pharm., 292/64, 206, 1958.

[965] S. S. Walkenstein, N. Chumakow, and J. Seifter, J. Pharm. Exp. Ther., 115, 16, 1955.

[966] S. S. Walkenstein, R. A. Corradino, R. Wiser, and C. H. Gudmundsen, Biochem. Pharmacol., 14, 121, 1965.

[967] S. S. Walkenstein, C. M. Knebel, J. A. Macmullen, and J. Seifter, J. Pharm. Exp. Ther., 123, 254, 1958.

[968] S. S. Walkenstein, J. A. Macmullen, C. Knebel, and J. Seifter, J. Am. Pharm. Assoc. Sci. Ed., 47, 20, 1958.

[969] S. S. Walkenstein and J. Seifter, J. Pharm. Exp. Ther., 125, 283, 1959.

[970] S. S. Walkenstein, R. Wiser, C. H. Gudmundsen, H. B. Kimmel, and R. A. Corradino, J. Pharm. Sci., 53, 1181, 1964.

[971] P. G. Waser and U. Lüthi, Helv. Physiol. Acta, 24, 259, 1966.

[972] T. Watabe, H. Yoshimura, and H. Tsukamoto, Chem. Pharm. Bull., 12, 1151, 1964.

[973] E. L. Way and T. K. Adler, Pharmacol. Rev., 12, 383, 1960.

[974] E. L. Way and T. K. Adler, Bull. O.M.S., 25, 227, 1961.

[975] E. L. Way and T. K. Adler, Bull. O.M.S., 26, 51, 1962.

[976] E. L. Way and T. K. Adler, Bull. O.M.S., 26, 261, 1962.

[977] E. L. Way, A. I. Gimble, W. P. McKelway, H. Ross, C. Sung, and H. Ellsworth, J. Pharm. Exp. Ther., 96, 477, 1949.

[978] E. L. Way, C. T. Peng, N. Allawala, and T. C. Daniels, J. Am. Pharm. Ass. Sci. Ed., 44, 65, 1955.

BIBLIOGRAPHY

[979] E. L. Way, B. T. Signorotti, C. H. March, and C. T. Peng, J. Pharm. Exp. Ther., 101, 249, 1951.

[980] E. L. Way, P. K. Smith, D. L. Howie, R. Weiss, and R. Swanson, J. Pharm. Exp. Ther., 93, 368, 1948.

[981] E. L. Way, C. Y. Sung, and W. P. McKelway, J. Pharm. Exp. Ther., 97, 222, 1949.

[982] J. L. Way, J. Pharm. Exp. Ther., 138, 258, 1962.

[983] J. L. Way, J. Pharm. Exp. Ther., 138, 264, 1962.

[984] J. L. Way, W. T. Brady, and H. S. Tong, J. Chromatog., 21, 280, 1966.

[985] J. L. Way, U. F. Feitknecht, C. N. Corder, and P. M. Miranda, Biochim. Biophys. Acta, 121, 432, 1966.

[986] J. L. Way, P. B. Masterson, and J. A. Beres, J. Pharm. Exp. Ther., 140, 117, 1963.

[987] J. L. Way, H. Tong, and R. Rabideau, Feder. Proc., 19, 276, 1960.

[988] C. J. Weber, J. J. Lalich, and R. H. Major, Proc. Soc. Exp. Biol. Med., 53, 190, 1943.

[989] J. H. Weikel and J. A. Labudde, J. Pharm. Exp. Ther., 138, 392, 1962.

[990] I. M. Weiner, A. E. Burnett, and B. R. Rennick, J. Pharm. Exp. Ther., 118, 470, 1956.

[991] I. M. Weiner and O. H. Müller, J. Pharm. Exp. Ther., 113, 241, 1955.

[992] H. Weinfeld, Feder. Proc., 10, 267, 1951.

[993] E. O. Weinman and T. A. Geissman, J. Pharm. Exp. Ther., 125, 1, 1959.

[994] R. M. Welch and A. H. Conney, Clin. Chem., 11, 1064, 1965.

[995] R. M. Welch, A. H. Conney, and J. J. Burns, Biochem. Pharmacol., 15, 521, 1966.

[996] J. S. Welles, R. C. Anderson, and K. K. Chen, Antibiotics Annual, 1954-1955, 291.

[997] J. S. Welles, R. E. McMahon, and W. J. Doran, J. Pharm. Exp. Ther., 139, 166, 1963.

[998] J. S. Welles, M. A. Root, and R. C. Anderson, Proc. Soc. Exp. Biol. Med., 101, 668, 1959.

[999] J. S. Welles, M. A. Root, and R. C. Anderson, Proc. Soc. Exp. Biol. Med., 107, 583, 1961.

[1000] G. B. West, J. Pharm. Pharmacol., 11, 595, 1959.

[1001] L. G. Whitby, J. Axelrod, and H. Weil-Malherbe, J. Pharm. Exp. Ther., 132, 193, 1961.

[1002] K. I. H. Williams, S. H. Burstein, and D. S. Layne, Arch. Biochem. Biophys., 117, 84, 1966.

[1003] K. I. H. Williams, K. S. Whittemore, T. N. Mellin, and D. S. Layne, Science, 149, 203, 1965.
[1004] R. T. Williams, Biochem. J., 35, 1169, 1941.
[1005] R. T. Williams, Biochem. J., 37, 329, 1943.
[1006] R. T. Williams, Biochem. J., 40, 219, 1946.
[1007] R. T. Williams, Biochem. J., 41, 1, 1947.
[1008] R. T. Williams, Detoxication Mechanisms. The Metabolism of Drugs, Toxic Substances and Other Organic Compounds (London: Chapman and Hall, Ltd., 1959.
[1009] V. Williams and F. E. Shideman, J. Pharm. Exp. Ther., 122, 84 A, 1958.
[1010] K. Willner, Arzneimitt. Forsch., 13, 26, 1963.
[1011] W. D. Winters, E. Spector, D. P. Wallach, and F. E. Shideman, J. Pharm. Exp. Ther., 114, 343, 1955.
[1012] E. H. Wiseman, J. Chiaini, and R. Pinson, J. Pharm. Sci., 53, 766, 1964.
[1013] E. H. Wiseman, E. C. Schreiber, and R. Pinson, Biochem. Pharmacol., 13, 1421, 1964.
[1014] R. Wiser and J. Seifter, Feder. Proc., 19, 390, 1960.
[1015] G. Wittenhagen and G. Mohnike, Deut. Med. Wschr., 81, 887, 1956.
[1016] G. Wittenhagen, G. Mohnike, and W. Langenbeck, Z. Physiol. Chem., 316, 157, 1959.
[1017] K. C. Wong and J. L. Spratt, Biochem. Pharmacol., 12, 577, 1963.
[1018] H. B. Wood and E. C. Horning, J. Am. Chem. Soc., 75, 5511, 1953.
[1019] L. A. Woods, Feder. Proc., 13, 419, 1954.
[1020] L. A. Woods, J. Pharm. Exp. Ther., 112, 158, 1954.
[1021] L. A. Woods and H. I. Chernov, Pharmacologist, 8, 206, 1966.
[1022] L. A. Woods, J. Cochin, E. J. Fornefeld, and F. G. McMahon, J. Pharm. Exp. Ther., 101, 188, 1951.
[1023] L. A. Woods, L. B. Mellett, and K. S. Andersen, J. Pharm. Exp. Ther., 124, 1, 1958.
[1024] L. A. Woods, H. E. Muehlenbeck, and L. B. Mellett, J. Pharm. Exp. Ther., 117, 117, 1956.
[1025] S. E. Wright, J. Pharm. Pharmacol., 14, 613, 1962.
[1026] J. B. Wyngaarden, L. A. Woods, and M. H. Seevers, Proc. Soc. Exp. Biol. Med., 66, 256, 1947.
[1027] K. Yamaguchi, Folia Pharmacol. Japan, 51, 631, 1955.
[1028] A. Yamamoto, H. Yoshimura, and H. Tsukamoto, Chem. Pharm. Bull., 10, 522, 1962.

[1029] A. Yamamoto, H. Yoshimura, and H. Tsukamoto, Chem. Pharm. Bull., 10, 540, 1962.

[1030] C. M. Yates, A. Todrick, and A. C. Tait, J. Pharm. Pharmacol., 15, 432, 1963.

[1031] H. Yoshimura, Pharm. Bull. (Japan), 5, 561, 1957.

[1032] H. Yoshimura, Chem. Pharm. Bull., 6, 13, 1958.

[1033] H. Yoshimura and H. Tsukamoto, Chem. Pharm. Bull., 10, 566, 1962.

[1034] M. Yoshimura and H. Tsukamoto, Chem. Pharm. Bull., 11, 689, 1963.

[1035] S. Yoshinari, Nippon Yakurigaku Zasshi, 61, 311, 1965.

[1036] J. A. Young and K. D. G. Edwards, J. Pharm. Exp. Ther., 145, 102, 1964.

[1037] R. L. Young and M. W. Gordon, J. Neurochem., 9, 161, 1962.

[1038] S. B. Zak, H. H. Tallan, G. P. Quinn, I. Fratta, and P. Greengard, J. Pharm. Exp. Ther., 141, 392, 1963.

[1039] V. Zamboni and A. Defranceschi, Biochim. Biophys. Acta, 14, 430, 1954.

[1040] L. Zathurecky, V. Krupa, and M. Rochova, Pharmazie, 21, 322, 1966.

[1041] H. L. Zauder, J. Pharm. Exp. Ther., 104, 11, 1952.

[1042] K. Zehnder, F. Kalberer, W. Kreis, and J. Rutschmann, Biochem. Pharmacol., 11, 535, 1962.

[1043] L. Zicha, F. Freytag, and F. Weist, Arzneimitt. Forsch., 15, 777, 1965.

[1044] G. R. Zins, J. Pharm. Exp. Ther., 150, 109, 1965.

AUTHOR INDEX

Underlined numerals refer to the pages in the text on which the author's name appears. Roman numerals refer to the reference numbers.

A

Aaron, R., 877
Abe, M., 193; 1
Abel, R. M., 301; 2
Abiko, Y., 195; 3
Acred, P., 291; 4
Adamson, R. H., 147, 251, 280; 5, 6, 412, 413, 559, 560, 692
Addison, E., 288; 7
Adler, T. K., 261, 262, 263; 8, 9, 10, 11, 973, 974, 975, 976
Agranoff, B. W., 90; 12
Ague, S. L., 5
Ahmad, P. M., 435, 436
Akagi, M., 242, 244; 13, 14
Akerman, B., 116; 15
Albert, A., 193; 16
Alburn, H. E., 789
Algeri, E. J., 127, 134; 17, 18
Alha, A. R., 147; 892
Al Hashimi, M. M., 702
Allawala, N., 9˙ˊ ˙
Allen, D. E., .15
Alleva, J. J., 21; 19
Allewijn, F. T., 206; 20
Allgen, L. G., 74; 21
Alpen, E. L., 2; 22
Alter, H., 318; 873
Alton, H., 46; 377, 378
Amundson, M. E., 26, 28; 23, 24
Anagnoste, B., 17; 375
Anders, M. W., 263; 25
Andersen, K. S., 1023
Andersen, M. L., 21
Anderson, R. C., 544, 545, 996, 998, 999
Angelucci, R., 246; 26
Anghileri, J., 234; 27
Appleton, P. J., 57
Archer, S., 784

Argus, M., 669
Arita, T., 473, 474, 475, 476
Ariyoshi, T., 925
Armstrong, M. D., 16, 41, 135, 153; 28, 29
Arnold, B., 141; 806
Aronow, L., 40, 123, 124
Artini, D., 26
Asami, T., 597
Asatoor, A. M., 15, 16; 30, 31, 32, 33
Ashley, J. J., 299; 34
Aström, A., 15
Aubert, J. P., 632
Axelrod, J., 15, 17, 25, 26, 31, 42, 43, 45, 108, 109, 110, 111, 113, 203, 231, 266, 273, 278, 294; 14, 35, 36, 37, 38, 39, 40, 41, 42, 43, 44, 45, 123, 124, 125, 126, 127, 128, 129, 130, 131, 209, 241, 520, 534, 842, 1001

B

Babbitt, B., 610
Bader, G., 844
Baer, J. E., 132, 138
Baggiolini, M., 201, 202; 46, 47, 85, 86, 87
Bailey, K. R., 831
Baker, E. M., 10, 597
Baladova, D., 123, 132; 542, 543
Baltes, B. J., 16; 48
Bami, H. L., 470
Banks, W. F., 574
Bärisch, R., 74
Barker, W. M., 555
Barnes, M. J., 295; 49
Barreto, R. C. R., 194; 50, 51, 52, 53
Bartley, W. C., 524

Baruel, J., 27
Basch, B. G., 751
Batchelor, F. R., 289; 782
Bauer, H., 276
Beattie, J. O., 153, 721
Beaven, M. A., 55, 56
Becher, A., 16; 54
Beck, J. L., 152, 153
Beckett, A. H., 17, 18, 48, 57, 58, 67, 116; 55, 56, 57, 58, 59, 60, 61, 62, 63, 64, 65
Beckmann, R., 155, 181, 182, 216; 66, 67, 68, 69, 70
Bedford, C., 5; 71
Beer, H., 79; 79
Beisenherz, G., 218, 249, 250; 72, 73, 74
Bell, S. A., 4; 536
Benakis, A., 135; 361
Benham, G. H., 54; 75
Benigni, J. D., 42; 76
Berberian, D. A., 784
Beres, J. A., 986
Beresford, J., 55; 717
Berg, D., 555
Berger, B. L., 155
Berger, F. M., 87, 89; 77, 270, 565, 768
Berger, M., 333
Bergman, W. L., 589
Bernhammer, E., 113; 78
Bernhard, H. V., 526
Bernhard, K., 79, 186, 214; 79, 80, 81, 82, 499
Bernstein, A., 252
Bernstein, E., 133, 135
Bernstein, J., 312; 906
Bertino, J. R., 459
Beyer, K. H., 14, 242; 83, 84, 905
Bickel, M. H., 80, 202; 46, 47, 85, 86, 87
Bieber, A. L., 254; 88
Bieber, S., 299, 300
Bieder, A., 209; 89, 90, 91, 92
Binder, B., 72, 73
Bird, H. L., 291; 93, 544
Birtley, R. D. N., 156; 94
Blake, M. W., 145; 95
Bleidner, W. E., 29; 96
Bletz, I., 638
Bloch, R., 525
Block, W., 70, 121, 142, 143; 97, 98
Bocchieri, S. F., 695
Bocher, C. A., 330

Bofinger, M., 290
Boissier, J. R., 213
Bolt, A. G., 66; 61, 99, 332
Boone, I. V., 194; 100, 101
Booth, A. N., 655
Booth, B. A., 459
Boothroyd, B., 295; 49
Borgstedt, H. H., 311; 262
Borner, K., 291; 102
Bösche, J., 215; 103
Boucherat, M., 593
Boura, A. L. A., 155; 104
Bowers, W., 911
Boxer, G. E., 762
Boyer, F., 161, 312; 105, 106, 632
Boyes, R. N., 57, 58
Bradley, R. M., 14
Bradley, W., 382
Brady, W. T., 984
Brattgard, S. O., 319; 107
Bratton, A. C., 113, 159, 213; 108, 368, 608
Brattsand, R., 107
Braun, G., 157; 109
Bray, H. G., 2, 107, 164, 166, 167; 110, 111, 112, 113, 114, 115, 116, 117, 118
Bridges, J. W., 167, 174; 119, 120
Broadley, K. J., 48; 121
Brobyn, R., 436
Brochmann-Hansen, E., 149, 122; 122, 663, 885
Brock, N., 428, 429
Brodie, B. B., 15, 22, 24, 31, 54, 55, 108, 109, 110, 111, 113, 121, 127, 129, 136, 137, 140, 178, 203, 212, 220, 224, 231, 238, 272, 276, 277; 40, 123, 124, 125, 126, 127, 128, 129, 130, 131, 132, 133, 134, 135, 136, 137, 138, 139, 155, 156, 157, 158, 159, 160, 219, 287, 357, 740, 791, 792, 842
Brooks, C. W. J., 17, 49; 140
Brooks, J. W., 25, 432
Brossi, A., 271; 141, 814, 815
Brown, B. T., 298, 299, 301; 34, 142, 143, 144
Brown, D. M., 4
Brown, H. D., 910
Brown, P., 708, 709
Browne, J. S. L., 614
Brownlee, G., 153; 380
Brubacher, G., 80, 82

AUTHOR INDEX

Bruce, R. B., 204, 205; 145, 146
Bruck, C. G. V., 172, 173; 147
Bruderer, H., 814, 815
Brunel, P., 89, 90
Brunner, H., 707
Büch, H., 111, 112; 148, 788
Buhler, D. R., 97, 98, 99; 149, 150, 151
Buhs, R. P., 38, 114, 115; 152, 153, 721, 910
Burchenal, J. H., 525
Burchfield, H. P., 65, 69; 154, 460, 461, 462, 463
Burnett, A. E., 990
Burnett, R., 373
Burns, J. J., 212, 234, 235, 260; 133, 155, 156, 157, 158, 159, 160, 217, 218, 245, 531, 995
Burstein, S. H., 1002
Bush, M. T., 121, 122, 132, 135, 143, 144; 161, 162, 163, 164, 165, 166, 167, 182
Bushby, S. R. M., 312; 168, 169
Bütikofer, E., 187; 170
Butler, T. C., 121, 132, 134, 135, 147, 151, 152, 204, 205, 317; 164, 165, 171, 172, 173, 174, 175, 176, 177, 178, 179, 180, 181, 182, 183, 184, 185, 963
Büttner, H., 186, 829
Buu-Hoï, N. P., 783
Buyske, D. A., 13, 177, 286, 287, 288; 216, 501, 502, 503, 704, 705
Buzard, J. A., 236; 187

C

Caffau, S., 288, 292; 188, 189
Cafruny, E. J., 441
Cain, C. K., 218; 711
Callahan, S., 300
Cambosos, N. M., 595
Campbell, A. D., 89; 190
Canas-Rodriguez, A., 258; 191
Cannon, P. J., 152
Capella, P., 49; 192
Carey, P. L., 571
Carlat, L. E., 304; 193
Carroll, N. V., 598
Carter, C. E., 254; 194
Cartoni, G. P., 19; 195
Cassano, G. B., 28; 196
Catanese, B., 241; 197, 846
Ceriotti, G., 192; 198

Cerna, M., 843
Cerny, A., 337
Chambers, J., 167
Chang, T., 259
Charalampous, K. D., 32, 35; 199, 200
Chase, G. D., 666
Chen, K. K., 544, 545, 996
Chenkin, T., 156
Chenoweth, M. B., 955, 956, 957
Chernov, H. I., 267; 1021
Chiaini, J., 1012
Childress, S. J., 165; 822
Choulis, N. H., 59
Christensen, C. B., 30; 201
Christensen, F., 259; 202, 203, 204, 205
Christensen, V. W., 911
Christman, A. A., 273; 222
Chulski, T., 667
Chumakow, N., 965
Ciocalteu, V., 110; 331
Clapp, J. W., 177; 206
Clare, N. T., 54; 207
Clark, B. B., 156; 552
Clark, L., 411
Clark, R. G., 288; 7
Clarkson, T. W., 314; 208
Clarmann, M. V., 290
Coburn, A. F., 1; 486
Cochin, J., 61, 132, 294; 209, 210, 211, 1022
Cohen, Y., 154, 242; 212, 213
Coles, F. K., 190, 439
Collier, H. B., 54; 214, 215
Colucci, D. F., 177; 216
Conaway, H., 874
Conklin, J. D., 187
Conn, H. L., 2
Conn, J. W., 563
Conney, A. H., 111, 218; 217, 218, 510, 994, 995
Conway, W. D., 574
Cooper, J. R., 136, 137; 219
Cope, O. B., 598
Coppi, G., 88; 220
Corbett, K., 178; 221
Corder, C. N., 985
Cordova, V. F., 393, 674
Cornish, H. H., 273; 222
Corradino, R. A., 966, 970
Corsi, G., 846
Corson, M., 409
Costerousse, O., 154; 212

Cottier, P., 170
Coutinho, C. B., 251, 252, 253, 257
Cox, E., 299, 302; 223, 224
Craig, A. W., 315; 225, 226, 227, 334
Craig, J. C., 57; 228
Craig, L. C., 277; 132, 229, 907
Cramer, J. L., 79; 230
Cresseri, A., 247; 26, 231
Crew, M. C., 251, 252
Crossley, M. L., 162; 778
Culp, H. W., 584, 586, 587
Culpan, R., 719
Cummings, A. J., 5, 8; 71, 232, 233, 234, 235
Curry, A. S., 134, 147; 236, 237, 238
Curry, J. H., 689
Curry, S. H., 67; 60, 61
Cuthbertson, W. F. J., 192; 239
Cutting, W. C., 609, 610

D

Dakhil, T., 961
Dalgliesh, C. E., 240
Daly, J., 35; 241
Daly, J. W., 61, 132; 210, 211
Daniels, T. C., 978
Danishefsky, I., 307; 242
d'Arconte, L., 81; 512, 845
d'Asaro, B. S., 835
Davidson, J. D., 251; 5, 692, 693
Davis, G. E., 404
Davis, M. A., 523
Davison, C., 7; 243
Dawson, J. M., 620
Day, M., 307; 244, 384
Dayan, R. J., 764
Dayton, P. G., 235; 245
Dearborn, E. H., 504
Debackere, M., 16, 17; 246, 289
De Carneri, I., 198
Dechavassine, M., 105
De Eds, F., 53; 247, 248, 655
Defranceschi, A., 192; 198, 249, 1039
Degwitz, E., 110; 250
Delfs, F. M., 147
Demis, D. J., 307; 572
Demoen, P. J., 206; 20
Denham, C., 412
Dennis, E. W., 784
Densen, P. M., 121; 166

Dequattro, V., 919
De Stefano, F., 19; 195
Dicarlo, F. J., 170, 318, 319; 251, 252, 253, 254, 255, 256, 257
Dickison, H. L., 165
Dietz, W., 142; 258
Dill, W. A., 190; 139, 259, 362, 363, 364, 365, 366, 367, 368
Diller, W., 260
Dingell, J. V., 261, 357
Dirschka, U., 316; 746
Distefano, V., 311; 262
Dixon, A. C., 597, 598
Dobson, F., 162; 263
Dobson, T., 523
Doenicke, A., 339
Dohrn, M., 226; 264
Domar, G., 312; 265
Doran, W. J., 997
Dorfmüller, T., 148; 266, 267, 268
Dost, F. N., 202; 749
Douglas, D. E., 116; 350
Douglas, J. F., 91, 95, 96, 106; 269, 270, 271, 272, 273, 296, 565
Douglas, J. S., 188; 274
Douglass, C. D., 199; 275
Drabner, J., 29, 81; 276
Dreiding, A., 830
Dresel, P. E., 89; 277
Dring, L. G., 16; 278
Driscoll, J. L., 279, 611
Dubnick, B., 20, 226; 280, 281
Duggan, D. E., 254; 282
Duhault, J., 20; 283
Duhm, B., 39; 284
Dulin, W. E., 852
Dumazert, C., 2; 285
Duncombe, W. G., 104
Duschinsky, R., 514
Dutton, G. J., 93; 286
Duvall, R. N., 336
Dvornik, D., 523
Dworak, G., 631

E

Earle, D. P., 178, 277; 287, 288
Easterday, D. E., 207; 304
Easton, N. R., 13; 585
Ebenezer, C. S., 722
Eberhardt, G., 340
Eberhardt, H., 16, 20, 62, 122; 289, 290, 291, 292

AUTHOR INDEX

Eberholst, I., 81; 293
Ebert, A. G., 70, 124; 294, 295
Ebigt, I., 142, 143; 98
Eddy, C. W., 247
Edelson, J., 272, 296
Edwards, K. D. G., 39; 1036
Edwards, S. A., 221
Egerton, J. R., 911
Eiber, H. B., 307; 242
Eidus, L., 959
Eisdorfer, I. B., 57; 297
Eisner, H. J., 287; 298
Elderfield, R. C., 907
Elias, A. L., 748
Elion, G. B., 252, 253, 254, 256; 299, 300, 301, 302, 394
Ellard, G. A., 310; 303
Ellenbogen, W. C., 15, 57; 297, 500
Ellin, R. I., 207; 304
Elliott, E. W., 262; 537
Elliott, H. W., 266; 601, 712, 713
Elliott, T. J., 382
Ellison, T., 16; 48, 305, 306
Ellsworth, H., 977
El Masry, A., 16; 307
El Ouachi, 12; 285
Emerson, K., 609, 610
Emmerson, J. L., 57, 90, 91; 308, 309
Enander, I., 207; 310, 311, 312
Engelman, K., 849
Epstein, W. L., 764
Esplin, D. W., 248; 313
Eubank, L. L., 190, 439
Euler, U. S. von, 47; 314
Evelyn, K. A., 614
Evertsbusch, V., 281; 315

F

Fabro, S., 183; 316
Faigle, J. W., 182; 317
Fajan, S. S., 563
Fancher, O. E., 548
Feitknecht, U. F., 985
Fenard, S., 20; 283
Ferlemann, G., 211; 318
Fickewirth, G., 281; 319
Finger, K., 236; 320
Fischer, K. 101; 321
Fisher, A. L., 268; 322
Fishman, J. K., 905

Fishman, V., 56, 57, 58, 61, 62, 67, 69, 79; 323, 324, 325, 326, 327, 371, 372, 373
Fishman, W. H., 250; 328, 329
Flanagan, D. F., 877
Flanagan, T. L., 55; 330
Folin, O., 110; 331
Font du Picard, Y., 213
Forist, A. A., 852, 853, 855, 856
Fornefeld, E. J., 1022
Forrest, F. M., 333
Forrest, I. S., 57, 66; 99, 332, 333
Fourneau, E., 151
Fouts, J. R., 280; 6
Fox, B. W., 239, 310; 225, 226, 334, 335
Foye, W. O., 5; 336
Franc, Z., 337
Francis, G. E., 110
Francova, V., 255; 337
Fratta, I., 1038
Fredga, A., 265
Frederickson, R. L., 695, 696
Freele, H., 784
Fretwurst, F., 230; 391, 392
Freundt, K. J., 290, 291, 292
Frey, H. H., 121, 129, 133, 136, 142; 338, 339, 340, 341, 342
Freytag, F., 1043
Friebolin, H., 734
Friedhoff, A. J., 34; 343, 344
Fröde, R., 74
Fujimoto, J. M., 265; 10, 345, 346, 347, 348
Fujimoto, M., 947
Fuller, A. T., 159; 349
Furner, R., 874
Furutani, M., 945

G

Gabourel, J. D., 381
Gabriel, T., 514
Gad, I., 14; 451
Gallo, D. G., 831
Galman, B. R., 32
Garattini, S., 22; 646
Garceau, A. J., 561
Garrod, J. M. B., 303
Gebauer, I., 74
Geddes, I. C., 116; 350
Geiling, E. M. K., 281; 315, 504, 688, 689

Geissman, T. A., 210; 993
Gerhards, E., 179, 311; 351, 352, 353, 354, 355
Gerritsen, G. C., 853
Gessner, P. K., 284; 356
Gibian, H., 351, 352, 353
Giefer, L., 198; 671, 672
Giles, W. G., 398, 399
Gillette, J. R., 55; 261, 357, 358
Gilligan, D. R., 166; 359
Gimble, A. I., 977
Ginsberg, T., 270
Giotti, A., 123; 360
Giraldi, P. N., 26, 231
Glasson, B., 135; 361
Glasko, A. J., 106, 275, 285; 259, 362, 363, 364, 365, 366, 367, 368
Gleason, H. L., 862
Gliniecke, G., 629
Godtfredsen, W. O., 293; 369
Gold, O., 35; 370
Goldenberg, H., 56, 57, 58, 61, 62, 67, 69, 79; 323, 324, 325, 326, 327, 371, 372, 373
Goldenthal, E. I., 656, 657
Goldman, A., 156
Goldschmidt, S., 123, 124; 371
Goldstein, M., 17, 34; 343, 375
Golumbic, C., 229, 907
Gönnert, R., 73; 376
Goodall, McC., 44, 46; 377, 378, 507,
Goodman, L. S., 676
Goodwin, S., 157
Gordis, E., 19; 379
Gordon, E., 520
Gordon, M., 680
Gordon, M. W., 16; 1037
Gorvin, J. H., 153; 380
Gosselin, R. E., 281; 381
Granstrand, B., 629
Grasso, A., 197
Graves, E. L., 88; 382
Green, J. P., 307; 244, 383, 384
Green, S., 250; 329
Greenberg, J., 470, 471
Greenberg, L. A., 107; 385, 551
Greenblatt, I. J., 805
Greene, F. E., 874
Greengard, P., 1038
Grimmer, G., 297; 386
Grishina, V. I., 841
Gross, E. G., 264; 387, 903

Grostic, M. F., 27; 854
Grunert, M., 449
Gudmundsen, C. H., 966, 970
Gudzinowicz, B. J., 63; 279, 388, 611, 737
Günther, H., 738
Gutman, A. B., 160
Gutzait, L., 305, 306
Guy, J. L., 3; 720
Gwilt, J. R., 31; 389

H

Haahti, E. O. A., 953
Hackenthal, E., 5; 390
Häfliger, O., 141
Halberkann, J., 230; 391, 392
Haley, T. J., 275; 590
Hall, C. R., 19; 393
Hamilton, L., 252; 394
Hamilton, R. E., 571
Hammar, C. H., 110; 395
Hampton, A., 254; 396
Handley, C. A., 315; 397
Handschumacher, R. E., 245; 790
Hansen, H. J., 398, 399
Hanshoff, G., 629
Hanson, S. W. F., 400
Hansson, E., 212; 196, 401
Harfenist, M., 510
Harley-Mason, J., 34; 402, 403
Harman, R. E., 216; 404, 910
Harmon, J. B., 96
Harpootlian, H., 151
Harris, S. C., 14; 405
Harthon, J. G. L., 107
Hartigan, J. M., 254, 255
Hartles, R. L., 163; 406, 407
Haskins, W. T., 222; 408
Hassler, H. M., 723, 724
Haüser, H., 148
Hausner, E. P., 259; 409
Haüssler, A., 178; 410
Haynes, L. J., 251, 257
Heaton, A., 327, 373
Heffter, A., 281; 319
Heller, A., 196; 411
Heming, A. E., 571
Henderson, E. S., 250; 412, 413, 692
Hennig, W., 213; 414
Henriksen, V., 55; 415
Hermann, B., 77, 78, 80; 416, 417, 418, 419

AUTHOR INDEX

Hermann, E. C., 96
Hertting, G., 46; 420
Hespe, W., 23, 282; 421, 422
Hess, S., 198; 423
Hess, S. M., 70; 5, 109, 294
Hewes, W. E., 96
Hewitt, M. I., 805
Hilbert, P., 107; 454
Hine, C. H., 153; 521
Hiratani, T., 9; 492
Hiroka, T., 659
Hirschmann, R., 876
Hirtz, J., 5; 424
Hitchings, G. H., 253; 299, 300, 301, 302, 795
Hitzig, W. H., 830
Hoffmann, E. A., 311; 437
Hoffmann, H., 233; 425
Hoffmann, I., 70; 426
Hoffmann, K., 427, 495, 496, 497, 498, 499
Hoffmann, P., 401
Hogan, R., 199; 275
Hogben, A. M., 842
Hogness, J. R., 786
Hohorst, H. J., 316; 428, 429
Hollister, L. E., 34; 344
Hollunger, G., 116; 430, 431
Holmstedt, B., 432
Horning, E. C., 17, 49; 140, 157, 192, 240, 432, 953, 1018
Horning, M. G., 240, 432
Howie, D. L., 980
Hübner, G., 145; 433
Hucker, H. B., 27, 311; 434, 435, 436, 437, 438
Huf, E. G., 88; 190, 439
Hug, C. C., 267, 268; 440, 441, 442
Huggins, R. A., 823
Hughes, H. B., 191; 443
Humphris, B. G., 111
Huus, I., 81; 21, 293, 415
Hynie, I., 92; 444

I

Ide, H., 920, 931, 932
Ikeda, G. J., 149; 900
Iliev, V., 513
Imanari, T., 491
Imhof, P., 170
Ings, R. M. J., 236; 445
Inscoe, J. K., 41, 44
Ireland, D. M., 239

Irrgang, K., 128; 446
Iveson, P., 305; 447
Ivy, A. C., 405
Iwainsky, H., 192, 210; 448, 449

J

Jackson, D., 458
Jackson, H., 315; 225, 226, 227, 334
Jackson, J. V., 129; 450
Jacobacci, S., 288, 292; 188, 189
Jacobsen, E., 14; 451
Jaffe, J. J., 738
Jaffe, M., 107, 230; 452, 453, 454
Jagenburg, O. R., 111; 455
Jäger, G., 339
Jahn, V., 105; 800
Jakobsson, C., 629
James, G. V., 160; 456
James, S. P., 112
Jänicke, G., 518
Jaques, L. B., 546
Jelinek, V., 337
Jelliffe, R. W., 302; 457, 458
Jermann, M., 222; 769
Jindra, A., 226; 847
Johns, D. G., 459
Johnson, D., 409
Johnson, D. E., 60, 63, 65; 460, 461, 462, 463, 464, 779, 780
Johnson, J. L., 563
Johnson, J. R., 32
Johnson, M. J., 289; 487
Johnson, M. L., 23
Johnston, R. L., 151
Jolles, G., 296; 465, 466
Jommi, G., 84; 467, 468
Jones, B. M., 236; 469
Jones, F. T., 655
Jönsson, B., 21
Josephson, E. S., 220; 470, 471
Jurovcik, M., 472, 743, 744
Just, M., 81, 82

K

Kacl, K., 444
Kakemi, K., 11, 196, 286, 288; 473, 474, 475, 476
Kalberer, F., 282; 477, 1042
Kalser, S. C., 23, 281; 381, 478, 479
Kamata, O., 929
Kamil, I. A., 2, 9, 32, 98, 138, 186, 188, 215, 224, 274, 313, 320; 480

Kamm, J. J., 55, 129; 358, 481
Kanda, M., 200; 576
Kane, P. O., 209; 482
Kaneda, K., 927
Kanegis, L. A., 503
Kapadia, G. G., 602
Kapadia, G. J., 602
Kapetanidis, I., 306; 483, 484
Kaplan, J., 845
Kaplan, S. A., 295; 485
Kapp, E. M., 1; 486
Karnovsky, M. L., 289; 487
Kashiwagi, K., 9; 493
Kasik, J. E., 411
Kaslander, J., 312; 488
Kass, D. J., 798
Kato, K., 921, 922
Katzung, B. G., 301; 489
Kaul, P. N., 226; 490
Kawai, S., 51; 491
Kawamata, J., 9; 492, 493
Kazenko, A., 364, 365, 366
Keasling, H. H., 855
Keberle, H., 183, 188, 321; 170, 317, 427, 494, 495, 496, 497, 498, 499, 707, 765
Keller, R. E., 15; 500
Kelly, J. M., 680
Kelly, R. G., 282, 283, 288; 501, 502, 503
Kelsey, F. E., 276; 504, 688
Kelvington, E. J., 478, 479
Kensler, C. J., 781, 848
Kerby, G. P., 307; 505
Kerr, D. N. S., 15; 33
Kerza-Kwiatecki, A., 627
Kety, S. S., 534
Khairallah, P. A., 356
Kibby, M. R., 119
Kido, M., 592
Kiese, M., 110, 113; 506
Kimmel, H. B., 970
King, C. T. G., 662
King, M. L., 5; 233
King, W. M., 44
Kinross-Wright, J., 199, 200
Kinross-Wright, V. J., 66; 742
Kiriyama, T., 475
Kirshner, N., 42; 507
Kiyomoto, A., 192; 698
Klatt, L., 72, 73
Kleinsorge, H., 70; 508
Klesczewski, S., 757, 758
Klicka, J., 208; 509

Klutch, A., 112; 510, 531
Knebel, C. M., 967, 968
Knox, K. L., 240, 432
Koe, B. K., 222; 511
Koechlin, B. A., 81, 198, 199, 240; 512, 513, 514, 515, 516, 818, 844, 871
Koepfli, J. B., 276; 628
Koga, T., 291; 517
Kolb, K. H., 311; 351, 352, 354, 355, 518
Komai, T., 884
Kondo, H., 946
König, J., 444
Konishi, R., 476
Kono, M., 173; 944, 945, 946
Kopf, R., 415
Köpf, R., 723, 724
Kopin, I. J., 43, 44, 45, 46; 42, 519, 520
Kopp, R., 725, 726
Kopsch, H. W., 145; 727
Korduba, C. A., 886
Koss, F. W., 72, 73, 74
Kovensky, A., 301
Kozelka, F. L., 153; 521
Krake, J. J., 633
Kramer, B., 207; 522
Kramer, M., 518
Kraml, M., 157; 523
Krantz, J. C., 320; 664
Krapcho, J., 109
Krause, D., 342
Krebs, H. A., 167; 524
Kreis, W., 115, 202; 477, 525, 526, 1042
Krisch, K., 113; 78
Kristerson, L., 401
Krol, G., 515, 818
Kruger, M., 272; 527, 528
Krüger-Thiemer, E., 260
Krupa, V., 1040
Kuehl, F. A., 404
Kuhn, R., 79; 529
Kum-Tatt, L., 104; 530
Kuntzman, R., 239; 357, 531
Kuribayashi, K., 174; 943
Kuroda, K., 220; 532
Kuroiwa, Y., 138, 139; 923
Kurokawa, Y., 839
Kushima, T., 947
Küssner, W., 386
Kutt, H., 135; 533

AUTHOR INDEX

L

Labrosse, E. H., 42; 534
Labudde, J. A., 203; 989
Lage, G. L., 301; 535
Laird, A. H., 402, 403
Lake, H. J., 113
Lalich, J. J., 988
Lambelin, G., 783
Landrau, M., 245
Langbein, J. W., 291
Lange, W. E., 4; 336, 536
Langenbeck, W., 640, 1016
Langham, W., 409
Langham, W. H., 786
Lannon, T. J., 798
Lannotti, A. T., 459
Laragh, J. H., 152
Larsen, E. R., 956
Lasagna, L., 721
Latham, M. E., 262; 537
Lauener, H., 550
Lauterbach, F., 303; 538, 539
Lavicka, J., 843
Law, G. L., 445
Lawrow, D., 230; 540, 541
Layne, D. S., 1002, 1003
Ledvina, M., 123, 132; 542, 543
Lee, C., 258, 291, 292; 544, 545, 546
Lee, G. E., 221
Lee, H. M., 26; 547, 588, 714
Lee, K. H., 485
Lee, M. L., 789
Leeling, J. L., 226; 548
Leeson, G. A., 280
Lehmann, K., 249; 549
Lehner, H., 74; 550
Leifer, E., 786
Lerbs, O. W., 292
Lester, B., 551
Lester, D., 107; 385
Leverett, R., 280
Levine, J., 719
Levine, R. M., 156; 552
Levy, L., 48
Lewis, H. B., 958
Lief, P. A., 133, 134, 135, 155
Lin, S. C. C., 213; 553
Lin, T. H., 56; 330, 554
Linderholm, H., 265
Lingner, K., 386
Link, K. P., 260; 555
Linkenheimer, W. H., 171; 556

Lipson, M., 54; 557
Litchfield, J. T., 608
Little, J. M., 642
Logemann, W., 26, 231
London, I., 88; 558
Long, J. P., 268; 322
Loo, T. L., 251, 252; 559, 560, 561
Loop, W., 172; 828
Lorenzetti, O. J., 320; 562
Losin, L., 621
Louis, L. H., 148; 563
Lowe, J. S., 309; 564
Lucas, R. C., 836
Luchi, R. J., 2
Ludwig, B. J., 88, 90, 93, 95; 270, 271, 272, 273, 565, 566
Lukas, S. D., 301; 567
Lundy, J. S., 568
Lüthi, U., 30, 322; 569, 570, 971
Luts, H., 566
Luttermoser, G. W., 222; 408
Lutz, A. H., 80, 81
Lynes, T. E., 96

M

Maas, A. R., 88; 571, 785
McAllister, B. M., 285; 572
McAuliff, J. P., 575
McBay, A. J., 127, 134; 129, 17, 18
McChesney, E. W., 163, 219, 220; 389, 573, 574, 575
McCoubrey, A., 104
McDowell, F., 533
Macek, K., 668
McGeer, P. L., 49; 827
McGrath, W. R., 182; 578
McIsaac, W. M., 3, 200; 356, 576, 577
McKelway, W. P., 977, 981
Mackensie, R. D., 182; 578
MacMahon, F. G., 579, 1022
McMahon, R. E., 13, 25, 28, 101, 104, 115, 150, 257; 580, 581, 582, 583, 584, 585, 586, 587, 588, 997
McMillan, A., 28
MacMullen, J. A., 967, 968
McNally, W. D., 15; 589
Maengwyn-Davies, G. D., 919
Magee, M., 100
Maggiolo, C., 275; 590

Magnussen, M. P., 129; 341
Mahaffee, C., 183, 184
Mahaffee, D., 183
Maître, L., 47; 591
Major, R. H., 988
Makisumi, S., 22; 592
Malament, S. G., 256
Mallein, R., 220; 593
Malorny, G., 318; 594
Mandel, H. G., 4, 6; 22, 595, 596
Manis, J. G., 3; 799
Manitto, P., 467, 468
Mann, J. D., 42
Mannering, G. J., 262, 263, 267, 270; 25, 597, 598, 599, 647
Manthei, R. W., 191, 198; 600, 787
Manthey, J. A., 28; 23, 24
March, C. H., 266; 601, 979
Marcus, F. I., 300; 602
Margraf, A. W., 193
Mark, L. C., 133, 135
Marks, V., 198; 603
Marme, W., 264; 604, 605
Marsh, D. F., 282; 606
Marshall, E. K., 113, 159, 160, 176, 213; 108, 607, 608, 609, 610
Marshall, F. J., 584, 586, 587
Martin, B. K., 15; 71, 234, 235
Martin, C. M., 912
Martin, H. F., 63; 611, 737
Martin, J. F., 299; 612
Maruyama, M., 286; 613, 687
Mary, N. Y., 228
Massart-Lëen, A. M., 17; 246
Masterson, P. B., 986
Masuda, Y., 659
Matzuk, A. R., 910
Maughan, G. B., 231; 614
Maul, W., 284
Mayer, E., 728, 729, 730
Maynert, E. W., 121, 123, 124, 126, 128, 129, 153; 360, 615, 616, 617, 618, 619, 620, 621, 622, 623, 624, 625, 626
Mazeau, L., 209; 89, 90, 91, 92
Mazel, P., 294; 166, 627
Mead, J., 276; 628
Medenwald, H., 284
Medler, E. M., 831
Meisinger, M. A. P., 404, 876
Melander, B., 155; 629

Mellett, L. B., 266, 267; 440, 441, 630, 648, 1023, 1024
Mellin, T. N., 1003
Mertel, H. E., 910
Messiha, F. S., 47
Metz, E., 301
Meyer, M., 565
Meyers, F. H., 301; 489
Meythaler, C., 236; 631
Michael, M. E., 561
Mighton, H., 229
Mighton, H. R., 907
Miksch, J., 54
Milhaud, G., 632
Millar, J. D., 461
Miller, E. A., 435, 436
Miller, K. S., 708, 709
Miller, L. D., 2
Miller, W. L., 148; 633
Miller, W. N., 587
Mills, G. T., 400
Milne, M. D., 32
Miranda, P. M., 208; 634, 985
Mirimanoff, A., 306; 484
Misra, A. L., 266, 267, 268; 635, 636
Mitsui, S., 932
Miya, T. S., 57, 72; 295, 308, 309, 710
Moerman, E., 303; 637
Mohler, W., 274; 638, 639
Mohnicke, G., 148; 640, 1015, 1016
Mohrschulz, W., 101; 641
Moller-Nielsen, I., 21
Moran, N. C., 54; 792
Morgan, A. M., 88; 642
Morgan, D. F., 280, 281
Morgan, S., 458
Mörner, K. A. H., 107, 110; 643, 644
Morrison, J. A., 205; 645
Morvillo, V., 22; 646
Moss, M. S., 129; 450
Muehlenbeck, H. E., 1024
Mueller, A. J., 831
Mueller, S., 302
Mule, S. J., 267, 268; 635, 636, 647, 648
Müller, J., 32; 850
Müller, O. H., 313, 314; 649, 991
Murata, T., 104; 650, 651, 652, 653, 654
Murray, C. W., 306; 655

N

Nadai, T., 474
Nadkarni, M. V., 315; 656, 657, 919
Nadler, S. B., 398, 399
Nagatsu, T., 491
Nair, V., 199; 658
Nakamura, E., 32; 838, 839, 840
Nakamura, K., 153; 659
Nakano, M., 475
Nakao, M., 10; 660, 661
Nakatsuji, K., 659
Nakazona, I., 3
Nannini, G., 26
Narrod, S. A., 238; 662
Nathan, H., 300
Nauckhoff, B., 21
Nauta, W. T., 421, 422
Nayak, K. P., 281; 663
Neale, F. C., 110, 114, 115
Needlemann, P., 320; 664
Neely, W. B., 318; 665
Neff, N., 33; 666
Nelson, E., 149; 667
Nelson, J. W., 562
Nemecek, O., 236; 668
Newell, M., 113; 669
Newman, J., 145, 146
Nicholls, P. J., 183, 188; 274, 670
Nielsch, W., 198; 671, 672
Nieschulz, O., 426
Nitz, D., 538
Niyogi, S. K., 131; 673, 674, 675
Noach, E. L., 153; 676
Nonkin, P., 252
Notation, A. D., 509
Novick, W. J., 330
Novotny, J., 248; 677
Nowak, H., 242; 678
Nowell, P. T., 679, 820
Numerof, P., 275; 680

O

Oberhansli, W., 515, 516
Oberst, F. W., 261, 264, 320; 681, 682, 683, 684, 685
Ogorzalec, S., 862
Ogston, A. G., 250; 686
Oguri, K., 924, 941
Ohara, A., 687
Ohashi, S., 473
Okasaki, Y., 175; 948

O'Keefe, E., 187
Oketani, Y., 12, 13
Okita, G. T., 299, 300; 34, 688, 689
Okumura, K., 169; 690
Okun, R., 48
Oldham, F. K., 504
Oliverio, V. T., 202, 251; 412, 413, 691, 692, 693
Olivet, A. J., 575
Onoue, K., 3
Opitz, K., 21; 694
O'Reilly, I., 667
Orengo, A., 199
Orwen, I., 47; 314
Osterberg, A. E., 568
Overby, L. R., 317; 695, 696, 697
Ozawa, H., 192; 698

P

Page, C. J., 458
Page, I. H., 356
Pala, G., 220
Palm, M., 629
Palmer, S., 514, 818
Papper, E. M., 133, 135, 155
Parke, D. V., 304; 447, 699, 700
Parker, J. R. B., 58
Parnell, E. W., 445
Paterson, A. R. P., 254; 396
Patzschke, K., 284
Paul, B. B., 317; 701
Paul, M. F., 187
Paul, W. D., 4; 702
Pechtold, F., 231; 703
Peckham, W. D., 886
Peets, E. A., 13, 14; 704, 705
Peng, C. T., 309; 706, 978, 979
Perel, J. M., 160
Perlman, P. L., 145; 95, 886
Peruzzotti, G., 784
Peskin, G. W., 2
Peters, G., 321; 707
Peters, J. H., 196, 198; 708, 709
Peterson, L., 259
Peterson, R. E., 301; 567
Pfeil, E., 145; 433
Pfleger, K., 148
Phillips, B. M., 72; 548, 710
Phillips, G. E., 253, 254, 255, 256, 257, 280, 281
Piepho, S. B., 526
Pinchbeck, F., 145
Pinson, R., 222; 511, 1012, 1013

Pitts, J., 146
Pitts, J. E., 145
Place, V. A., 705
Plampin, J. N., 218; 711
Plotnikoff, N. P., 212; 712, 713
Plummer, A. J., 835
Poet, R., 134
Poet, R. B., 88; 558
Pohland, A., 25; 547, 714
Polin, D., 153
Polli, J. F., 589
Pollock, S., 699
Popendicker, K., 426, 639
Porter, C. C., 36, 213; 438, 715, 716
Porter, G. S., 55; 717
Portwich, F., 186, 829
Posner, H. S., 59, 67; 718, 719
Postma, E., 82; 818, 819
Potter, G. D., 3; 720
Powell, L. S., 565
Prellwitz, W., 110; 395
Prescher, K., 538
Prescott, L. F., 39; 721
Preuss, F. R., 102, 103, 145, 151, 232, 244; 722, 723, 724, 725, 726, 727, 728, 729, 730, 731, 732, 733, 734
Pribilla, O., 82, 86; 735, 736
Price, S., 60; 737
Prien, E. L., 336
Prusoff, W. H., 239; 335, 738
Pryor, M. Z., 751
Pugh, C. T., 291; 93
Pulver, R., 77, 259; 418, 419, 739

Q

Queisnerova, M., 668
Quinn, G. P., 224; 740, 1038

R

Raaflaub, J., 200; 741
Rabideau, R., 987
Rabinowitz, J. L., 666
Ragland, J. B., 66; 742
Ramachandran, K., 959
Rambacher, P., 54
Randolph, M. M., 478, 479
Ranger, D., 298; 143
Rapoport, H., 248
Raska, K., 247, 248; 472, 743, 744

Raskova, H., 677
Raspe, G., 353, 518
Ratcliffe, J., 34; 745
Ratcliffe, R. J. M., 469
Rauen, H. M., 316; 746
Raventos, J., 121, 140; 747
Ray, A. P., 471
Ray, F. E., 669
Raz, K., 337
Rebstock, M. C., 362
Recchi, N., 27
Redetzki, H. M., 251; 748
Redfield, B. G., 423
Reed, D. J., 202; 749
Reeder, E., 871
Rees, C. W., 193; 16
Reichenthal, J., 273; 43, 131, 157
Reid, J. C., 561
Reineke, L. M., 633
Reiser, M., 638, 639
Remy, C. N., 253; 750
Renner, G., 110, 113; 506
Rennick, B. R., 44; 751, 990
Renton, R., 235
Repke, K., 297, 300; 539, 752, 753, 754, 755, 756, 757, 758
Reynolds, S. W., 554
Richards, R. K., 121; 759, 897
Richardson, A. P., 908
Richarz, G., 232, 234; 760
Richter, D., 14; 761
Rickards, J. C., 24; 762
Ridolfi, P., 846
Rieder, J., 86; 763, 814, 815
Rieders, F., 393, 674, 675
Riegelman, S., 295; 485, 764
Riess, W., 171; 170, 317, 497, 498, 765
Rietbrock, N., 594
Riley, R. F., 87, 88, 89; 766, 767, 768
Rist, N., 106
Ritter, P., 222; 769
Ritterband, A., 160
Roberts, D. J., 48; 121, 770, 771
Roberts, J. B., 156; 94, 772, 773, 774
Roberts, J. J., 310; 775
Robertson, A., 389
Robinson, A. E., 65; 55, 56, 776
Robinson, D., 2; 11, 777
Robinson, E. J., 162; 778
Robinson, H. J., 912
Robinson, J. D., 244

AUTHOR INDEX

Robson, R. D., 104
Rochova, M., 1040
Rodriguez, C. F., 65; 462, 463, 464, 779, 780
Rodwell, V. W., 22, 596
Rogers, B. S., 101
Rogers, J. E., 574
Rogers, W. I., 115; 781, 848
Rolinson, G. N., 288; 782
Roller, K., 233; 425
Roncucci, R., 321; 783
Rondelet, J., 593
Rondish, I. M., 330, 554
Roos, A. M. de, 421, 422
Root, M. A., 998, 999
Rose, J. C., 877
Rose, R. K., 156, 157
Rosen, L., 378, 507
Rosenblum, C., 911, 912
Rosi, D., 73; 784
Rösner, K., 508
Ross, H., 977
Ross, J. J., 56; 785
Ross, S., 15
Rossi, G. V., 666
Roth, L., 757, 758
Roth, L. J., 127, 191; 411, 786, 787
Rothstein, A., 208
Routh, J. I., 702
Rovenstine, E. A., 135
Rowland, M., 17; 62, 63, 64
Roxburgh, G., 302; 223
Roy, S. K., 228
Rubinstein, D., 317; 701
Rubio, F., 514
Rüdiger, W., 111; 148, 788
Ruelius, H. W., 84; 789
Rundles, S. W., 300, 301
Rupp, W. D., 245; 790
Rustemeyer, J., 340
Rutschmann, J., 477, 1042
Ruyle, W. V., 876
Ryman, B. E., 116

S

Sabino, S. O., 53
Sakamato, Y., 839, 840
Salzman, N. P., 54, 55; 791, 792
Sammons, H. G., 160; 793
Santomenna, D. M., 479
Sarcione, E. J., 253; 794
Sarett, H. P., 831

Sartorelli, A. C., 254; 88, 459
Saviard, M., 105
Scales, B., 303
Scannell, J. P., 253; 795
Schachter, D., 3; 796, 797, 798, 799
Schäfer, A., 54
Schanker, L. S., 270; 599
Schatz, F., 105; 800
Schaumann, O., 25; 801
Schayer, R. W., 2; 802
Scheline, R. R., 803
Scherman, R., 533
Schindler, W., 77; 419, 804
Schlameus, W., 464
Schlosser, A., 271, 272, 296
Schmerzler, E., 263; 805
Schmid, E., 631
Schmid, H. J., 427
Schmid, K., 170, 317, 498, 499, 707, 765
Schmidt, G., 141, 215; 103, 806, 961
Schmidt, P., 272; 527, 528
Schmutz, J., 550
Schneider, M., 594
Schneider, W. C., 252; 807
Schnider, O., 141
Schoetensack, W., 760
Schorre, G., 678
Schreiber, E. C., 1013
Schuchardt, G. S., 905
Schuhmacher, H., 182; 316, 808, 809, 810
Schüle, A., 74
Schulert, A., 156
Schulze, P. E., 355, 518
Schüppel, R., 232; 811, 812
Schwartz, D. E., 200, 201, 224; 741, 813, 814, 815
Schwartz, M. A., 82, 199; 515, 516, 816, 817, 818, 819, 845
Schwerd, W., 276
Scott, B., 79; 230
Scott, C. A., 156; 679, 820
Scott, E. G., 547
Scudi, J. V., 164, 165; 821, 822, 954
Searle, M. L., 405
Seevers, M. H., 1026
Sehrt, I., 449
Seibert, R. A., 265, 315; 397, 823
Seifter, J., 66, 90; 965, 967, 968, 969, 1014
Seiler, N., 33; 824, 825, 826
Sekine, Y., 951

Sekules, G., 220
Sen, N. P., 49; 827
Senoh, S., 41
Serick, E., 172; 147, 828
Serra, M. T., 99, 332
Sestanj, K., 523
Seydel, J., 172; 186, 829
Sezaki, H., 474, 475
Shafer, C. L., 409
Shahidi, N. T., 112; 830
Shannon, J. A., 288
Shaw, F. H., 262; 11
Shaw, K. N. F., 28, 29
Sheffner, A. L., 318; 831
Shekosky, J. M., 574
Shelswell, J., 160; 793, 832
Shen, T. Y., 876
Shepheard, E. E., 298; 142, 833
Shepherd, D. M., 47; 834
Sheppard, H., 186, 275; 835, 836, 837
Shibasaki, J., 118; 838, 839, 840
Shideman, F. E., 140, 142, 144; 869, 870, 1009, 1011
Shlyakhman, A. L., 22; 841
Shnol, S. E., 841
Shofer, R., 44
Shore, P. A., 269; 740, 842
Siblikova, O., 235; 843
Sicam, L. E., 245
Signorotti, B. T., 979
Silanos, M. A., 467, 468
Silva, J. A. F. de, 84; 844, 845
Silvestrini, B., 238; 197, 846
Simanis, J., 627
Simon, F., 252
Simon, M. J., 783
Simson, G., 158
Sipal, Z., 226; 847
Siwak, A., 115; 848
Sjoerdsma, A., 39; 44, 849
Sjöstrand, S. E., 196
Skinner, J. T., 14; 84
Sklow, N. J., 251, 252
Slater, I. H., 89; 277
Sliver, N. J., 257
Slotta, H. K., 32; 850
Smetana, R., 677
Smith, C. C., 221; 762, 851
Smith, D. L., 27, 150, 180, 229; 852, 853, 854, 855, 856
Smith, F. D., 689
Smith, J. L., 152, 153

Smith, J. N., 108, 109, 110, 111, 166, 320; 307, 480, 857, 858, 859, 860, 861
Smith, K., 19; 902
Smith, L. B., 688
Smith, N., 273
Smith, P., 34; 745
Smith, P. K., 1; 22, 243, 595, 596, 656, 657, 862, 919, 980
Smith, R. L., 182; 278, 316, 803, 808, 809, 810, 863, 864, 865, 866
Smythies, J. R., 403
Snow, G. A., 309; 303, 867
Snyder, F. H., 320; 685
Soehring, K., 142, 232; 258, 812
Sörbo, B., 310, 311, 312
Sorm, F., 472, 743, 744
Sormova, Z., 472, 743, 744
Specht, W., 101; 321
Spector, E., 32, 142, 144; 868, 869, 870, 1011
Speth, O. C., 152, 153, 721
Spinks, J. W. T., 546
Spratt, J. L., 301; 535, 1017
Stadler, R., 71; 899
Staehelin, M., 47; 591
Stagg, R. B. L., 808
Stanier, J. E., 250; 686
Staudinger, H., 110; 250
Stefanovic, J., 845
Sternbach, L. H., 871
Stevens, S. G. E., 469
Stier, A., 318; 872, 873
Stitzel, R. E., 268; 874
Stöffler, G., 898
Stoll, C. G., 862
Stolnikow, D., 264; 875
Stolzenberg, S. J., 556
Storey, I. D. E., 93; 286
Stormann, H., 35; 370
Strachan, R. G., 876
Strang, V. G., 101
Straub, W. H., 234; 877
Straube, L., 317; 697
Strömme, J. H., 313; 878, 879
Struck, W. A., 563
Strufe, R., 73; 880
Struller, R., 678
Stubbs, A. L., 163; 881
Stutzman, L., 253; 794
Sudendey, F., 342
Suga, T., 12
Sullivan, H. R., 714

AUTHOR INDEX

Sulser, F., 261, 357
Sundwall, A., 310, 311, 312
Sung, C., 977
Sung, C. Y., 116; 882, 981
Surrey, A. R., 575
Sutherland, R., 208
Suzuki, T., 292; 883
Suzuki, Y., 286, 289; 613, 884
Svendsen, A. B., 49, 122; 122, 885
Swales, W. E., 215
Swanson, R., 980
Sweeney, W. M., 705
Sykes, W. O., 524
Symchowicz, S., 70, 295; 886, 887, 888

T

Tabern, D. L., 897, 954
Tabone, J., 106
Tait, A. C., 1030
Takabatake, E., 889, 890, 920, 925, 926
Takada, M., 12
Takeda, M., 891
Takenouchi, T., 138; 915
Takeuchi, T., 884
Talbot, M. H., 336
Tallan, H. H., 1038
Talso, P. J., 688, 689
Tamminen, V., 147; 892
Tamura, Z., 491
Tanaka, F., 259; 883
Tanaka, I., 231; 893, 894
Tanaka, K., 592
Tatsumi, K., 921, 933, 934
Tatum, A. L., 895, 896
Tauchert, E., 426
Taylor, D. J., 470, 471
Taylor, J. D., 141; 759, 897
Telc, A., 15
Terlain, B., 465, 466
Tesarek, B., 843
Thalmann, K., 508
Thauer, R. K., 162; 898
Thiemer, K., 71; 899
Thiriaux, J., 783
Thomas, B. H., 94, 772, 773, 774
Thomas, J. O., 247, 248
Thomas, J. P., 465, 466
Thomas, R. C., 149; 900
Thomas, R. E., 301; 901
Thompson, G. F., 19; 902
Thompson, V., 264; 387, 903

Thorp, R. H., 298; 833
Thorpe, W. V., 174, 306; 110, 111, 112, 113, 114, 115, 116, 117, 118, 904
Threlfall, T. L., 221
Tillson, E. K., 178; 905
Titus, D. C., 36; 716
Titus, E., 88, 127, 219, 254, 312; 229, 282, 906, 907, 908, 909
Tocco, D. J., 240, 241; 910, 911, 912
Tochino, Y., 137; 913, 914
Toczko, K., 111; 455
Todrick, A., 1030
Toki, K., 916, 917
Toki, S., 138; 915, 916, 917, 927, 935, 936, 937, 938, 939, 940
Tomchick, R., 45
Tompsett, S. L., 135; 918
Tong, H. S., 984
Tosolini, G., 231
Towne, C. A., 281
Trams, E. G., 310; 919
Trefouel, J., 151
Trenner, N. R., 152, 153, 721, 910
Trevoy, L. W., 546
Triggs, E. J., 64
Tripathy, S. P., 959
Troestler, J., 106
Trousof, N., 218
Truant, A. P., 116; 882
Truitt, E. B., 642
Trujillo, T., 409
Tsai, I., 531
Tsien, W. H., 836, 837
Tsukamoto, H., 2, 3, 9, 92, 130, 136, 138, 139, 144, 154, 169, 189, 278, 280; 891, 916, 917, 920, 921, 922, 923, 924, 925, 926, 927, 928, 929, 930, 931, 932, 933, 934, 935, 936, 937, 938, 939, 940, 941, 942, 972, 1028, 1029, 1033, 1034
Turnbull, L., 146
Turner, D. H., 4
Turney, D. F., 100
Tye, A., 562

U

Udenfriend, S., 137, 224; 136, 137, 138, 139, 423
Ueda, M., 174; 943, 949, 950
Uehleke, H., 898

AUTHOR INDEX

Ulick, S., 908
Ungar, F., 509
Uno, T., 165, 170, 173, 174, 175; 944, 945, 946, 947, 948, 949, 950, 951, 952

V

Vachek, G., 843
Valzelli, G., 26, 231
Vanden Heuvel, W. J. A., 63; 432, 953
Vanderbrook, M. J., 633
Vandevoorde, J. P., 399
Van Dyke, H. B., 121, 123, 126, 128; 622, 623, 624, 625, 626, 954
Van Dyke, R. A., 317; 955, 956, 957
Vangedal, S., 293; 369
Van Loon, E. J., 129; 305, 306, 330, 481, 554
Van Poznak, A., 957
Vecchio, T. J., 856
Vejdelek, Z. J., 668
Vendsalu, A., 849
Venkataraman, A., 8; 958
Venkataraman, P., 198; 959
Venkataraman, P. R., 958
Verbiscar, A. J., 42; 76
Vidic, E., 24, 25; 960
Vitulova, O., 843
Vogel, M., 760
Vogt, W., 211; 318, 961
Voigt, K. M., 232; 731
Von Kaulla, K. N., 259; 739
Vuilleumier, J. P., 81, 82, 499

W

Waddell, W. J., 130; 184, 185, 962, 963
Waldi, D., 271; 964
Walkenstein, S. S., 20, 66, 85, 90, 91, 214; 965, 966, 967, 968, 969, 970
Walker, K. E., 199, 200
Wall, P. E., 29
Wallach, D. P., 1011
Wangler, J., 243
Warkentin, D. L., 525
Warwick, G. P., 310; 775

Waser, P. G., 30, 322; 569, 570, 971
Watabe, T., 279; 924, 941, 972
Way, E. L., 8, 24, 213, 261, 265; 10, 346, 347, 348, 553, 663, 712, 713, 973, 974, 975, 976, 977, 978, 979, 980, 981
Way, J. L., 206, 208, 212; 634, 982, 983, 984, 985, 986, 987
Weathers, H. H., 193
Weber, C. J., 164; 988
Weder, H. J., 87
Wegner, L. A., 284
Wehr, R., 123, 124; 374
Weichselbaum, T. E., 193
Weikel, J. H., 203; 989
Weiler, H., 231; 414
Weil-Malherbe, H., 45, 1001
Weiner, I. M., 313, 314; 649, 990, 991
Weiner, M., 158
Weinfeld, H., 272; 992
Weinman, E. O., 210; 993
Weischer, M. L., 21; 694
Weiss, H., 127; 909
Weiss, R., 980
Weissbach, H., 423
Weist, F., 1043
Welch, R. M., 31, 111, 113; 994, 995
Welch, W. J., 288
Welles, J. S., 133, 149, 150, 292; 588, 996, 997, 998, 999
Wempe, E., 260
Wempen, I. M., 907
Wepierre, J., 213
Werner, G., 825
West, G. B., 47; 834, 1000
West, W. A., 566
Wexler, S., 159
Wheeler, R. J., 65; 154
Whitby, L. G., 43; 1001
White, H. J., 608
White, K., 118
Whittemore, K. S., 1003
Wicha, H., 178; 410
Wiechmann, M., 825, 826
Wilk, A. L., 662
Wilkinson, G. R., 18; 65
Williams, C. E., 823
Williams, K. I. H., 311; 1002, 1003
Williams, R. A. D., 863

AUTHOR INDEX

Williams, R. T., 2, 3, 108, 109, 110, 111, 159, 160, 161, 162, 163, 166, 174, 182, 200, 320; 119, 120, 263, 278, 307, 316, 400, 406, 407, 447, 480, 577, 699, 700, 777, 793, 803, 808, 809, 810, 832, 857, 858, 859, 860, 861, 863, 864, 865, 866, 881, 904, 1004, 1005, 1006, 1007, 1008
Williams, V., 140; 1009
Willig, G., 732, 733, 734
Willner, K., 271; 1010
Wills, J. H., 381
Wilson, A., 94, 679, 772, 773, 774, 820
Wilson, M. J., 4
Winters, W., 533
Winters, W. D., 149; 1011
Wiseman, E. H., 151, 222; 1012, 1013
Wiser, R., 90, 91; 966, 970, 1014
Witkop, B., 41, 241
Wittenhagen, G., 148; 640, 1015, 1016
Wodinsky, I., 848
Woiwod, A. J., 312; 168, 169
Wolf, L. M., 363, 364, 365, 366, 367, 368
Wolf, V., 147
Wolff, W., 239
Wollack, A., 155
Wong, K. C., 301; 1017
Wong, K. K., 295; 887, 888
Wood, H. B., 1018
Woodbury, D. M., 248; 313, 676
Woods, L. A., 32, 115, 262, 264, 265, 266, 267, 268, 270; 442, 579, 630, 635, 636, 648, 1019, 1020, 1021, 1022, 1023, 1024, 1026
Woods, P. B., 111
Wozniak, L. A., 556
Wright, J. B., 563
Wright, S. E., 297, 298, 299, 301, 302; 34, 142, 143, 144, 223, 224, 612, 833, 901, 1025

Wulf, R. J., 287; 298
Wyler, H., 830
Wyngaarden, J. B., 87; 1026

Y

Yamaguchi, K., 193; 1027
Yamamoto, A., 9, 92; 928, 929, 1028, 1029
Yamamura, Y., 3
Yamane, S., 13
Yamasaki, M., 687
Yamashina, H., 476
Yanagisawa, I., 10; 661
Yarger, K., 240
Yasuda, H., 951
Yates, C. M., 79; 1030
Yesair, D. W., 781
Yim, G. K. W., 295, 309
Yoshida, K., 922
Yoshida, T., 3
Yoshimura, H., 92, 136, 137, 138; 891, 924, 926, 930, 931, 932, 933, 934, 935, 936, 937, 938, 939, 940, 941, 972, 1028, 1029, 1031, 1032, 1033
Yoshimura, M., 154, 189; 942, 1034
Yoshinari, S., 10; 1035
Young, J. A., 39; 1036
Young, R. L., 16; 785, 1037
Yu, T. F., 160
Yu, W., 805
Yutaka, S., 952

Z

Zak, S. B., 250; 1038
Zamboni, V., 192; 198, 249, 1039
Zampaglione, N. G., 634
Zathurecky, L., 304; 1040
Zauder, H. L., 264; 1041
Zehnder, K., 71; 1042
Zicha, L., 236, 320; 320, 1043
Ziemann, A., 428, 429
Zins, G. R., 245; 1044

SUBJECT INDEX

A

7-Acetamidodiazepam, 86
S-(1-Acetamido-4-hydroxphenyl)-cysteine, 112
4-Acetamido-2-morpholino-1,3,5-triazine, 246
p-Acetamidophenol (see also: paracetamol), 107-113
p-Acetamidophenol glucuronide, 108, 110, 112
p-Acetamidophenol sulfate, 108, 110, 112
Acetanilide, 107, 110-112
Acetohexamide, 150, 151
Acetone, 259
Acetone isonicotinoylhydrazone, 195
p-Acetylaminophenylacetic acid, 213
4-Acetylamino-1-phenyl-2,3-dimethyl-5-pyrazolone, 230, 231, 333
N-Acetyl-p-aminosalicylic acid, 7, 11
N-Acetyl p-aminosalicylic acid glucuronide ester, 9, 11
N-Acetyl anileridine, 213
Acetylecysteine, 318
Acetyldidesmethylchlorpromazine, 64
Acetyldidesmethylchlorpromazine-sulfoxide, 64
Acetylidigitoxin, 299
N-Acetyl-3,4-dimethoxyhydroxyphenethylamine, 35
N-Acetyl-S-ethylcysteine, 310
N-Acetylhydralazine, 200, 201
Acetylhydrazine, 195, 200
Acetyl-7-hydroxychlorpromazine, 64
Acetylisoniazid (1-isonicotinoyl-2-acetylhydrazine), 193
N-Acetylmescaline, 35
Acetylmetanilamide, 162
Acetylmethadol, 25
Acetylmonodesmethylchlorpromazine, 64
Acetylmonodesmethylchlorpromazine sulfoxide, 64
N-Acetylnortriptyline, 28

Acetylsalicylic acid, 4, 5
N-Acetylsulfadiazine, 166
N-Acetylsulfadimethoxine, 169
N-Acetylsulfaethoxypyridazine, 171
Acetylsulfafurazol, 173
Acetylsulfaguanidine, 176
N-Acetylsulfamethizol, 176
N-Acetylsulfametomidine, 171
Acetylsulfamoxol, 172
N-Acetylsulfanilamide, 160, 161
N-Acetylsulfasomizol, 176
Acetylsulfathiazol, 175
N-Acetylsulfisomidine, 171
Adrenaline (epinephrine), 40, 42-51
1-(4-Aldoximinopyridinium)-3-(4-cyanopyridinium)propane, 208
Allantoin, 255
Alloferin, 322
Allopurinol, 256
Alloxanthine, 256
N-Alkyl-4-bromobenzenesulfonamide, 180
5-Allyl-5-(1-hydroxy-1-methylbutyl) barbituric acid, 132
5-Allyl-5-(3-hydroxy-1-methylbutyl) barbituric acid, 130
5-Allyl-5-(1-methyl-3-carboxypropyl) barbituric acid, 130, 145
5-Allyl-5-(1-methyl-3-carboxypropyl)-2-thiobarbituric acid, 145
5-Allyl-5-(1-methyl-3-hydroxybutyl)barbituric acid, 130, 145
5-Allyl-5-(1-methyl-3-hydroxybutyl)-2-thiobarbituric acid, 145
Amantadine, 29
Amidopyrine (see: aminophenazone)
p-Aminobenzoic acid, 22, 23
2-Amino-5-chlorobenzophenone, 84
4-Amino-7-chloroquinoline, 220, 221
7-Aminodiazepam, 85
α-Aminodiphenylacetic acid, 153
α-Aminoglutarimide, 184
Aminoglutethimide, 188
p-Aminohippuric acid, 23
2-Amino 5-hydroxyl-4,6-dimethylpyrimidine, 168

Aminophenazone (amidopyrine), 230-233
Aminophenazone glucuronide, 230
m-Aminophenol, 8, 9, 11
m-Aminophenol glucuronide, 10
m-Aminophenol sulfate, 10
p-Aminophenol, 106-109, 111, 160
p-Aminophenol glucuronide, 107-109
4-Amino-1-phenyl-2,3-dimethyl-5-pyrazolone, 230, 231, 233
Aminopropylone, 231
p-Aminosalicylglutamine, 10
p-Aminosalicylic acid, 8-11
p-Aminosalicylic acid glucuronide ester, 10, 11
p-Aminosalicylic acid glucuronide ether, 10, 11
p-Aminosalicylic acid sulfate, 11
p-Aminosalicyluric acid, 9, 11, 12
Amitriptyline, 27, 28, 29
Amobarbital, 128
Amphetamine, 14-21, 50
Ampicillin, 289
Anhydrochlortetracycline, 290
Anhydrodemethylchlortetracycline, 290
Anhydrotetracycline, 289, 290
Anileridine, 213
Aniline, 110, 111
Antazonite, 206
Antipyrine (see: phenazone)
Arsanilic acid, 317
Atropine, 281, 282
Aureominic acid, 288
AY-8682, 157
5-Azacytidine, 247
6-Azacytidine, 248
5-Azauracil, 245
6-Azauridine, 245

B

Barbital, 123, 124, 141-144
Benzoic acid, 31, 36, 106, 274
Benzomethamine, 156
2-Benzothiazolemercapturic acid, 177
2-Benzoxazolone, 105-107
Benzquinamide, 222, 223
Benzydamine, 163, 241
Benzyl-N-benzylcarbethoxyhydroxamate, 106
N-Benzyl-N,N'-dimethylguanidine, 155

Benzylhydrazine, 199
Benzylpenicillin, 291
Bisacodyl, 211
N,N-Bis-(2-chloroethyl)-O-(3-aminopropyl)phosphoric acid amido ester, 316
4',4''-Bis-(2-imidazolin-2-yl)terephthalanilide, 115
4',4''-Bis-(1,4,5,6-tetrahydro-2-pyrimidinyl)terephthalanilide, 115
4-Bromobenzenesulfonamide, 180
2-Bromo-2-ethyl-3-hydroxybutyryl urea, 147, 148
Brucine, 278-280
Bufotenine, 284
Busulfan, 309, 310
Butamoxane, 257
1,4-Butanediol, 310
Buthalital, 142
Butobarbital, 124
p-Butoxyphenylacetamide, 321
p-Butoxyphenylacethydroxamic acid, 321
p-Butoxphenylacetic acid, 321
N-Butylbiguanide, 155
N-Butyl-4-bromobenzenesulfonamide, 180
1-Butyl-3-(p-carboxyphenyl)sulfonylurea, 147
1-Butyl-3-(p-hydroxymethylphenyl)sulfonylurea, 148
9-Butylmercaptopurine, 254
9-Butylmercaptopurine glucuronide, 254
N-Butylnoradrenaline, 48
Butynamine, 13, 91

C

Caffeine, 273
Carbadrine, 48
O-Carbamoylphenoxyacetic acid, 8
Carbenoxolone, 305
Carbenoxolone glucuronide, 305
2-Carboxamido-N-methylpyridinium, 207
5-(3-Carboxyallyl)-5-ethylbarbituric acid, 132
p-Carboxybenzenesulfonamide, 163
6-(4-Carboxybutyl)thiopurine, 253
2-Carboxy-8-ethoxy-1,4-benzodioxane, 258
6-(2-Carboxyethyl)thiopurine, 255

SUBJECT INDEX

Carboxymeprobamate, 93
2-Carboxy-N-methylpyridinium, 207
4- O-Carboxyphenylamino)-7-chloroquinoline, 220
3-Carboxypropylallyl acetylthioureide, 145
1-(3'-Carboxypropyl)-3,7-dimethylxanthine, 274
Carbromal, 147
Carisoprodol, 95
Caronamide, 178
Chloramphenicol, 285, 286
Chloramphenicol glucuronide, 286
Chloramphenicol monosuccinate, 286
Chlorcyclizine, 239
Chlordiazepoxide, 81, 82
Chlorethomoxane, 258
Chlormerodrine, 314
5-Chloro-8-acetoxyquinoline, 222
7-Chloro-2-amino-5-phenyl-3H-1,4-benzodiazepine-4-oxide, 82
p-Chlorobenzenesulfonamide, 149
p-Chlorobenzenesulfonylurea, 149
N-(2-Chloroethyl)aziridine, 316
Chloroform, 317
5-Chloro-2-(3'-hydroxycyclohexyl)-1-oxo-6-sulfamoylisoindoline, 179
2-Chloro-10-(3'-hydroxypropyl)phenothiazine, 62
5-Chloro-8-hydroxyquinoline, 222
p-Chlorophenol, 98-100
2-Chlorophenothiazine, 62, 64, 65
2-Chlorophenothiazine-N-propionic acid, 65
2-Chlorophenothiazine sulfoxide, 63, 65
p-Chlorophenoxyacetic acid, 97, 99
p-Chlorophenoxylactic acid, 97, 101
p-Chlorophenylthiourea, 117
6-Chloropurine, 254, 255
6-Chloropurine ribonucleotide, 254
Chloroquine, 219, 221
4-(7-Chloro-4-quinolylamino)-1-methyl-1-butanol, 221
4-Chloro-5-sulfamoylanthranilic acid, 178
6-Chlorouric acid, 255
Chlorphenesine carbamate, 97, 98
Chlorphenesine carbamate glucuronide, 97, 98
Chlorphentermine, 20
Chlorpromazine, 54-66
Chlorpromazine-N-oxide, 57, 62-65

Chlorpromazine-N-oxide sulfoxide, 62, 64, 65
Chlorpromazine sulfoxide, 55-58, 61-63, 65
Chlorpropamide, 149
Chlorprothixene, 74
Chlorprothixene sulfoxide, 74
Chlorthalidone, 218
Chlortetracycline, 287, 290, 291
Chlorzoxazone, 218
Chondocurarine, 282
Cinchonidine, 277
Cinchonine, 277
Cinchophene, 226, 227
Clioquinol, 222
Clorexolone, 178
Cocaine, 282
Codeine, 261, 262, 270
Convallatoxin, 303
2-Cyano-N-methylpyridinium, 207, 208
Cyclizine, 239
Cyclobarbital, 138
5-Cyclohexenonyl-5-ethylbarbituric acid, 138
Cyclohexenylmethyl-N-methylacetylurea, 137
Cyclophosphamide, 316
Cymarin, 303
Cymarol, 303

D

4-Deamino-4,7-dihydroxydichlor-omethotrexate, 251
Decamethonium, 30
Deferoxamine, 321
5a-(11a)-Dehydrochlortetracycline, 290
Demethylchlortetracycline, 287, 290
Deptropine, 282
Desethylhydroxychloroquine, 221
Desmethylamitriptyline, 28
N-Desmethylbutynamine, 13
N-Desmethyldeptropine, 283
6-Desmethyl-6-deoxytetracycline, 290
N-Desmethyldiazepam, 83, 84
N-Desmethylerythromycin, 287, 292
Desmethylgriseofulvin, 295
Desmethylhexobarbital, 136
Desmethylimipramine, 77, 79, 80
Desmethylmephentermine, 20
N-Desmethylmethixene, 74, 75

SUBJECT INDEX

N-Desmethylmethixenesulfone, 75
N-Desmethylmethixenesulfoxide, 75
N-Desmethylorphenadrine, 23
N-Desmethylpethidine, 212, 213
N-Desmethylproheptazine, 214
Desmethylpropoxyphene, 26
N-Desmethyltropine, 283
Dextromethorpan, 271, 272
N,N'-Diacetylsulfanilamide, 153
Diallylmelamine, 245, 246
2,4-Diamino-7-hydroxy-6-pteridine-carboxylic acid, 251
Diaphenylsulfone, 312
Diaphenylsulfone glucuronide, 312
Diazepam, 83-85
Dibenamine, 26
Dibenzosuberol, 283
Dibenzylamine, 27
N,N-Dibenzyl-β-hydroxyethylamine, 26
3',5'-Dibromosulfanilanilide, 163
3',5'-Dibromosulfanilanilide glucuronide, 155
p,p'-Dichlorocarbanilide, 117
Di-(2-chloroethyl)amine, 316, 317
Dichloromethotrexate, 251
3,5-Dichloro-4-methylaminobenzoyl-glutamic acid, 251
p,p'-Dichlorothiocarbanilide, 117
Dicoumarol, 258, 259
Didesethylchloroquine, 221
Didesethyllidocaine, 116
Didesmethylaminoaureominic acid, 288
Didesmethylchlorpromazine, 60-64
Didesmethylchlorpromazine sulfoxide, 55-63
Didesmethylimipramine, 78-80
N,N-Didesmethylorphenadrine, 23, 24
4,4'-Diethoxythiocarbanilide, 118
Diethylaminoacetic acid, 116
Diethylaminoethanol, 23
Diethyl disulfide, 309
Diethyldithiocarbamate glucuronide, 312, 313
Diethyldithiocarbamic acid, 312, 313
Digitoxigenin, 300
Digitoxin, 298-302
Digitoxose, 301
Digoxigenin, 299-302
Digoxigenin digitoxoside, 299, 302
Digoxin, 298-300, 302
Dihydrodigoxigenin, 267, 302

1-(4',5'-Dihydrohexyl)-3,7-dimethylxanthine, 274
Dihydromorphine, 267-269
Dihydromorphinone, 269
Dihydronormorphine, 269
Dihydrostreptomycin, 292
5,6-Dihydro-6-(2-thienyl)imidazo-[2,1-b]-thiazole, 206
2,3-Dihydroxybenzoic acid, 1
2,5-Dihydroxybenzoic acid, 1
p,p'-Dihydroxycarbanilide, 117
Dihydroxycinchophene, 227
2,5-Dihydroxy-4,6-dimethylpyrimidine, 67
3,4-Dihydroxymandelic acid, 42, 43, 46
3,4-Dihydroxy-5-methoxyphenethylamine, 34
3,5-Dithydroxy-4-methoxyphenethylamine, 34
3,4-Dihydroxy-5-methoxyphenylacetic acid, 34
3,4-Dihydroxynorephedrine, 50
3,4-Dihydroxyphenylacetic acid, 306
3,4-Dihydroxyphenylacetone, 37, 38, 41
3,4-Dihydroxyphenylglycol, 42, 43
Dihydroxyphenylglycol sulfate, 46
1,4-Dihydroxyphthalazine, 201
1,4-Dihydroxyphthalazine glucuronide, 201
5-(2,3-Dihydroxypropyl)-5-(1-methylbutyl)barbituric acid, 130, 131
Dihydroxyquinine, 277
1,2-Dihydroxysulfonamidobenzene, 161
Dihydroxytetrabenazine, 225
p,p'-Dihydroxythiocarbanilide, 117
Diiodohydroxyquinoline, 222
N,N'-Diisonicotinoylhydrazine, 186, 187, 194, 195, 198
3,5-Dimethoxy-4-hydroxyphenethylamine, 35
(3,4-Dimethoxyphenyl)ethylamine, 34
N,N-Dimethylbiguanide, 154
N,N-Dimethyl-2-hydroxymethylpiperidinium, 216
3,5-Dimethylhydroxypyrazole, 230
1,3-Dimethyl Nirvanol, 152
3,5-Dimethylpyrazole, 229
Dimethylsulfide, 311
Dimethylsulfone, 311

SUBJECT INDEX

Dimethylsulfoxide, 311
1,3-Dimethyluric acid, 272, 273
2,4-Dioxo-3,3-diethyl-5-hydroxymethyltetrahydropyridine, 215
2,4-Dioxo-3,3-diethyl-5-methyltetrahydropyridine, 215
4,6-Dioxo-5,5-diethyltetrahydronicotinaldehyde, 215
4,6-Dioxo-5,5-diethyltetrahydronicotinic acid, 215
2,4-Dioxo-3,3-diethyltetrahydropyridine, 215
Diphenylhydantoic acid, 154
Diphesatine, 211
Dipyridamole, 249, 250
Dipyridamole glucuronide, 249, 250
Disulfiram, 312
Ditophal, 310, 311
Dolcental, 104, 151
Dopa, 36, 40
Dopamine, 36, 40, 50, 51
Doxapram, 204

E

Enoxolone, 304
Enoxolone glucuronide, 304
Enoxolone sulfate, 304
Ephedrine, 15, 18, 50, 51
4-Epianhydrochlortetracycline, 290
4-Epianhydrodemethylchlortetracycline, 290
4-Epianhydrotetracycline, 290
4-Epichlortetracycline, 290
5a-Epichlortetracycline, 290
4-Epidemethylchlortetracycline, 290
6-Epidesmethyldeoxytetracycline, 290
3-Epidigoxigenin, 301
4-Epiisochlortetracycline, 290
Epinephrine (see: Adrenaline)
4-Epitetracycline, 287, 289-291
Erythromycin, 291, 292
Ethambutol, 13, 14
Ethanalisonicotinoylhydrazone, 195
Ethenzamide, 7
Ethinamate, 101
Ethionamide, 209, 210
Ethionamidesulfoxide, 200, 209, 210
Ethoheptazine, 213, 214
Ethomoxane, 257, 258
Ethopabate, 114
Ethosuximide, 190
2-Ethoxy-4-acetamidobenzoic acid, 114
2-Ethoxy-4-aminobenzoic acid, 114
2-Ethoxy-4-amino-5-hydroxybenzoic acid, 5-O-sulfate, 114
p-Ethoxyaniline, 111
Ethoxybenzoic acid, 7
3-Ethoxycarbonylphenytoin, 154
2-Ethoxy-N-methylpyridinium, 208
5-Ethyl-6-azauracil, 245
5-Ethylbarbituric acid, 123
Ethylbiscoumacetate, 259, 260
N-Ethyl-4-bromobenzenesulfonamide, 180
Ethylbutylthiobarbital, 142, 143
2-Ethylbutyrylurea, 147, 148
2-Ethyl-4-carbamoylpyridine, 209
5-Ethyl-5-(1-cyclohexenyl)-4,6-dioxohexahydropyrimidine, 138
5-Ethyl-5-(1-cyclohexenyl)hydantoin, 154
S-Ethylcysteine, 310
2,2'-Ethylenediimunodibutyric acid, 15
Ethyl ether, 317
5-Ethyl-5-(3-hydroxybutyl)barbituric acid, 124
5-Ethyl-5-β-hydroxyethylbarbituric acid, 124
5-Ethyl-5-(3-hydroxyisoamyl)barbituric acid, 128
5-Ethyl-5-(3-hydroxy-1-methylbutyl)barbituric acid, 127-129
5-Ethyl-5-(3-hydroxy-3-methylbutyl)barbituric acid, 129
5-Ethyl-5-p-hydroxyphenylhydantoin, 152
2-Ethylisonicontinic acid, 209
Ethylmercaptan, 309
9-Ethylmercaptopurine, 254
Ethyl methanesulfonate, 309
Ethyl-(1-methylbutyl)malonuric acid, 126
5-Ethyl-5-(1-methyl-3-carboxypropyl)barbituric acid, 141
5-Ethyl-5-(1-methyl-3-carboxypropyl)thiobarbituric acid, 141
Ethylmorphine, 268
3-Ethyl Nirvanol, 152
N-Ethylnoradrenaline, 48
N-Ethylphenylisopropylamine, 21
Ethyl trichloramate, 106
Ethyl trichloromate glucuronide, 106
1-Ethynyl-3-acetoxy-1-cyclohexylcarbamate, 103

1-Ethynyl-4-acetoxycyclohexylcarbamate, 103
1-Ethynyl-1,2-cyclohexanediol, 104
Ethynylcyclohexanol, 102
1-Ethynyl-1-hydroxy-1-cyclohexylcarbamate, 102
1-Ethynyl-1-hydroxy-2-cyclohexylcarbamate, 103
1-Ethynyl-2-hydroxy-1-cyclohexylcarbamate, 102, 103
1-Ethynyl-3-hydroxy-1-cyclohexylcarbamate, 102
1-Ethynyl-4-hydroxy-1-cyclohexylcarbamate, 101, 102
Etosalamide, 8
Etymide, 157

F

Fenfluramine, 20, 21
5-Fluorocytosine, 240
Fluphenazine, 70
Fluphenazine sulfoxide, 70
Formaldehyde, 318
2-Formyl-N-methylpyridinium, 207
Furosemide, 178
Fusidic acid, 293

G

Gentisamide, 6, 7
Gentisamide glucuronide, 6
Gentisic acid, 4-7, 200
Glafenine, 220
Glucose isonicotinoylhydrazone, 195, 196
Glucuronic acid, 1, 2, 16, 54, 94, 169, 170
Glucuronolactone, 94, 169
Glutamine, 184
Glutamic acid, 184
Glutarimide, 181
Glutethimide, 183, 186-188
Glutethimide glucuronide, 186-188
Glyceryl nitrates, 320
Glyceryl trinitrate, 320
Glymidine, 179
Griseofulvin, 295
Guanidine, 258
Guanoxan, 258
Guanylurea ribonucleoside, 247

H

Halothane, 317
Helveticoside, 304
Heparin, 306, 307
1-(1H-hexahydro-1-azepinyl)-3-(p-acetylphenylsulfonyl)urea, 150
Hexetal, 124
Hexobarbital, 135-139
Hexobendine, 35
1-Hexyl-3,7-dimethylxanthine, 274
Hippuric acid, 16, 22, 105, 106
Homofenazine, 71
Homofenazine sulfoxide, 71
Homovanillylamine, 50
Homoveratrylamine, 50
Hordenine, 50
Hydralazine, 199-201
Hydralazine glucuronide, 201
Hydrogen iodide, 207
Hydroxyacetanilide, 111
Hydroxyamobarbital, 128
p-Hydroxyamphetamine, 15, 16, 21
p-Hydroxyamphetamine glucuronide, 24
p-Hydroxybenzenesulfonamide, 160
p-Hydroxybenzenesulfonamide glucuronide, 160
p-Hydroxybenzenesulfonamide sulfate, 160
p-Hydroxybenzoic acid, 105
8-Hydroxybutamoxane, 257
3-Hydroxybutylbiguanide, 156
Hydroxycarbamide, 147
p-Hydroxycarbanilide, 117
Hydroxychloroquine, 220, 221
7-Hydroxychlorpromazine, 57, 60, 61, 63, 64
7-Hydroxychlorpromazine glucuronide, 60
7-Hydroxychlorpromazine sulfate, 60
6-Hydroxychlorzoxazone, 218
2-Hydroxycinchonine, 277
Hydroxycinchophene, 227
p-Hydroxycinnamic acid, 105
5-(3-Hydroxy-1-cyclohexenyl)-3,5-dimethylbarbituric acid, 137
2-Hydroxydesmethylimipramine, 78-80
2-Hydroxydismethylimipramine glucuronide, 78

SUBJECT INDEX

10-Hydroxydesmethylimipramine, 80
p-Hydroxydesmethylmephentermine, 20
3-Hydroxydiazepam, 83
7-Hydroxydichloromethotrexate, 251
7-Hydroxydicoumarol, 259
3-Hydroxy-4,5-dimethoxyphenethylamine, 34
4-Hydroxy-3,5-dimethoxyphenethylamine, 35
2-Hydroxy-3,6-dimethylpyrimidine, 167, 168
p-Hydroxyephedrine, 18
Hydroxyethinamate, 101, 102, 104
Hydroxyethinamate glucuronide, 101
Hydroxyethoheptazine, 214
5-(1-Hydroxyethyl)-6-azauracil, 245
1-[p-(1-Hydroxyethyl)phenylsulfonyl]-3-cyclohexylurea, 150
Hydroxyhexamide, 151
3-Hydroxyhexobarbital glucuronide, 137
1-(5'-Hydroxyhexyl)-3,7-dimethylxanthine, 275
Hydroxyhydralazine, 201
Hydroxyhydralazine glucuronide, 201
10-Hydroxyiminobenzyl, 80
2-Hydroxyimipramine, 77-80
2-Hydroxyimipramine glucuronide, 78
10-Hydroxyimipramine, 80
4-Hydroxyindolacetic acid, 283, 284
5-Hydroxyindolacetic acid, 283, 284
p-Hydroxylaminobenzene sulfonic acid, 160
p-Hydroxymephentermine, 20
4-Hydroxymepivacaine, 212
4-Hydroxymetanilamide, 162, 163
2-Hydroxymethixene, 75
7-Hydroxymethotrexate, 251
α-Hydroxy-β-(o-methoxy-p-propionylphenoxy)propionic acid, 89
2-Hydroxy-3-methoxystrychnine, 279
5-Hydroxy-5(1-methylbutyl)barbituric acid, 130
Hydroxymethylmeprobamate, 90, 91, 93
Hydroxymethylmeprobamate glucuronide, 90, 91, 93

3-Hydroxy-N-methylmorphinan, 271, 272
2-Hydroxymethyl-2-propyl-1,3-propanediol dicarbamate, 90
3-Hydroxy-N-methylpyridinium, 156
2-Hydroxymethyl-3-o-tolyl-(3H)-4-quinazolinone, 243
3-Hydroxymorphinan, 271, 272
p-Hydroxynorephedrine, 15, 17
10-Hydroxynortriptyline, 29
4'-Hydroxypapaverine, 278
3-Hydroxyphenacetin, 111
2-Hydroxyphenacetin glucuronide, 112
4-Hydroxyphenazone, 230
2-Hydroxyphenetidine glucuronide, 112
p-Hydroxyphenobarbital, 123, 134, 135
3-Hydroxyphenothiazine, 70
p-Hydroxyphentermine, 21
p-Hydroxphenylacetic acid, 105
β-Hydroxyphenylethylamine, 50, 51
1-(p-Hydroxyphenyl)-2-phenyl-3,5-dioxo-4-(3-hydroxybutyl)pyrazolidine, 237
1-(p-Hydroxyphenyl)-2-phenyl-3,5-dioxo-4-(3-oxobutyl)pyrazolidine, 237
m-Hydroxyphenyltrimethylammonium bromide, 156
1-Hydroxyphtalazine, 201
1-Hydroxypromazine, 68
2-Hydroxypromazine, 68
3-Hydroxypromazine, 67, 68
4-Hydroxypromazine, 68
β-Hydroxypropylcarisoprodol, 95
β-Hydroxypropylmeprobamate, 91, 93, 95, 96
β-Hydroxypropylmeprobamate glucuronide, 93
β-Hydroxypropyltybamate, 96
2-Hydroxyquinine, 276, 277
2'-Hydroxyquinine, 276, 277
2-Hydroxystrychnine, 280, 281
Hydroxysulfadimerazine, 167
2-Hydroxysulfanilamide, 168
3-Hydroxysulfanilamide, 162, 163, 167, 168
N-Hydroxysulfanilamide, 160, 162, 163
2-Hydroxysulfanilic acid, 168

3-Hydroxysulfanilic acid, 168
5'-Hydroxysulfapyridine, 165
Hydroxysulfinpyrazone, 235
Hydroxytetrabenazine, 224
p-Hydroxythiocarbanilide, 117
p-Hydroxythiocarbanilide glucuronide, 117
5-Hydroxytiabendazole, 240, 241
5-Hydroxytiabendazole glucuronide, 240
5-Hydroxytiabendazole sulfate, 240
Hydroxytoluic acid, 5, 6
5-Hydroxytryptamine, 51
6-Hydroxyzoxazolamine, 218
Hypoxanthine, 255

I

Idoxuridine, 239
Iminobenzyl, 80
Imipramine, 77, 79
Imipramine-N-oxide, 79
Indomethacin, 216, 217
Indomethacin glucuronide, 216, 217
Inosityl nicotinate, 319
4-Iodophenazone, 234
Iodouracil, 239
Iproniazid, 198
N-Isobutylnoradrenaline, 48
Isocarboxazid, 199
Isochlortetracycline, 290
Isoglutamine, 184
Isoniazid, 191-199
Isonicotinamide, 194
Isonicotinic acid, 191-198
Isonicotinic acid N-oxide, 186
1-Isonicotinoyl-2-acetylhydrazine (acetylisoniazid), 192, 194, 195
N-Isonicotinoylglycine, 192, 193, 194, 196, 198
1-Isonicotinoyl-2-glycylhydrazine, 194, 195
Isoprenaline, 46-48
N-Isopropyl-4-bromobenzenesulfonamide, 177
N-Isopropylterephtalamic acid, 202
Isopyrine, 232, 233
Isotetracycline, 290

K

Kalypnon, 132
Kanamycin, 292, 293

α-Ketoglutaric acid hydralazine hydrazone, 192, 201
α-Ketoglutaric acid isonicotinoyl hydrazone, 193, 195
Ketomeprobamate, 93
Ketophenylbutazone, 235, 236, 237

L

Lanatoside C, 298, 299
Levallorphan, 270
Levorphanol, 268, 269
Levulic acid, 2
Lidocaine, 115
Lucanthone, 73
Lucanthone sulfoxide, 75

M

Mafenide, 163
Mebutamate, 197
Meclozine, 238
Melitracene, 81
Mephenesine, 87, 88
Mephenesine carbamate, 89
Mephenoxalone, 205
Mephentermine, 20
Mephenytoin, 152
Mepivacaine, 212
Meprobamate, 89-96
Meprobamate glucuronide, 90, 91
Meprobamate-N-monoglucurone, 94
Meprophendiol, 88, 89
Meralluride, 315
2-Mercaptobenzothiazole, 177
2-Mercaptobenzothiazole glucuronide, 177
Mercaptopurine, 252, 253
Mersalyl, 314
Mescaline, 32-35, 50
Metadrenaline, 41, 43, 44, 46, 48-51
Metanilamide, 162
Metaraminol, 47
Metaxalone, 205
Methadone, 24, 25
Methanesulfonic acid, 310
Methanesulfonic acid ethyl ester, 309
Methaqualone, 242
Metharbital, 132
Methitural, 142

SUBJECT INDEX

Methixene, 74, 75
Methixene-N-oxide, 75
Methixene sulfone, 75
Methixene sulfoxide, 75
Methocarbamol, 88, 89
Methocarbamol glucuronide, 89
Methohexital, 133
Methotrexate, 250
7-Methoxychlorpromazine, 58
7-Methoxydicoumarol, 259
Methoxyflurane, 317
3-Methoxy-4-hydroxymandelic acid, 41, 43, 44, 46
3-Methoxy-4-hydroxyphenylacetone, 37, 38, 41
β-(3-Methoxy-4-hydroxyphenyl)ethylamine, 51
3-Methoxy-4-hydroxyphenylglycol, 42, 44
3-Methoxy-4-hydroxyphenylglycol sulfate, 46
2-Methoxy-3-hydroxystrychnine, 278, 280
3-Methoxyisoprenaline, 47
3-Methoxymethyldopa, 37-41
3-Methoxy-α-methyldopamine, 37, 39, 41
3-Methoxy-α-methyldopamine glucuronide, 37, 41
3-Methoxymorphinan, 271, 272
β-(4-methoxyphenyl)ethylamine, 50, 51
o-Methoxy-p-propionylphenoxyacetic acid, 88
p-Methoxy-o-sulfonyloxyaniline, 112
3-Methylacetylsalicylic acid, 6
5-(1-Methylallyl)-5-(1-methyl-4-hydroxy-2-pentinyl) barbituric acid, 133
2-Methylamino-5-chlorobenzophenone, 84
4-Methylamino-2,3-dimethyl-1-phenyl-5-pyrazolone, 233
p-Methylaminophenol, 32
Methyl-N-(o-aminophenyl)-N-(3-dimethylaminopropyl) anthranilate, 27
2-Methylbenzhydrol, 23
N-Methyl-4-bromobenzene sulfonamide, 176
5-(1-Methylbutyl) barbituric acid, 130, 132, 133
5-(3-Methyl-5-carboxyphenoxymethyl)-2-oxazolidinone, 205

2-Methyl-2-(1-cyclohexenyl) glutarimide, 188, 189
5-Methyl-5-(1-cyclohexenyl) hydantoin, 154, 155
3-Methyl-3,4-dihydroxy-4-phenyl-1-butyne, 320
Methyldopa, 36-42
Methyldopa mono-O-sulfate, 38
α-Methyldopamine, 34, 39-41
α-Methyldopamine glucuronide, 41
Methylene chloride, 317
6-Methyleneoxtetracycline, 290
Methylephedrine, 18, 19
1-Methyl-2-ethyl-4-carbamoyl-6-oxodihydropyridine, 210
1-Methyl-2-ethyl-S-oxo-4-thiocarbamoyl-6-oxodihydropyridine, 210
Methylethylsulfone, 309, 310
Methylethylsulfoxide, 309, 310
1-Methyl-2-ethyl-4-thiocarbamoyl 6-oxodihydropyridine, 210
1-Methyl-2-ethyl-4-thiocarbamoyl-pyridinium, 210
N-Methylglutethimide, 187, 188
Methylglyoxal-bis-guanylhydrazone, 202
Methylhydrazine, 202
2-Methyl-2-(β-hydroxyl-α-methylpropyl)-1,3-propanedioldicarbamate, 97
2-Methyl-3-(2'-hydroxymethylphenyl)-(3H)-4-quinazolinone, 243
2-Methyl-3-(3'-hydroxy-2'-methylphenyl)-(3H)-4-quinazolinone, 243
2-Methyl-3-(4'-hydroxy-2'-methylphenyl)-(3H)-4-quinazolinone, 243
2-Methyl-3-(5'-hydroxy-2'-methylphenyl)-(3H)-4-quinazolinone, 243
2-Methyl-3-(6'-hydroxy-2'-methylphenyl)-(3H)-4-quinazolinone, 243
β-(2-methyl-4-hydroxyphenoxy)lactic acid, 88
N-Methylisonicotinic acid, 197
6-Methylmercaptopurine, 253
N-Methylmethadone, 28
1-Methyl Nirvanol, 152
2-Methyl-5-nitro-1-imidazolyl-acetic acid, 238

α-Methylnoradrenaline, 46
Methylpentynol, 320, 321
N-Methylphenacetin, 112
N-Methylphenetidine, 112
Methylphenobarbital, 135
1-Methyl-4-phenylisonipecotic acid, 212
5-Methylpyrazole-3-carboxylic acid, 229, 230
1-Methylpyridinium-4-carboxyhydrazine iodide, 195
N-Methyl-α-pyridone, 207, 208
Methylrubazonic acid, 232, 233
3-Methylsalicyluric acid, 6, 7
N-Methylthiocarbanilide, 118, 119
6-Methylthiopurine, 253
2-Methyl-3-o-tolyl-5-hydroxy-(3H)-4-quinazolinone, 243
2-Methyl-3-o-tolyl-6-hydroxy-(3H)-4-quinazolinone, 243
2-Methyl-3-o-tolyl-7-hydroxy-(3H)-4-quinazolinone, 243
2-Methyl-3-o-tolyl-8-hydroxy-(3H)-4-quinazolinone, 243
1-Methyluric acid, 272
Methyprylone, 214
Meticillin, 289
Metofoline, 224, 225
Metronidazole, 236
Metyrapone, 208
Monochloracetanilide, 113
Monodesethylchloroquine, 219, 221
Monodesethyllidocaine, 115
Monodesmethylchlorpromazine, 61-64
Monodesmethylchlorpromazine sulfoxide, 56, 58-64
Monodesmethylprocarbazine, 201
Mondesmethylpromazine, 66
Monodesmethylpromazine sulfoxide, 64
Moroxydine, 155
Morphine, 264-267, 270, 271
Morphine glucuronide, 264-266
Morphine sulfate, 267

N

Nalorphine, 268
Nalorphine glucuronide, 268
Neostigmine, 156
Nicotinamide, 193, 279
Nicotine, 18
Nicotinic acid, 193

Nifenazone, 233
Nirvanol, 152
Nitrazepam, 86
5-Nitro-2-aminobenzophenone, 85
Nitrofurantoin, 236
Noradrenaline, 41, 43, 44, 49-51
Norcodeine, 263
Norcyclizine, 239
Norephedrine, 15, 18, 19, 50, 51
Normetadrenaline, 44, 49-51
Normetadrenaline glucuronide, 46
Normetadrenaline-O-sulfate, 46
Normorphine, 266, 267
Norpseudoephedrine, 18
Nortriptyline, 28, 29
Noscapine, 281

O

Octopamine, 50, 51
Opipramol, 29, 80
Orphenadrine, 23, 24
Orthanilamide, 163
Ouabaine, 302
Oxazepam, 83-85
Oxazepam glucuronide, 85
Oxedrine, 48
5-(3-Oxo-1-cyclohexen-1-yl)-3,5-dimethylbarbituric acid, 136, 137
5-(3-Oxo-1-cyclohexen-1-yl)-5-methylbarbituric acid, 136, 137
1-(5'-Oxohexyl)-3,7-dimethylxanthine, 274
Oxolamine, 238
Oxonazine, 246
Oxychloroquine, 219
Oxyphenbutazone, 236
Oxytetracycline, 288, 290, 291

P

Pamaquine, 220
Papaverine, 278
Paracetamol (See also: p-Acetamidophenol), 31, 107
Paramethadione, 205
Paromomycin, 292
Pecazine, 70
Pecazine sulfoxide, 70
Penicillamine, 291
Penicillamine disulfide, 291
Penicilloic acid, 291
Pentaerythritol, 320
Penaerthrityl nitrates, 319, 320

SUBJECT INDEX 393

Pentaerythrityl tetranitrate, 318-320
Pentaquine, 221
Pentetrazole, 248
Pentobarbital, 126, 127, 134, 142, 179
Perphenazine, 70
Perphenazine sulfoxide, 70
Pethidine, 212
Phenacetin, 31, 107, 110, 111, 113
Phenacetin glucuronide, 110, 113
Phenazone (antipyrine), 230
Phenetidin, 109-111
Phenobarbital, 134-138
Phenothiazine, 53-55, 60, 61, 66
Phenothiazine sulfoxide, 70
Phenothiazone, 53, 54
Phenoxybenzamine, 27
N-Phenoxyisopropyl-N-benzylamine, 27
[2-(o-Phenoxyphenyl)amino]-4-amino-6-hydroxy-1,3,5-triazine, 247
[2-(o-Phenoxyphenyl)amino]-4-amino-1,3,5-triazine, 247
Phenprobamate, 105
Phentermine, 20
Phenylacetic acid, 35, 105
Phenylacetone, 16
Phenylbutazone, 234, 237
2-Phenyl-2-(1-cyclohexenyl)glutarimide, 189
Phenylephrine, 50, 51
β-Phenylethylamine, 18, 50, 51
α-Phenyl-α-ethylglutaconimide 186-188
α-Phenylglutarimide, 184, 186-188
5-Phenyl-5-hydroxyphenylhydantoin, 153, 154
5-Phenyl-5-p-hydroxyphenylhydantoin glucuronide, 152, 153
4-Phenylisonipecotic acid, 212
1-Phenyl-5-methyl-4,5-pyrazolinedione, 232
1-Phenyl-3-methyl-5-pyrazolone, 232
1-Phenylpropan-2-ol, 18
Phenylpropionic acid, 105
4-Phenylthioethyl-1,2-diphenyl-3,5-pyrazolinedione, 235
1-Phenyl-2-thiourea, 117
Phenytoin, 152-154
Phthalazine, 201
Phthalimide, 184
Phthalic acid, 184

N-Phthaloyl glutamine, 184
N-Phthaloyl glutamic acid, 184
N-Phthaloyl glutarimide, 184
N-Phthaloyl isoglutamine, 184
N-Phthalyl glutamine, 183, 184
N-Phthalyl glutamic acid, 183, 184
N-Phthalyl isoglutamine, 184
Piribenzile, 216
Pralidoxime, 206, 208
Prilocaine, 116
Pristinamycin, 296
Probenecid, 178
Procaine, 22, 23
Procarbazine, 200-202
Prochlorperazine, 72
Prodilidine, 203
Proheptazine, 214
Promazine, 60, 66, 67
Promazine sulfoxide, 66
Pronarcon, 133
Prontosil, 159
Propoxyphene, 26
N-Propylalanine, 116
N-Propyl-4-bromobenzenesulfonamide, 169
Pseudoephedrine, 18
Pseudomorphine, 266
Psilocin, 282, 283
Psilocybin, 284
Puromycin, 294
Pyridostigmine, 156
Pyridoxal isonicotinoylhydrazone, 195, 196
Pyruvic acid hydralazine hydrazone, 192, 200, 201
Pyruvic acid isonicotinoylhydrazone, 193, 195

Q

Quercetine, 306
Quindonium, 226
Quindonium glucuronide, 226
Quinidine, 277
Quinine, 1, 276
2-Quinolyl-1-piperazine, 226

R

Reserpine, 275, 277
Reserpic acid, 277
Rubazonic acid, 232, 233
Rutoside, 306

S

Salicylamide, 6, 7
Salicylamide glucuronide, 3, 7
Salicylamide sulfate, 7
Salicylglucuronic acid, 1
Salicylic acid, 1-7, 220
Salicylic acid glucuronide, 7
Salicylic acid glucuronide ester, 213
Salicylic acid glucuronide ether, 2, 3
Salicylohydrazide, 200
Salicyluric acid, 1-7, 200
Secbutabarbital, 126
Secobarbital, 130, 145
Streptomycin, 292
Strophantidine, 303
Strychnine, 280, 281
Sulfadiazine, 165, 166
Sulfadiazine glucuronide, 166
Sulfadiazine sulfonate, 166
Sulfadimerazine (sulfadimidine), 166-168
Sulfadimethoxine, 167-169
Sulfadimethoxine glucuronide, 167-169
Sulfdimidine (see: Sulfadimidine)
Sulfaethoxypyridazine, 171
Sulfafurazol, 173, 174
Sulfafurazol glucuronide, 173
Sulfafurazol sulfonate, 173
Sulfaguanidine, 176
Sulfamethizol, 175, 176
Sulfamethizol glucuronide, 176
Sulfamethizol sulfate, 176
Sulfamethoxazol, 174
Sulfamethoxazol glucuronide, 174
Sulfametomidine, 170
Sulfamoxol, 172
Sulfanilamide, 160-166, 168, 172, 173
Sulfanilcarbamide, 173
Sulfanilic acid, 167-169, 176
Sulfaphenazol, 171, 172
Sulfaphenazol glucuronide, 172
Sulfapyridine, 164, 165
Sulfasomizol, 174, 176
Sulfasomizol glucuronide, 175, 176
Sulfasomizol sulfate, 175, 176
Sulfathiazol, 174, 175
Sulfathiazol glucuronide, 174
Sulfathiazol sulfonate, 174
Sulfinpyrazone, 235
Sulfisomidine, 170

2-Sulfonamidobenzothiazole, 177
Synephrine, 50

T

Tacrine, 226
Tetrabenazine, 224, 225
Tetracycline, 286, 287, 290, 291
4-(1, 4, 5, 6, -Tetrahydro-2-pyrimidinyl)-4'-[p-(1, 4, 5, 6-tetrahydro-2-pyrimidinyl)phenyl] carbamoylbenzanilide, 115
Thalidomide, 181-185
Theobromine, 273
Theophylline, 272
Thiambutosine, 118
Thiamylal, 144, 145
Thiobarbital, 140, 143, 144
Thiocarbanilide, 116
Thioguanine, 254
Thioguanine mononucleotide, 254, 256
Thioguanosine, 254
Thioguanosine deoxyriboside, 252
Thionol, 53
Thiopental, 140-143
Thioridazine, 71, 72
Thioridazine disulfone, 72
Thioridazine disulfoxide, 72
Thioridazine monosulfoxide, 72
Thiourea, 141
2-Thiouric acid, 251
6-Thiouric acid, 251, 252, 256
8-Thiouruc acid, 251
Tiabendazole, 240
Tolazoline, 238
Tolbutamide, 148, 229
o-Toluidine, 116
β-(o-Tolyoxy) lactic acid, 87-89
p-Tolylthiourea, 117
Tranylcypromine, 21
Triamterene, 249
Trichlorethanol, 106
Triethylenephosphoramide, 315
Triethylenethiophosphoramide, 315
Trofluoroacetic acid, 318
m-Trifluoromethylphenylisopropylamine, 21
2, 3, 5-Trihydroxybenzoic acid, 2
Trihydroxyethylrutoside, 306
3, 4, 5, -Trihydroxyphenethylamine, 34

SUBJECT INDEX

Trimedoxime, 208
Trimethadione, 204
3,4,5,-Trimethoxybenzoic acid, 35, 256
3,4,5,-Trimethoxymandelic acid, 34
3,4,5,-Trimethoxyphenylacetic acid, 32-35
3,4,5,-Trimethoxyphenethylamine, 34
3,4,5,-Trimethoxyphenylethanol, 34
2,4,6-Trioxo-3,3-diethyl-5-methyl-piperidine, 215
Tripelennamine, 210
Tropine, 283
Tubocurarine, 282
Tubocurarine dimethyl ether, 282
Tybamate, 96
Tyramine, 50, 51

U

Uracil, 239
Urea, 255

4-Ureidoaminophenazone, 230
Uric acid, 255, 272, 273
Uroheparin, 307

V

Vanillic acid, 46
Vanillylamine, 50
Vinylethylbarbituric acid, 124

W

Warfain, 260

X

3,5-Xylenol, 206
Xylidine, 116

Z

Zoxazolamine, 218